Musical Instruments in the 21st Century

Till Bovermann · Alberto de Campo
Hauke Egermann · Sarah-Indriyati Hardjowirogo
Stefan Weinzierl
Editors

Musical Instruments in the 21st Century

Identities, Configurations, Practices

Springer

Editors
Till Bovermann
Institut für zeitbasierte Medien
Universität der Künste Berlin
Berlin
Germany

Alberto de Campo
Institut für zeitbasierte Medien
Universität der Künste Berlin
Berlin
Germany

Hauke Egermann
Department of Music
University of York
York
UK

Sarah-Indriyati Hardjowirogo
Audio Communication Group
Technische Universität Berlin
Berlin
Germany

Stefan Weinzierl
Audio Communication Group
Technische Universität Berlin
Berlin
Germany

ISBN 978-981-10-2950-9 ISBN 978-981-10-2951-6 (eBook)
DOI 10.1007/978-981-10-2951-6

Library of Congress Control Number: 2016954600

Printed on acid-free paper

This Springer imprint is published by Springer Nature
The registered company is Springer Nature Singapore Pte Ltd.
The registered company address is: 152 Beach Road, #22-06/08 Gateway East, Singapore 189721, Singapore

Contents

Introduction

Till Bovermann, Alberto de Campo, Hauke Egermann, Sarah Hardjowirogo and Stefan Weinzierl

Abstract This book started as a number of notes attached to a wall, with eight people from different academic backgrounds sticking little dots on them. The notes had several keywords written on them, "electronic music", "live performance", "improvisation" and the like. The points were used to vote for a keyword that would set the thematic focus of an upcoming workshop, which was meant to prepare the ground for the work on this book. There was a lively debate on which keyword represented the most promising topic in the context of contemporary musical instruments that would be of interest not only to scholars from diverse academic fields, but also to practitioners both from musical instrument design and artistic practice. Eventually, the winning note was the one that read "*instrumentality*". There had been a lot of discussion around that term beforehand, and it seemed to offer an interesting anchor for a book that was intended to juxtapose a variety of perspectives related to contemporary musical instruments.

Anyone dealing with (the design of, the study of, the performance with, the production of, …) electronic and digital instruments knows that they are fundamentally different from traditional ones in many respects. These differences affect not only

T. Bovermann (✉) · A. de Campo
Institut für zeitbasierte Medien, Universität der Künste Berlin, Berlin, Germany
e-mail: t.bovermann@udk-berlin.de

A. de Campo
e-mail: decampo@udk-berlin.de

H. Egermann
Department of Music, University of York, York, UK
e-mail: hauke.egermann@york.ac.uk

S. Hardjowirogo · S. Weinzierl
Audio Communication Group, Technische Universität Berlin, Berlin, Germany
e-mail: hardjowirogo@tu-berlin.de

S. Weinzierl
e-mail: stefan.weinzierl@tu-berlin.de

T. Bovermann et al. (eds.), *Musical Instruments in the 21st Century*,
DOI 10.1007/978-981-10-2951-6_1

the way we design, build, and play instruments, but they also influence the way we think about them. Along with new instrumental configurations and practices, the identity of the musical instrument has changed, too. In this context, the concept of instrumentality, taken as a specific quality of musical instruments, may serve as a starting point for observations from the most diverse backgrounds that seek to comprehend how exactly our interaction with and our conception of musical instruments has changed through digital technology—and what follows from that. Thinking about what makes a sounding metal box, or a piece of software and a sensor interface a musical instrument and why constitutes an important first step towards understanding the meanings and functions of musical instruments in the 21st century.

In the past three years, we have been dealing with a multitude of questions, challenges, and approaches in the context of contemporary musical instruments within a project on the "Design, Development and Dissemination of New Musical Instruments" (3DMIN, http://www.3dmin.org/). As a collaboration between the Technical University (TU Berlin) and the University of the Arts (UdK Berlin), we approached the topic both from an artistic perspective—creating new musical instruments and performance setups together with concrete musical projects with mixed groups of students (composers, perfomers, engineers)—and from a scientific perspective, investigating different artistic and technical concepts for the development of musical interfaces by performers/composers and the ways these are perceived by the audience.

The collection of articles in this book presents some of the results of the 3DMIN project, and puts them in context with invited articles by artists, designers and musicologists, thus aiming to create a wider discourse on the role of musical instruments in contemporary forms of electronic music and sound art. We divided the contributions into four parts corresponding to different sets of processes in which instrumentality plays a key role; these terms are not intended as categories to keep these aspects separate, but as attractors which pull articles with more direct crosslinks closer to each other. The fact that many articles could fit in more than one section is a clear indication of the degree of interdisciplinarity achieved in current artistic and scientific research on contemporary musical practice.

In part I, **Think Know Reflect, Sarah Hardjowirogo** prepares the playing field by introducing the concept of *instrumentality*, considering the differences, but also the similarities between traditional and electronic musical instruments. By identifying a number of criteria for instrumentality, she develops a theoretical framework for musical instruments that connects contemporary ones with those known for centuries.

A second approach to the concept of instrumentality is undertaken by **Caroline Cance**, who draws on evidence gathered from a linguistic study on the designations of digital musical devices. Her findings suggest that, more than the object itself, it is the action patterns the object is integrated into that decide whether something is referred to as a musical instrument or not.

Bernd Enders, in his pointed summary of the history of music technology, outlines the technological development of the virtual musical instrument from prehistory to the digital age. According to his explanations, only ten stages of development separate the first simple instruments from the latest software devices, yet in their technological complexity and their musical functionality they are worlds apart.

Paul Théberge resorts to Gilles Deleuze's notion of "assemblage" to take account of the increased complexity of contemporary musical instruments, which almost always consist of more than just one component. Such configurations, he argues, become musical instruments only when placed within a particular network of relationships.

The concept of assemblage is also central to the argument put forward by **Deniz Peters**. His fascinating case study discusses the phenomenon of an unexpected additional musical voice emerging from a installatory assemblage which requires being played by three musicians jointly, and his reflections lead to the conclusion that this an example of an instrumentality that is distributed, personal, and self-agential.

Jin Hyun Kim and Uwe Seifert discuss the shortcomings of using historical classification systems for analyzing contemporary music instruments. Trying to identify new potential classification criteria, they present interactivity and agency as important emerging theoretical concepts.

Part II, **Design Make Create,** is opened by **Amelie Hinrichsen** and **Johanna Schindler** reporting on their views on integrating embodied musician-instrument relations into musical instrument design. As product designer respectively ethnographer, they approach the idea of musicality by proposing a design process where musical instrument prototypes are developed inspired by improvisation practices originating in contemporary dance.

While Hinrichsen and Schindler look directly at the design process, **Giuseppe Torre** and **Kristina Andersen** are more interested in the sustaining impression and unfolding of instrumentality. Their article discusses how the act of perceiving a digital object as a musical instrument can be considered as directly proportional to the amount (and quality) of time invested in its development and refinement to suit individual needs rather than generic ones. They support their arguments by a case study on one of the pioneers and developers of digital musical instruments: Michael Waisvisz and his work on The Hands.

Rebecca Fiebrink discusses years of artistic practice and research with machine learning applied to instrument design. Her program Wekinator allows for a highly interactive approach to creating complex mappings from controllers to sound processes. She argues convincingly that such approaches can be creatively very satisfying by making design work less technical and more intuitive, which the wide variety of artists who work with her software confirm.

Thor Magnusson investigates how sound is represented visually in software instruments. Starting out from a knowledgeable historical summary, he describes the challenge of creating instruments in the digital domain and provides a number of examples that vividly illustrate how decisively different strategies of software design shape the identity of the resulting instrument.

Marten Seedorf and **Christof Martin Schultz** present *loop* as a software toolbox meant to introduce techniques and tools for sound field synthesis to beginners already at school level. They highlight the importance of musical instruments, in this case in the form of software instruments, not only for the public performance, but also for a general introduction to the forms of music they are used for, as we can learn from the traditional practices of classical music.

As an instrument builder and lecturer, **Jeff Snyder** looks into the five year adventure of designing, prototyping and constantly rethinking of the Birl, an instrument that morphed from a large, strange electromechanical contraption into a miniature wind controller. His notion on instrumentality involves keywords like ideas emerging from accidents, surprise, intense prototyping, and, last not least, letting professional instrumentalists extensively test and perform with his prototypes.

Constant testing of and performing with an ever-changing instrument is also at the heart of **Hans Tammen**'s work. Over 15 years of development, practice and performance and play, his "Endangered Guitar", a hybrid interactive instrument meant to facilitate live sound processing underwent lots of adaptions and transitions, reflecting on the artist's personal viewpoints towards performance.

In the first article in Part III, **Compose Play Perform, Marije A.J. Baalman** argues that, since electronics and code have become essential parts of current musical practice, boundaries between composition, instrument design and performance are blurring. She questions common concepts like composition, instrument design, and improvisation and investigates questions on the influences of design decisions, new materialities and the maker's skills.

Antye Greie-Ripatti and **Till Bovermann** look at instrumentality from outside: Their concept of **sonic wild{er}ness interventions** takes musical practice away from conventional performance sites. In a kaleidoscope of short text fragments, they define a sonic variant of the notion of "wilderness", they consider practical aspects and artistic consequences of playing music in remote areas, and they discuss implications of this practice on the notion of instrumentality.

Long-time collaborators and friends **Bjørnar Habbestad** and **Jeff Carey** interview each other about what it means to play an instrument in a computer music context, about insights on their respective personal instrument development process, and those stemming from their collaborative work. Their projects were based on evolving software frameworks that led to the initiation of the Modality project, which brought together several authors in this book and 3DMIN associates.

Andreas Pysiewicz and **Stefan Weinzierl** look upon interfaces for sound spatialisation rather than sound generation. Based on an inventory of controllers for the real time spatialisation of sound as part of musical performances, they discuss to what extent these can be considered as musical instruments in light of the theoretical concepts discussed throughout this book.

Comparing B.B. King's play with his guitar *Lucille* and Mari Kimura's play with a robotic instrument called *GuitarBot*, **Philip Auslander** analyses the role of instrumentality and agency in musical performance. He discovers parallels between instrumental and ventriloquist performances and stresses the importance of dramatizing in both genres.

Being trained as a flutist, specializing in New Music, and being active in a multitude of musical domains, **Bjørnar Habbestad** argues that instrumentality is located between the instrument-object and the player-subject, and proposes that transgression plays a central role in defining and expanding instrumentalities in contemporary musical practice.

Hernani Villaseñor Ramírez discusses an interesting 'corner case' of instrumentality: His group LiveCodeNet performs by means of live coding where all code is shared between players by network. What exactly is the instrument being played here, and where is instrumental agency located?

Hildebrand Marques Lopes, Alberto de Campo, and **Hannes Hoelzl** begin the complex account of their **Trio Brachiale** with a "many-festo" of their artistic aims. They propose a new concept, *Second Order Virtuosity*, for describing their notions of contemporary musical practice, which they base on rich conceptual background informing their approach, sources of inspiration they find in other artists' work, and a choice of representative details of their personal and shared artistic practices.

In the fourth and last part, **Listen Perceive Feel**, **Gina Emerson** and **Hauke Egermann** adopt the perspective of the audience by investigating the effect of different mapping strategies and in particular the perceived causality of gesture and sound on the experience of musical instruments.

After that, **Kai Siedenburg** presents an even closer insight into the cognitive mechanisms of auditory instrument recognition and its implications for the design of new musical instruments. While timbre research previously focused on merely sensory phenomena, his contribution examines how familiarity with a musical instrument timbre changes its perception.

Looking into historical archive data, **Song Hui Chon** tries to identify which design parameters of musical instruments predict its success in being used as a solo instrument. Here, she reports that acoustical parameters like higher median pitch, higher salient timbre, and more potential musicians increase the probability that a certain instrument will be used in a solo role in Western classical concertos.

In the last article of this volume, **Dafna Naphtali** discusses the implications and consequences that the invisibility of her instrument, voice and electronics, had on the forms of instrumentality and the communication strategies she chose over the course of her artistic practice.

The editors wish to thank all authors for their diverse, thoughtful and inspiring contributions to this collection. We also wish to thank all artists and scholars who have participated in the 3DMIN project, as well as the Einstein Stiftung Berlin for its generous support of the project. And we wish to thank the Springer Verlag, in particular Mr. Christoph Baumann, for a good and understanding cooperation and for making this publication possible.

Part I
Think Know Reflect

Instrumentality. On the Construction of Instrumental Identity

Sarah-Indriyati Hardjowirogo

Abstract The musical instruments of the 21st century and those of earlier times differ in many respects, be it their appearance, their technical functionality, their playing technique, or their sounds. And as they have changed, so too have our understandings of what a musical instrument is. The lacking precision of the current notion of the instrument and its incompatibility with contemporary instrumental forms are consequences of a technocultural process that raises fundamental questions about the identity of the musical instrument: When (and why) is something a musical instrument—and when (and why) is it not? In order to grasp the slight differences between the yet-to-be-defined instrumental and the assumed other, it seems reasonable to speak of instrumentality when denoting this particular specificity that instruments are supposed to feature. The present contribution seeks to prepare the ground for a reflective discussion on the concept of instrumentality and the underlying theoretical problem by considering not only the differences, but also the similarities between traditional and electronic musical instruments. Using a couple of different approaches to and views on the concept and defining a number of criteria of instrumentality, it eventually yields a picture of musical instruments that connects the contemporary ones with those known for centuries.

1 Introduction

> If a traditional and relatively precise definition of 'instrument' excludes large areas of contemporary musical practice from our field of study, we might be better off with less precise alternatives. (Kvifte 2008, p. 56)

This contribution is based upon work supported by the Einstein Foundation Berlin.

S.-I. Hardjowirogo (✉)
3DMIN Project, Audio Communication Group, Technische Universität Berlin, Berlin, Germany
e-mail: hardjowirogo@tu-berlin.de

T. Bovermann et al. (eds.), *Musical Instruments in the 21st Century*,
DOI 10.1007/978-981-10-2951-6_2

The past century has witnessed a number of technological changes which resulted in far-reaching consequences for all realms of musical practice. In the context of music production, the processes of phono-graphy, electrification, digitalisation, and interconnectedness gave rise to a huge number of new musical instruments which differ significantly from those known previously. The usage of recorded sounds, the synthesis of sounds that are physically irreproducible, playing instruments that are purely virtual, or having instruments communicate among each other via a network, identify milestones in the history of musical instruments. Yet, at the same time, they blur the boundaries between something we are prone to call 'instrument' and other categories such as 'medium', 'system', 'configuration', 'machine'.

Many contemporary sound producing devices, and in particular those that consist of a whole set of different functional parts, some of which may be software or based on other kinds of media technology, raise the question—to spectators as well as to organologists and other theoreticians, and maybe even to some musicians themselves—of whether they are (still or already) musical instruments. Complaints about the rather boring appearance of laptop performers, for instance, are known well enough, and frequently they are combined with this admittedly simple, but by no means trivial question. What, then, *are* musical instruments in the 21st century, and how can they be recognised as such? What do they have in common with instruments such as a violin, a piano, or a trumpet, and what are the differences between them? What is their relation to other sound producing devices? What defines contemporary musical instruments as musical instruments?

In order to grasp that specific quality musical instruments are assumed to feature as distinguished from other sound producing devices (or 'non-instruments' in general) the concept of instrumentality has been used ever more frequently over the past couple of years. The present contribution explores the usage of this concept in some pertinent works, discusses its use for the study of contemporary musical instruments and works out a number of criteria that appear to be crucial for the construction of instrumental identity.

2 Musical Instruments versus Other Things

Any project that involves introducing a new concept or sharpening an introduced one must start out with one question: Why? If the term instrumentality shall denote something like the 'essence of the musical instrument', as that which defines a musical instrument as such, it will first have to be explained why such a concept should be needed. Why should it not be sufficient to define a musical instrument as, say, "*any object that produces sound*", just like several (musicological and general) encyclopaedias do, following Hornbostel's statement that "[f]or purposes of research everything must count as a musical instrument with which sound can be produced intentionally" (Hornbostel 1933, p. 129)?

There are (at least) two answers to this question that are both surprisingly simple. The first answer is: *Musical instruments are not the only things that are used to produce sound.* On the contrary, our world today is full of things that are used to produce sound but are no musical instruments. An iPod, for example, is something that is undoubtedly used to produce sound but normally isn't referred to as a musical instrument but rather as a playback device. A violin, however, is something that at least people familiar with Western music culture immediately recognise as an item belonging to the class of musical instruments.[1]

But it is not always equally easy to tell whether something is a musical instrument or not. Consider, for instance, the cases of other sound media, such as turntables, radios, or tape machines. If it were only about them being used to produce sound, then all of them would clearly be musical instruments. But then, how are we to categorise smartphones, tablets, and laptops? And finally, what about objects like saws, combs, and oil drums? All of these things are or have been used more or less often to produce sound in a musical context. However, all of them have originally been designed with another purpose: While the mentioned sound media have the original purpose of playing back previously recorded or receiving broadcast sound, smartphones, tablets, and laptops have multiple purposes and can, among other things, also be used to produce sound, and everyday devices such as saws, combs, and oil drums have an original purpose that has nothing to do with sound at all.

What can be immediately learned from these examples is that instrumentality, or simply being a musical instrument must not be understood as a property an object as such has or has not. Rather, it seems to result from using something in a particular way which we think of as instrumental. Consequently, an object is not per se a musical instrument (ontological definition) but it becomes a musical instrument by using it as such (utilitarian definition).

But there is something else that can be learned from these examples—and this is where we get to the second answer: *Musical instruments are more than only sound-producing devices.* As the above examples should have made clear, there are some objects we immediately recognise as musical instruments, while we can surely say of others that they are no musical instruments and of yet others that they are used as musical instruments more or less regularly. This means that we are able to order all of those objects according to their 'degree of instrumentality', and this is to say that, apparently, there are some objects that, to us, are more 'instrumental' than others. Why is that so?

One could assume that it might have to do with the different purposes these objects have been designed for and that, for instance, we recognise the violin immediately as a musical instrument because it has never been anything else than that for centuries, while a saw might be used for sawing much more often than for

[1]Interestingly, this clear distinction between musical instruments and playback devices is relatively new: as recently as in the 1930s, gramophones and phonographs, the playback devices of the time, were referred to as musical instruments (cf., e.g., Straebel 1996, p. 219).

making music. However, there are a lot of examples which suggest that the original purpose of the objects is only partly relevant in this regard. The original purpose of a radio, for instance, could perhaps best be described as receiving and displaying broadcast audio signals. But in the context of a composition like John Cage's *Imaginary Landscape No. 1* this very purpose is being used for another purpose, namely that to make music. As will be described later, the *intention* with which an object is used is undoubtedly something that plays a major role in the construction of instrumentality. But its *purpose* is something that is dependent on the intention of the person using it, and thus it is situational.

What is probably equally important, here, is the fact that some of these objects have undergone a long process of culturalisation as musical instruments, while others have not (yet). Culturalisation in this regard means that they have been used for the purpose of making (a more or less specific kind of) music regularly and for a long time in the context of a particular culture. The importance of this aspect becomes even clearer when considering some other examples. Most electronic instruments, for instance, are objects that are designed for the single purpose of being used as musical instruments. Still, probably only few people will recognise The Hands (Fig. 1, above) as a musical instrument (even if the earliest version of it is more than 30 years old) because it has never been sold commercially and for that reason hasn't been used widely.

The other two images show examples for instruments that are very common in other parts of the world but scarcely known to most people in Europe. They don't have any other purpose than being used to make music, still to many Europeans they could probably be just about anything: they don't *mean* anything to them because they are not *culturalised* as musical instruments in the cultural context that they are familiar with.

Subsuming, the answer to the question of why a concept denoting what defines a musical instrument as such should be needed is that the traditional definition of musical instruments as sound-producing devices is not sufficient any more—first, because musical instruments are not the only things that are used to produce sound and second, because they are more than only sound-producing devices. And this is to say that it is not at all easy to define what a musical instrument essentially is and that, in order to do so, we need to be able to tell what the difference between musical instruments and other sound-producing devices is.

3 Musical Instruments and Musical Instrument Concepts

This specificity of musical instruments as distinguished from other sound-producing devices is expressed by the concept of instrumentality, which, as the above considerations suggest, seems to be a graduable and dynamic concept that is not tied to an object per se but is rather a matter of cultural negotiation. Yet, another important question remains to be answered: Why should we *want* to define what a musical instrument is? This entirely legitimate question is often

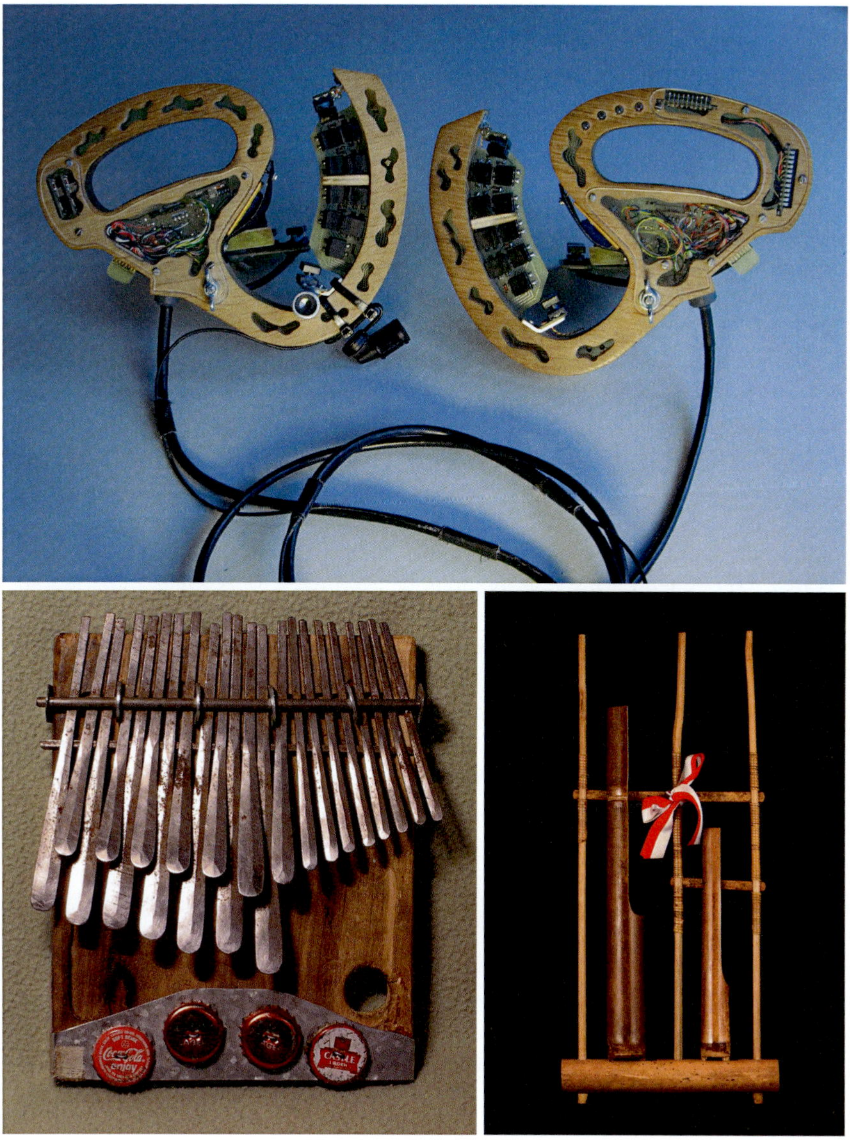

Fig. 1 Whether something is recognised as a musical instrument or not is not least a matter of culturalisation. The Hands (*above*) by Michel Waisvisz are a prominent example for an early gestural controller. The Mbira (*left*), widely distributed in Africa, is played by plucking its tines with the thumbs. Angklungs (*right*) are single-pitch instruments made from bamboo that are used in the context of traditional music in Indonesia

accompanied by the comment: *A musical instrument doesn't become one by calling it an instrument but by using it as such.* But what, then, does it *mean* to use something as a musical instrument? What are the actions typically associated with musical instruments? And what, other than that, constitutes a musical instrument as such?

Answering these questions may contribute to a better understanding of contemporary musical practice in general and of the way technocultural processes like electrification, digitalisation, virtualisation and the like have influenced the design and use of musical instruments in our culture. The repeated questions of whether something is a musical instrument or not indicate that fundamental cultural concepts are in transition—once again. Taking a look at the many different musical instrument concepts to be found in the recent literature, this becomes all the more visible.

It doesn't take much effort to find as many as six different musical instrument concepts already in a small selection of sources, which particularly show the degree of disagreement on the precise extension of the notion of musical instrument. Roughly a century ago, von Hornbostel and Sachs (1914) have established the traditional organological definition of musical instruments as *sound generators*. Recently, Harenberg (2012) has applied that concept to virtual instruments and claims that, consequently, in configurations of a *software sound generator* and a hardware controller interface only the former one is the instrument. Bense (2012), in contrast, argues that, in virtual instruments, it is the *interface* that is equivalent with the instrument. This view is also supported by the title of the NIME (New Interfaces for Musical Expression) conference, which deals with topics centred on digital musical instruments. A common definition of digital musical instruments conceives them in accordance with Malloch et al. (2006) as *tripartite systems* consisting of a sound generator, a control interface, and the mapping that defines how one is connected to the other. Enders (1987) has described musical instruments as *quadripartite systems* consisting of discrete modules for the generation, control, modification, and storage of sound and explicitly includes automatically controlled systems. Accordingly, Großmann (2010) discusses the status of *reproduction media* as musical instruments.

This list is, of course, only exemplary, but it illustrates quite appropriately why re-negotiating the concept of musical instrument should matter: There is anything but a consensus on what a musical instrument actually is, and the situation gets particularly complicated when it comes to contemporary instruments. Consequently, a definition referring to both traditional and electronic or digital musical instruments is yet to be made.[2]

In this context, a concept that is able to capture the common essentials of musical instruments could be of use. And that is where the notion of instrumentality comes into play.

[2]Tellef Kvifte's 1989 book has made a promising attempt in this regard, but in the meantime the situation has changed significantly through the advent of digital musical instruments (Kvifte 1989).

4 Previous Approaches to Instrumentality

In his 1987 article entitled "Instrumentalities", Burrows (1987) discusses the relation between the musical instrument and its player, thereby addressing some of what he considers characteristic, if not defining features of the musical instrument. Instead of denoting a specific instrumental quality, the concept of instrumentality, here, is used to describe the purpose of musical instruments, yet his ideas reflect exactly what the present reading of the term is supposed to stand for: the elements that constitute a musical instrument.

The most important feature of musical instruments, in his opinion, is their role as *mediators* between the performer's body and the sound they produce, or, between the inside and the outside of the human body (ibid., p. 117). He is interested in the transitions between the physicality of the human body and the sounding body of the instrument on the one hand and the volatility of the realms of sound and music on the other and seeks to apply the concept of "transitional object" to musical instruments (p. 120ff).

Burrows' understanding of instrumentality (in the sense intended here) is very clearly dominated by the function of musical instruments as mediators between apparently contrary realms, namely between corporeality and fluidity, between the inside and the outside and between the material and the immaterial.

To him, musical instruments are both part of the human body and external to it, they are literally means of physical expression, and this exactly is what Burrows regards as their purpose or—in his sense of the term—instrumentality.

His considerations are taken up by Philip Auslander[3] who opposes to Burrows' idea of the instrument having "its own agency with which the musician must negotiate" the image of the ventriloquist's dummy which, similar to musical instruments, needs to be acted upon by a human in order to make a sound, while the illusion of it having some kind of agency of its own is crucial to the performance. Following an argument by Godlovitch (1998), he stresses the importance of the specific circumstances under which instrumental sounds are produced: other than the mere production of particular sounds, he claims, instrumental performance involves techniques of producing them that are supposed to appear difficult to outside observers. This relates to the popular idea that *effort* be a key feature of instrumentality—perceived effort, that is, not actual effort, as Auslander concedes.

Apparently, then, instrumentality is not so much a matter of actual playing skills, but rather of the demonstration or, as he puts it, "dramatizing" of such skills.

The subjects of effort and of demonstrated instrumentality are also present in John Croft's 2007 paper "Theses on Liveness" (Croft 2007), in which a number of "conditions of instrumentality" are defined—conditions that must be fulfilled so that an audience would recognise a given setup of live electronics as an instrument. These conditions can be roughly summarised as the claim that the relationship between a performer's actions and the resulting sound be as transparent as possible

[3]Cf. his contribution reprinted in the present volume.

to the audience. Interestingly, then, Croft identifies the perception of the system as an instrument with the perception of the performance as live. This sort of perceived liveness is closely related to what Auslander describes as the perceived difficulty of an instrumental performance. For this reason, Croft argues, it should be of interest to any musician to achieve the highest possible degree of instrumentality, and thus of liveness—because "there is nothing inherently interesting about the fact that a computer can generate a sound in response to a person's action; this is why the triggering of sounds using sensors is often dull – or, at best, merely interesting" (ibid., p. 65).

His main argument is that instruments are simply more interesting to watch for the audience and that that should be motivation enough for musicians to optimise the instrumentality of their live-electronics setup.

Apart from the transparency of the relationship between playing action and resulting sound, Croft emphasizes the importance of physical effort and expressivity, of a "unified expressive persona normally associated with a solo performance" (p. 63), for his concept of instrumentality, which he sees threatened by the disembodied sounds coming out of a loudspeaker.

Although his sound-aesthetical ideal might be a very specific one, his approach provides a good example for the prominent role that is assigned to effort in the context of instrumentality concepts.

In contrast, the argument made by Philip Alperson in his 2007 paper "The Instrumentality of Music" (Alperson 2007) takes an entirely different direction. Even though he doesn't use the term instrumentality with regard to instruments, but rather to music itself, he still makes some interesting points concerning his concept of musical instrument and thereby contributes to the present definition of the term. He starts by defining what he calls the commonsense view of musical instruments:

> Typically, we think of instruments as discrete, self subsisting material objects, intentionally crafted for the purpose of making music by performing musicians. (ibid., p. 38)

Discussing what role the aspect of intention or purpose plays for instrumentality, he finds that there are numerous examples for instruments that haven't been designed as musical instruments originally but still involve some kind of human intention, namely "the intention to use the object as a musical instrument." (ibid.)

Another characteristic of musical instruments Alperson questions is their being external to the human body. Many musicians are so intimately tied to their instruments, he argues, that "it is difficult to know where the instrument ends and the rest of the body begins" (p. 46). Finally, he stresses the importance of what he calls the immaterial features of musical instruments. Being "musically, culturally, and conceptually situated" (p. 42), they cannot be fully understood if they are reduced to mere material objects—as it is usually done in traditional organology. Instead, they have to be studied in the context of their cultural and historical embeddedness.

In a recent study on the instrumentality of "new digital musical devices", Cance et al. (2013) have combined a linguistic analysis of the concept of musical instrument with an interview study, in which a number of experts had to give their personal definitions of musical instruments. The authors summarise their findings in the statement that "it appears that "*instrument*" does not actually refer to a device [...] but rather qualifies its interaction with users [...]" (ibid., p. 297). In their opinion, instrumentality is not so much dependent on the properties of a device itself, but rather on the actions and meanings it is embedded in. This view again turns the focus away from the instrument as a material object and upon what Alperson calls its immaterial features.

5 Criteria of Instrumentality: A Preliminary Inventory

At the beginning of this chapter, instrumentality has roughly been defined as 'that which defines a musical instrument as such', as 'the essence of the musical instrument', and as a 'specific instrumental quality'. More precisely, it denotes the potential for things to be used as musical instruments or, yet differently, their instrumental potential as such. Instrumentality in this sense represents a complex, culturally and temporally shaped structure of actions, knowledge, and meaning associated with things that can be used to produce sound. However, as also suggested by the findings of Cance et al., the term must not be understood as denoting a property an object per se has or has not, but it is rather intended as a means of capturing the instrumental potential of a given artefact. Also, it must not be conceived as a constant, but rather a graduable, dynamic term which means that an object may be more or less instrumental, according to its expression of the characteristics associated with instrumentality.

A brief analysis of the above-presented works may serve as a starting point for the identification of those characteristics or criteria that are crucial for the construction of instrumentality. The following list represents only a first, rough approximation to those cornerstones of instrumentality. However, the cited references show that there are numerous examples in contemporary musical practice and current research suggesting that the mentioned criteria do actually matter for the construction of an instrument's identity. Those criteria that appear repeatedly and thus presumably play a major role are the following:

1. **Sound Production**

Obviously, musical instruments necessarily have to be able to produce sound in some way. This criterion represents the traditional musicological notion of instrument originating from von Hornbostel and Sachs (1914) and is—quite reasonably—neither questioned nor emphasized by any of the mentioned works. In digital musical instruments, however, the instrument's sound is not an immediate result of the sonic characteristics of a material object anymore, as is the case with traditional instruments. This means that, in the design of novel digital instruments, the instrument's sonic identity and its physical appearance have to be designed independently from each other. While there is quite a large amount of works dealing with physical interface design, most prominently represented by the NIME community,[4] relatively little attention has been given so far to questions regarding the sound design of digital musical instruments.

2. **Intention/Purpose**

As already suggested by the original meaning of the Latin *instrumentum* ('device' or 'tool'), intention and purpose are quite decisive features for the construction of instrumentality in that playing a musical instrument always requires both the intention to do so and the purposeful use of something (that may also have a different original purpose) as a musical instrument. This criterion is particularly mentioned by Alperson, who addresses the relevance of intention for the process of instrument building.

Furthermore, as McCaleb (2014, p. 83) points out, it is also important on a performative level: with regard to ensemble performance, he states that "performers' musical intentions influence, to varying degrees, the way they [...] operate their instruments. In performance, there is a correlation between intention [...] and action [...]." This correlation becomes particularly obvious when considering instrumental borderline cases such as the turntable, that allow for both an instrumental and a non-instrumental use. Here, it is primarily the performer's intention that makes the difference between the two.[5]

3. **Learnability/Virtuosity**

Both learnability and virtuosity involve the opportunity to improve one's playing skills through exercise. In a broader sense this means that the higher the impact of practising an instrument, the higher its degree of instrumentality. The idea of developing specific instrumental techniques over time is also congruent with the idea expressed by Auslander that, at least in professional instrumental performance, playing an instrument should appear more difficult than pressing a play button.

Such a demonstration of playing skills can directly be connected to Cohen's (2008, p. 58) idea of virtuosity, which he defines as "the exhibition of something difficult done without apparent effort." Monteiro (2007, p. 316) takes it even further and declares, "[v]irtuosity also means the possibility to bypass some kind of impossibility [...], to go beyond reality, to cheat triviality."

This moment of bypassing the impossible is, according to Hegel (1975, p. 958), the very moment the instrument comes to life: "In this sort of execution we enjoy the topmost peak of musical vitality, the wonderful secret of an external tool's becoming a perfectly animated instrument [...]."

In order to make this happen, however, the instrument has to be learned first. In a paper investigating possible reasons for the success or failure of newly designed digital instruments, Jordà (2004) has identified learnability and playability, but also

[4]A good overview is given in Marshall (2008).

[5]Such cases also exemplify the relevance of a specific instrumental sort of acting on something in order to make it an instrument.

effort, as determining an instrument's "efficiency"—a term he uses to express the correlation between the time needed to learn an instrument and the acquired playing skills.

But especially when it comes to digital instruments, the learning process can be challenging and quite different from that known from traditional instruments. This is, on the one hand, due to the fact that electronic instruments usually lack visual or haptic feedback, which makes them more difficult to play. On the other hand, their learning procedures and playing techniques are not yet standardized and often must be developed first.

Several efforts have been made to facilitate the learning of such instruments by means of technical innovation. Jordà (2003), for instance, has shown how the implementation of visual feedback can improve the learnability of an interactive music system. Merrill and Paradiso (2005) have gone further by transferring part of the learning task to the instrument itself, so as to teach the instrument the desired mappings by example. However, novel instruments are still inadequately integrated into institutional music education, although students, teachers, and instrument designers would probably benefit alike.

4. **Playability/Control/Immediacy/Agency/Interaction**

Although these are actually five quite different notions, they share some common features that are mentioned both by Auslander and Croft as well as by Cance et al. The requirement that a musical instrument be playable may be seen as a somewhat broader expression for the traditional idea of the instrumentalist controlling the instrument. Both playing and controlling an instrument involve immediacy regarding the connection between the instrumentalist's actions and the instrument's sound, but they differ in the degree of agency they ascribe to the instrument. In this regard, interaction can be understood as a concept of instrumental play that ascribes as much agency to the instrument as it does to the performer.

The question of how electronic and digital musical instruments should best be controlled has been—and still is—one of the key issues in the pertaining academic discourses for quite a while now. As early as 1991, Joel Ryan from the Studio for Electro-Instrumental Music (STEIM) in Amsterdam has problematised the "mediating distance which confronts each composer when encountering the computer" and proclaimed a "quest for immediacy in music" (Ryan 1991, p. 3) for both aesthetical and practical reasons. A few years later, Levitin et al. (2002) were among many who, in a similar way, expressed their displeasure with the persistent dominance of the keyboard metaphor in electronic musical instruments.

There was not exactly a lack of suggestions of how to solve the interface problem—but many things had to be considered that were unknown in traditional lutherie.

The interaction with musical instruments had to be thought anew, lessons had to be learned from HCI (Holland et al. 2013), from the other performative arts (cf., e.g., Benford 2010), as well as from the newly formed discipline of interaction design (e.g., Franinovic and Visell 2007).

Although playability did (and does) play a role, for instance, in violin making in the sense of how to improve responsivity and ease of play through particular constructional measures, in the context of HCI, playability issues gain a whole new meaning. Thus, for example, in a playability evaluation of a virtual bowed string instrument (a virtual violin interface, that is), playability means that “the acoustical analysis of the waveforms produced by the model fall within the region of the multidimensional space given by the parameters of the model” (Young and Serafin 2003). Analogously, numerous PM-modeled instruments have been subjected to playability evaluations in order to allow for a latency-free, reliable, and authentic-sounding real-time play (cf., e.g., Vergez and Tisserand 2006).

Ever since Joel Ryan’s call, there have been innumerable approaches to establish alternate forms of control in musical instruments, ranging from gestural and biosignal control (see Miranda and Wanderley 2006 for an overview) over feedback control (e.g., Berdahl et al. 2012) to shared control (e.g., Gurevich 2014), where part of the control is transferred to the instrument itself.

In this context, the idea of musical instruments having their own agency (Bates 2012) has become a popular and much-discussed topic, in artistic programs (Jenkinson 2004; de Campo 2014) as well as in theoretical discourses (Kim 2007; Magnusson 2009).[6]

5. **Expressivity/Effort/Corporeality**

These three, too, represent fairly different concepts all of which, however, address the physical aspect of instrumental performance. The claim that playing an instrument require physical action or even effort, mentioned by Burrows, Auslander, Croft, and Alperson, goes back to the romanticistic idea of the (both physically and aesthetically) expressive play of the virtuoso and is becoming ever more popular again in the context of contemporary instrument building.

The variety of current works investigating the function and meaning of gestures, tactility, ergonomics and the like in the context of musical instruments (e.g., Wanderley and Battier 2000; Godøy and Leman 2010) shows how the physical aspects of instrumental practice are being brought back to the fore after having been ignored in the study of musical instruments for quite some time.

That a musical instrument is a means of musical expression and should therefore enable an expressive play is largely uncontested. Therefore, one of the main goals of contemporary musical instrument design is to find ways of creating instruments that inherently allow expressivity, for instance, by means of transparent mappings (e.g., Fels et al. 2002). However, as Malloch et al. (2006) note, “[e]xpressivity is commonly used to discuss the virtue of an interaction design in absolute terms, yet expressive interfaces rely on the goals of the user and the context of output perception to generate information.” This problem is also addressed by Arfib et al. (2005), who explore how expressiveness can be obtained by performing specific gestures.

[6]See also Philip Auslander’s contribution in this volume.

The idea of expressivity is often mentioned in connection with effort, a pairing that is known from traditional instruments. This appears to be in contrast to the effortless play of electronic instruments, as D'Escriván (2006, p. 183) notes, not without stating, however, that "a young generation seems content to accept that there may be no apparent correlation between input effort and sound output."

6. **"Immaterial Features"/Cultural Embeddedness**

The cultural embeddedness of an instrument or its "immaterial features" are particularly emphasized by Alperson and Cance et al. Whereas Alperson stresses the significance of an established instrument's cultural situatedness for its instrumental status, Cance et al. especially refer to the importance for a new instrument to take up existing aesthetical practices.

In a similar way, Dawe (2003, p. 274) has pointed out that the "value and meaning [of musical instruments is] negotiated and contested in a variety of cultural arenas" and that, apart from studying its physical functionality and its location in the organological system, an instrument's identity cannot be fully understood without studying the cultural contexts in which it is embedded. Despite this being a key issue in ethnomusicology, as is impressively demonstrated, e.g., by Kartomi (1990), still it is all too often forgotten in the study of contemporary musical instruments.

7. **Audience Perception/Liveness**

Meeting the audience's expectations, be it with regard to the difficulty of the performance, its liveness or its expressiveness, seems to be a criterion that should not be underestimated. Following the arguments of both Auslander and Croft, instrumentality in the sense of a category that legitimates instrumental performance is highly dependent on audience perception.

To date, the role of audience perception has not received much attention in the study of contemporary instrumental performance. Only recently have some works addressed the connection between audience perception and the evaluation of novel digital instruments (e.g., Barbosa et al. 2012; Brown et al. 2013). Following this idea, Gina Emerson's contribution to the present volume illustrates how much transparent mappings matter for the audience's perception of instrumentality.

As also stated by Croft, the perception of instrumentality is directly connected with the perception of liveness. Ever since Philip Auslander's 1999 book on liveness (Auslander 1999), the term has become increasingly popular and still inspires a significant amount of works in the field. Lately, there has been a number of attempts to capture the perceived liveness of digital musical instruments, e.g. Marshall et al. (2012), Bown et al. (2014) and Berthaut et al. (2015).

6 Conclusion

The aim of this chapter was to introduce the concept of instrumentality, discuss its use for the study of contemporary instruments, and define a number of criteria that, based on a literature review, appear to be crucial for the construction of instrumentality.

As such, they also identify those fields of research that scholars and designers need to pay particular attention to when studying and creating electronic and digital musical instruments that are not only technically appealing, but also artistically versatile, culturally meaningful and visually intriguing artefacts.

The study of contemporary instruments confronts us with a number of fundamental issues regarding the way of how instrumental identity is being constructed that cannot be answered without taking into account that musical instruments are a lot more than just arbitrary objects that produce sound. They are complex, culturally freighted artefacts allowing for particular ways of interaction that result in particular sounds. Their identity as musical instruments—their instrumentality—is constructed in the interplay of various criteria, among the most relevant of which seem to be those mentioned above. If the underlying principles of this interplay were better understood, they could inform the design process of new musical instruments and thus contribute to the development of instruments with a characteristic and coherent identity. But above all, they would provide general insights about how processes of culturalisation work: how arbitrary objects turn into meaningful things with a well-determined function—such as, for example, musical instruments.

References

Alperson, P. (2007). The instrumentality of music. *Journal of Aesthetics and Art Criticism, 66*, 37–51.

Arfib, D., Couturier, J.-M., & Kessous, L. (2005). Expressiveness and digital musical instrument design. *Journal of New Music Research, 34*(1), 125–136.

Auslander, P. (1999). *Liveness. Performance in a mediatized culture.* London, New York: Routledge.

Barbosa, J., Calegario, F., Teichrieb, V., Ramalho, G., & McGlynn, P. (2012). Considering audience's view towards an evaluation methodology for digital musical instruments. In *Proceedings of the New Interfaces for Musical Expression (NIME) 2012.*

Bates, E. (2012). The social life of musical instruments. *Ethnomusicology, 56*(3), 363–395.

Benford, S. (2010). Performing musical interaction: Lessons from the study of extended theatrical performances. *Computer Music Journal, 34*(4), 49–61.

Bense, A. (2012). *Musik und Virtualität. Digitale Virtualität im Kontext computerbasierter Musikproduktion.* Osnabrück: epOs.

Berdahl, E., Smith, J. O., III, & Niemeyer, G. (2012). Feedback control of acoustic musical instruments. *Journal of the Acoustic Society of America, 131*(1), 963–973.

Berthaut, F., Coyle, D., Moore, J., & Limerick, H. (2015). Liveness through the lens of agency and causality. In *Proceedings of the New Interfaces for Musical Expression (NIME) 2015.*

Bown, O., Bell, R., & Parkinson, A. (2014). Examining the perception of liveness and activity in laptop music: Listeners' inference about what the performer is doing from the audio alone. In *Proceedings of the New Interfaces for Musical Expression (NIME) 2014.*

Brown, A. R., Gifford, T., & Voltz, B. (2013). Factors affecting audience perceptions of agency in human computer musical partnerships. In *Proceedings of the 9th ACM Conference on Creativity & Cognition.*

Burrows, D. (1987). Instrumentalities. *Journal of Musicology, 5*(1), 117–125.

Cance, C., Genevois, H., & Dubois, D. (2013). What is instrumentality in new digital devices? A contribution from cognitive linguistics & psychology. In M. Castellengo & H. Genevois (Eds.), *La Musique et ses instruments* (pp. 283–297). Paris: Delatour.

Cohen, T. (2008). *Thinking of others: On the talent for metaphor.* Princeton: Princeton University Press.

Croft, J. (2007). Theses on liveness. *Organised Sound, 12*(1), 59–66.

Dawe, K. (2003). The cultural study of musical instruments. In M. Clayton, T. Herbert, & R. Middleton (Eds.), *The cultural study of music: A critical introduction* (pp. 274–283). New York, London: Routledge.

D'Escriván, J. (2006). To sing the body electric: Instruments and effort in the performance of electronic music. *Contemporary Music Review, 25*(1–2), 183–191.

de Campo, A. (2014). Lose control, gain influence—Concepts for metacontrol. In A. Georgaki & G. Kouroupetroglou (Eds.), *Proceedings of ICMC/SMC 2014* (pp. 217–222). Athens: National and Kapodistrian University of Athens.

Enders, B. (1987). Instrumentenkunde – Form, Funktion und Definition des Musikinstruments im Spannungsfeld zwischen Musik und Technik. In A. Edler, S. Helms, & H. Hopf (Eds.), *Musikpädagogik und Musikwissenschaft* (pp. 306–345). Wilhelmshaven: Noetzel.

Fels, S., Gadd, A., & Mulder, A. (2002). Mapping transparency through metaphor: Towards more expressive musical instruments. *Organised Sound, 7*(2), 109–126.

Franinovic, K., & Visell, Y. (2007). New musical interfaces in context: Sonic interaction design in the urban setting. In *Proceedings of the New Interfaces for Musical Expression (NIME) 2007.*

Godlovitch, S. (1998). *Musical performance: A philosophical study.* London: Routledge.

Godøy, R. I., & Leman, M. (Eds.). (2010). *Musical gesture: Sound, movement, and meaning.* New York, Milton Park: Routledge.

Großmann, R. (2010). Distanzierte Verhältnisse? Zur Musikinstrumentalisierung der Reproduktionsmedien. In M. Harenberg & D. Weissberg (Eds.), *Klang (ohne) Körper. Spuren und Potenziale des Körpers in der elektronischen Musik* (pp. 183–200). Bielefeld: transcript.

Gurevich, M. (2014). Distributed control in a mechatronic musical instrument. In *Proceedings of the New Interfaces for Musical Expression (NIME) 2014.*

Harenberg, M. (2012). *Virtuelle Instrumente im akustischen Cyberspace. Zur musikalischen Ästhetik des digitalen Zeitalters.* Bielefeld: transcript.

Hegel, G. W. F. (1975). *Aesthetics. Lectures on fine art* (trans: Knox, T. M.). Oxford: Clarendon Press.

Holland, S., Wilkie, K., Mulholland, P., & Seago, A. (Eds.). (2013). *Music and human-computer interaction.* London: Springer.

Jenkinson, T. (2004). Collaborating with machines. *Flux Magazine, 3*(2004).

Jordà, S. (2003). Interactive music systems for everyone: Exploring visual feedback as a way for creating more intuitive, efficient and learnable instruments. In *Proceedings of the Stockholm Music Acoustic Conference 2003.*

Jordà, S. (2004). Digital instruments and players: Part I—Efficiency and apprenticeship. In *Proceedings of the new interfaces for musical expression (NIME) 2004.*

Kartomi, M. J. (1990). *On concepts and classifications of musical instruments.* Chicago, London: The University of Chicago Press.

Kim, J. H. (2007). Toward embodied musical machines. In C. Lischka & A. Sick (Eds.), *Machines as agency. Artistic perspectives* (pp. 18–35). Bielefeld: Transcript.

Kvifte, T. (1989). *Instruments and the electronic age. Towards a unified description of playing technique*. Solum Forlag: Oslo.

Kvifte, T. (2008). What is a musical instrument? *Svensk Tidskrift för Musikforskning, 1*(2008), 45–56.

Levitin, D. J., McAdams, S., & Adams, R. L. (2002). Control parameters for musical instruments: A foundation for new mappings of gesture to sound. *Organised Sound, 7*(2), 171–189.

Magnusson, T. (2009). Of epistemic tools: Musical instruments as cognitive extensions. *Organised Sound, 14*(2), 168–176.

Malloch, J., Birnbaum, D., Sinyor, E., & Wanderley, M. M. (2006). Towards a new conceptual framework for digital musical instruments. In *Proceedings of the 9th International Conference on Digital Audio Effects (DAFx) 2006*.

Marshall, M. T. (2008). *Physical interface design for digital musical instruments*. Ph.D. thesis. McGill University, Montreal, QC, Canada.

Marshall, M. T., Fraser, M., Bennett, P., & Subramaniam, S. (2012). Emotional response as a measure of liveness in new musical instrument performance. In *Proceedings of the Conference on Human Factors in Computing Systems (CHI) 2012*.

McCaleb, J. M. (2014). *Embodied knowledge in ensemble performance*. Farnham, Burlington: Ashgate.

Merrill, D. J., & Paradiso, J. A. (2005). Personalization, expressivity, and learnability of an implicit mapping strategy for physical interfaces. In *Proceedings of the Conference on Human Factors in Computing Systems (CHI) 2005*.

Miranda, E. R., & Wanderley, M. M. (2006). *New digital musical instruments: Control and interaction beyond the keyboard*. Middleton, WI: A-R Editions.

Monteiro, F. (2007). Virtuosity: Some (quasi phenomenological) thoughts. In *International Symposium on Performance Science*, 315–320.

Ryan, J. (1991). Some remarks on musical instrument design at STEIM. In P. Nelson & S. Montague (Eds.), *New instruments for the performance of electronic music. Live electronics. Contemporary Music Review*, *6*(1), 3–18. Reading: Harwood Academic Publishers.

Straebel, V. (1996). Klangraum und Klanginstallation. Klanginstallation zwischen elek-troakustischer Technik, Performance und Skulptur. In Akademie der Künste Berlin/Helga de la Motte-Haber (Eds.), *Klangkunst: erschienen anlässlich von Sonambiente, Festival für Hören und Sehen, Internationale Klangkunst im Rahmen der 300 Jahr-Feier der Akademie der Künste*, August 9–September 8, 1996 (pp. 219–221). München, New York: Prestel.

Vergez, C., & Tisserand, P. (2006). The Brass project, from physical models to virtual musical instruments: Playability issues. In: R. Kronland-Martinet, T. Voinier, & S. Ystad (Eds.), *Computer music modeling and retrieval* (pp. 24–33). Berlin, Heidelberg: Springer.

von Hornbostel, E. M. (1933). The ethnology of African sound-instruments. In C. Sachs (Ed.), *Comments on 'Geist und Werden der Musikinstrumente'. Africa, 6*(2), 129–157.

von Hornbostel, E. M., & Sachs, C. (1914). Systematik der Musikinstrumente. *Ein Versuch. Zeitschrift für Ethnologie, 46*(4–5), 553–590.

Wanderley, M. M., & Battier, M. (Eds.). (2000). *Trends in gestural control of music*. Paris: IRCAM.

Young, D., & Serafin, S. (2003). Playability evaluation of a virtual bowed string instrument. In *Proceedings of the New Interfaces for Musical Expression (NIME) 2003*.

From Musical Instruments as Ontological Entities to Instrumental Quality: A Linguistic Exploration of Musical Instrumentality in the Digital Era

Caroline Cance

Abstract The development of electricity, sound technology, electronics and computer science during the last 150 years has allowed the emergence of new kinds of musical devices. This paradigm shift from traditional to digital instruments has strong consequences for instrument identity and for the relationship between the musician and her/his instrument. Grounded in a situated cognitive linguistics perspective, this contribution first explores various definitions of the instrument (from general dictionaries and musicology literature) before analysing how members of the computer music community name and define their *instrument/interface/device*, etc. Analysing the different strategies of instrument naming used by designers and users of digital instruments and by authors in computer music literature allows us to study the on-going construction and negotiation of a new terminology. By highlighting the instability, the fuzziness but also the diversity of what an instrument is to these different speakers, these analyses contribute to a better understanding of the conditions of instrumentality in the digital era. More than just referring to a device, the notion of *instrument* rather qualifies the interaction with the users, thus allowing a new shift from the instrument as an ontological entity to an instrumental quality.

1 Introduction

For the last 150 years the technological development of electricity, sound recording and reproduction, electronics and computer science has allowed the emergence of new kinds of musical devices. Among others, Cadoz (1999) has extensively studied the strong consequences that electric and digital decoupling has for musician–instrument interactions compared to what happens with traditional instruments. This paradigm shift from mechanical to digital instruments raises new questions

C. Cance (✉)
Laboratoire Ligérien de Linguistique (LLL) – UMR 7270,
Université d'Orléans, Orléans, France
e-mail: caroline.cance@univ-orleans.fr

T. Bovermann et al. (eds.), *Musical Instruments in the 21st Century*,
DOI 10.1007/978-981-10-2951-6_3

about the identity of instruments: What is an instrument? What makes an instrument an instrument? How does an interface get its "instrumentality"?

Grounded in linguistics within a situated cognition approach, this contribution explores the ways members of the computer music community name and define the *instrument/interface/device*, etc. they develop, play and/or study, extending the reflection initiated in Cance et al. (2013). After a brief explanation of our theoretical and methodological framework (Sect. 2) we examine definitions of *instrument* in dictionaries and in academic discourses (Sect. 3) before diving into users' discourses about their practices concerning musical digital devices (Sect. 4). To conclude, we analyse how members of the computer music community name their devices in the titles of their communications (Sect. 5).

2 A Linguistic and Cognitive Perspective on Musical Instrumentality

This work was initiated and mainly developed within a collaborative research project[1] revolving around two computer music devices: the Meta-Instrument (de Laubier and Goudard 2006) and the Meta-Mallette (de Laubier and Goudard 2007). In addition to the study of technological developments carried out by the other partners, my main contribution as a linguist and cognitive scientist focused on the discourses and practices that emerged from the development and use of these devices.

2.1 A Multidisciplinary Framework…

Each new technology, each new artefact brings about new practices, including language practices allowing the developers and users to communicate, exchange, define, and also build and share know-how and knowledge, therefore constituting a *community of practices* as proposed by Wenger (1998) and developed in linguistics by Eckert (2006). Language is not transparent: in each language every speaker has the possibility of choosing between different ways of saying, expressing and referring to something. This non-trivial choice (i) relies on both individual (psychological) and collective (linguistic, cultural, sociological and historical) strategies and constraints and (ii) contributes to the construction of shared meaning in discourse. Analysing how people talk about the new musical devices they develop and play can therefore contribute to a better understanding of the underlying conceptualisations expressed and co-constructed in discourse. It allows us to highlight the

[1]The ANR-2PIM Project in 2006–2009, involving music associations (Puce Muse, Grande Fabrique) and research laboratories (Labri, LAM, LIMSI, IRCAM, McGill).

dynamic negotiation of new definitions and conceptualisations emerging from these new practices.

In order to observe and examine these language practices as cognitive but also social, historical and cultural practices, we adopt a multidisciplinary perspective of discourse analysis grounded in linguistics (Dubois 2009; Rastier 1991; Temmerman 2000) and psychology (Clark 1996; Te Molder and Potter 2005) within a situated cognition framework (Hutchins 1995; Barsalou 2008; Croft 2009). In this chapter I focus on the speakers' activities of defining, naming, referring to and categorising in discourse (Cance and Dubois 2015; Clark and Wilkes-Gibbs 1986; Mondada 1997) as productive indicators of the construction, negotiation and stabilisation of meaning.

2.2 *… for a Heterogeneous Corpus*

To observe both the emerging and un-stabilised practices in discourse and the circulation of discourses, I built a heterogeneous corpus by assembling different kinds of discourses on musical instruments, digital instruments and instrumentality:

- lexicographic definitions of *instrument* in French (fr) and English (en) general dictionaries (Sect. 3.1);
- academic definitions and discussions (fr & en) about instruments in musicology, ethnomusicology and organology literature (Sect. 3.2);
- discourses (fr) about instruments and instrumentality provided by different kinds of computer music users when interviewed about their practices (Sects. 4.1–4.2);
- terminologies (fr & en) used in the titles of papers at the main computer music conferences (Sect. 5).

In order to contrast lexicographic and academic discourses with users' spontaneous productions (among the French community of computer music), this linguistic work was mainly carried out in French. Nevertheless, a brief analysis of English terminologies and definitions is also provided to highlight the similarities and differences between the two languages, and to provide more evidence of the crucial role of language diversity in conceptualisation. All French quotations are translated and reproduced in English[2] in the body of the text, while the original French is given in endnotes.

[2]My own translation.

3 From Lexicographic and Academic Instrument Definitions…

3.1 Lexicographic Definitions

Various conclusions[3] can be drawn from the analysis of *instrument* definitions in two French dictionaries:

- *Le Petit Robert* (PR) (Rey-Debove and Rey 2007)
- *Le Trésor de la Langue Française informatisé* (TLFi) (ATILF 2016)
  and two English dictionaries:
- *Webster's New World Dictionary* (WD) (Neufeldt and David 1994)
- *New Oxford American Dictionary* (OD) (McKean 2007).

Dictionaries distinguish between a generic and a specific definition of *instrument* in relation to music. First, *instrument* is defined as a "concrete thing that allows the user to act on the physical world[i]" and as a "manufactured object […][ii]". The PR contrasts *instrument* with *tool*: "more general and less concrete than tool […][iii]". English dictionaries also link *instrument* and *tool*, but in a hyponymic relation, defining the former as a "specific tool used for specific purposes (scientific or artistic) and delicate work". In this respect, Simondon (1958), known as a specialist in the epistemology of technology, put these two concepts in perspective by defining both as prolongations of the body for the purpose of either performing a gesture (*tool*) or getting a better perception (*instrument*).[iv] In addition, the TLFi specifies the domains in which the instrument is involved, namely technology as well as science and art, and also emphasises the creative aspect.

Second, *musical instrument* is defined by the TLFi as an "object entirely made or prepared from another natural or artificial object, the former being conceived to produce sounds and to serve as an expressive means for composers and performers".[v] This focus on sound production as a medium for the users' expressivity does not appear in the other definitions. In the English dictionaries *musical instrument* is defined as "an object or device for producing musical sounds" without any explicit mention of agency as in the TLFi definition. As for the PR, it only exemplifies musical instruments in a typological enumeration based on organological classification (*areophone*, *chordophone*, […] *orchestral instruments*).

This brief overview of various definitions already illustrates a plurality of points of view for a given concept. While *instrument* is defined as a thing, an object, a tool or a device specifically conceived for a purpose, the relationships between all these concepts are not unequivocal. Moreover, *musical instrument* can either be defined intensionally (focusing on the sound production property) or extensionally (in an organological taxonomy), and the relation between the instrument and the user is barely taken into account (TLFi).

[3]For an extensive analysis see Cance et al. (2013).

3.2 (Ethno-) Musicology Definitions

Echoing the previous section, the ethnomusicologist Dournon (1996) warns against the reduction of instruments to their capacity to produce sounds, stressing on the contrary the importance of the symbolic meaning conveyed and societally situated:

> A musical instrument is not an object as others are; it produces sounds and carries meaning. It includes an additional aspect, due to its functional and symbolic role in society.[vi]

After expressing how difficult it is to define a musical instrument, the music philosopher Sève (2011) proposes a definition based on J.-C. Risset's works insisting on the cultural inscription of musical sounds and instruments as musical qualities that can only be considered (and are meaningful) within a specific culture:

> The musical instrument is a machine that is separable from the human body, that can be repaired piece by piece and that allows the transformation of the energy produced by the body of the person who plays into sounds considered as musical within the culture in which the instrument is used.[vii]

Concerning the musical quality of sounds, the music historian Michaud-Pradeilles (1983) recommends focusing on the function or the use of the instrument

> […] without any restrictive criterion, such as whether it is or is not made by a human being. In this way, the useless distinction between musical sounds and noise can be avoided.[viii]

This overview of instrument definitions by academics in musicology and the philosophy of the arts shows how perilous this undertaking is. If musical instruments are characterised by their capacity to produce sounds, one cannot neglect their symbolic and functional role. Whether or not an instrument is designed by a human is not a distinguishing criterion; however, it is necessary to consider the social, historical and cultural inscription of an instrument within a specific music system situated in history and culture.

Therefore an ontological definition of the instrument (as an entity with essential properties per se) remains very reductive, even *useless* according to Schaeffner (1994, p. 9), who suggests it would be better to consider the musical instrument in relation to a musical quality that can be acquired:

> Can we define the term 'musical instrument'? It is impossible, just as we cannot give a precise definition of music that would be valid in every situation, every period, and every use of this art. The problem of the instrument is linked to the question of the boundaries of music. An object can be sonorous; what is it that allows us to describe it as musical? What are the qualities that allow music to promote it to the same position as other instruments?[ix]

If it is so hard to define a musical instrument within the "traditional" history of music, how can this be done in the context of new digital musical technologies that do not fall within such a long history? Only a few researchers have investigated this issue, such as Brunner (2009), who has worked on

> the cultural implications embedded in the use and concept of the notion of the instrument in the field of computer music research,

and Hardjowirogo (this volume), whose work focuses on the increasing use of the concept of instrumentality to

> grasp that specific quality musical instruments are assumed to feature as distinguished from other sound producing devices

in the study of contemporary musical instruments, in order to identify

> relevant criteria for the construction of instrumental identity.

The next sections dedicated to analysing the discourse practices of computer music community members will allow us to put into perspective these two concepts (their use, their meaning) from a linguistic point of view.

4 … to the Emergence of a Terminology in a Computer Music Community of Practices

As described in Sect. 1, the work presented here was a contribution to the 2PIM project and consisted in the study of practices and discourses of computer music users in relation to two specific devices: the Meta-Instrument (MI) and the Meta-Mallette (MM). Developed by Puce Muse, both devices are generally described as controllers mapped with sound processing software and visual synthesis. Whereas the Meta-Instrument is an ad hoc device principally assigned to individual and expert musical practice (Fig. 1), the Meta-Mallette uses commercial interfaces (mainly joystick but also gamepad, Wii, etc.) and is conceived to be played immediately (without any necessary prerequisite skills or knowledge) and collectively (for example within a joystick orchestra, see Fig. 2).

To collect, document and analyse such practices, we adopted an ethnographic approach[4] including interviews, participant observation during project meetings and workshops dedicated to the MM, and participation in the Meta-Orchestre (Joystick Orchestra). The following section focuses on the results of these interviews.

[4]Quite innovative in this particular domain, with the exception of Booth and Gurevich (2012), who proposed an ethnographic approach of ensemble laptop performance, and Stowell et al. (2008), who developed a qualitative evaluation of digital musical interfaces through discourse analysis.

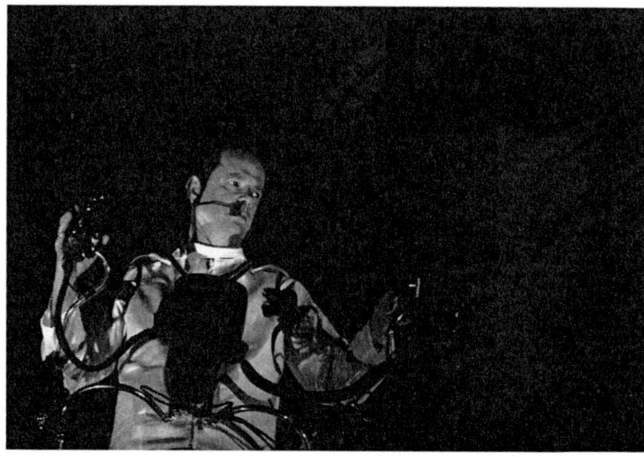

Fig. 1 S. de Laubier, Meta-Instrument creator. © Puce Muse

Fig. 2 The Meta-Orchestre playing the Meta-Mallette. © Puce Muse

4.1 Strategies for Naming and Categorising Digital Devices in Users' Interviews

Ten users of digital musical devices[5] were interviewed using a semi-directive methodology. They were asked non-directive questions (e.g.: *What is the Meta-Instrument in your opinion? How would you characterize it?*[x]) ranging from general to specific topics. After a full transcription, we conducted linguistic analyses on the lexical, morphosyntactic, semantic and discursive level, paying particular attention to how speakers refer to, name, categorize and define in their discourse the devices (MI and/or MM) they developed and/or played.

[5]All interviewees knew both the MI and the MM and had already used the MI and/or the MM before. As creators, developers, composers and/or performers, researchers, teachers or students from the conservatory, some of them combine different competencies and also use other devices they have or have not created themselves.

First of all we noticed a variety of lexical forms used to refer to the MI and the MM. Interviewees talked about *instrument*, *dispositif* (device), *outil* (tool), *objet* (object), *machine* and *engin* (machine), but also about *interface*, *logiciel* (software) and *programme*, which are specific to the computing domain.[6]

They modified some of these nouns with attributive adjectives, such as *instrument virtuel* (virtual instrument), *dispositif musical et visuel* (musical and visual device), *dispositif instrumental* (instrumental device), *machinerie portable* (portable machinery), *logiciel très souple* (very flexible software) and *instrument logiciel* (software instrument), to locate the reference either in the digital domain (*instrument virtuel*, ~ *logiciel*) or in the musical domain (*dispositif musical et visual*, ~ *instrumental*), depending on the reference domain of the head noun.

Generally speaking, *instrument* was more frequently used to refer to the MI, and *device* to refer to the MM. *Interface(s)* appeared to be used either to refer to all the devices that can be connected to the MM software (joystick, gamepad, graphic tablet, wii, etc.) or as a generic term (including MI, MM, joysticks and so on).

Nevertheless, the use of these different forms depends on each interviewee and her/his conceptualisations of the MM, MI and other devices. While some interviewees considered the MI and/or the MM as instruments:

IN1 So for me **the MI and the MM are instruments** that enable one to generate/manage sound and image in real-time,[xi]

or "like an instrument":

IN7 It's [MM] **a little bit like an instrument** but it works with a computer. But apart from that for me it's **like an instrument**,[xii]

another interviewee (IN6) described the MI and the MM as tools rather than instruments, using hedges[7] to attenuate his statement:

IN6 I describe them as **tools more than instruments** for the moment, maybe not yet,[xiii]

explaining:

The MM and MI devices are not mature enough yet to have a relation the same relationship with the user that an instrument has with its instrumentalist.[xiv]

Others (such as IN3) almost never used *instrument* preferring *interface*, *device*, *tool* and *machinery* instead:

[6]*Interface* is defined in the field of computing as "a device or program for connecting two items of hardware or software so that they can be jointly operated or communicate with each other" (OD).

[7]Hedges in linguistics refer to all the markers of uncertainty used in discourse. In the following example they are underlined.

IN3 That's really the idea, to have **a lighter device** (MM), because **the MI is a pretty heavy device.** It's **big machinery with several machines** that calculate image and sound in real-time.[xv]

All these interfaces are are are still very poor […]; when I play it does not vibrate as a guitar soundbox.[xvi]

Re-categorisation strategies can also be observed for instance with IN5 (music teacher that uses the MM with his students) who reformulated *instrument* with *object* (several times in his interview):

IN5 I wanted to create something with the Mozart samples and they they had to interpret it […], which was not easy moreover with **an instrument an object** a little bit new[xvii]

or:

IN6 There is a really big potential with these **instrum/interfaces.**[xviii]

Here, IN6 started by referring to *instrum*—but stopped before the end of the word replacing it by *interfaces*. This type of speech disfluency[8] indicates some uncertainty and fuzziness in the categorisation with an unclear distinction between interface and instrument.

Finding the appropriate term was also a challenge for the interviewer. Over the interviews none of the different denominations appeared to be neutral or generic enough, and one can observe retrospectively the different strategies developed to solve this matter, as in the following example:

to IN3 Can you describe to me the different / **so again I say interfaces with inverted commas I don't know how do you call them** / but describe the different **tools instruments interfaces devices** that are usable with the MM?,[xix]

which shows first a metalinguistic digression in order to anticipate the possible incompatibility of the word "interfaces" and then a list containing no less than four distinct denominations (!), from which the interviewee (IN3) could pick.

This list strategy used frequently could also be combined with an explicit naming question:

to IN4 If you need to describe explicitly to someone who has no clue about these kinds of **tools in/instruments** / how would you describe these kinds of of **devices**, **instruments**, **tools**?

[8]All the "irregularities" in speech, such as hesitations, disruptions and false starts (truncated words, repeated words or syllables, etc.).

> First we can even start first by **discussing how you name these kinds of devices** and how you would describe them,[xx]

which led IN4 to ratify the proposition:

> IN4 So actually the term def/device is generic enough to adapt to all scenarios. So a musical instrument is a device. It's also a tool, something we will work, with which we're gonna work.[xxi]

These examples illustrate the reflexive activity of the interviewer and her implication in this progressive construction and stabilisation of meaning.

4.2 *Defining Instrumentality*

In the course of each interview a network of semantic interdependencies was forged, showing various relations (opposition, complementarity, etc.) between these different linguistic forms that contribute to progressively build definitions of the MI and the MM and implicitly of what an instrument is. For instance, Meta-Instrument and Meta-Mallette were defined by successive oppositions, as illustrated by IN1, who after defining both MI and MM as instruments, contrasted later on in her interview

- a **meta-instrument**:

> Well, as I told you, for me it's [MI] **an instrument** that I rather refer to as **a musical instrument** [...] because there is really a gestural relationship in this instrument, so that it means it's really close to an acoustic instrument [...] and it's really the instrument that becomes one with the instrumentalist,[xxii]

- and a **meta-device**:

> So as for the Meta-Mallette, unlike the Meta-Instrument, I would say that it's **less of a musical instrument and more of a musical and visual device**.[xxiii]

Moreover, the boundaries of the instrument category were not fixed and signs of negotiation could be noticed. Although the MI was generally considered as an instrument by the interviewees, depending on the context it could be:

- excluded from the category of instrument (considered less than an instrument[9]):

[9]Echoing the notion of "voiceless instrument" proposed by Bricout (2011) to take into account this particularity of the digital devices not generating sound by themselves but needing the computer and a specific algorithm to do so.

IN6 No, in itself **I wouldn't put it in the instrument category,** it's the Meta-Instrument plus erm what we called the software instrument […], it's the couple made of the Meta-Instrument and the interpretation algorithm,[xiv]

- or included in a super-instrument category (considered more than an instrument):

IN9 Yes, I think they are instruments, **it's more than a simple instrument** because in fact inside they have a memory […] because even Serge's instrument has a memory, too.[xv]

Reciprocally, whereas the MM was mostly considered as a device composed of software and various possible interfaces (such as joysticks), in the context of a collective musical practice the joystick could become the user's instrument:

IN1 A **musical and visual device made of software and instruments** […] which are joysticks for the time being.

In fact when I'm talking about it I'm gonna say instrument for joystick meaning… meaning **you take your instrument to play it** […] so you're gonna take the object joystick to play […],[xvi]

while inside the small community of the MM, *instrument* usually referred to software (*program*, *games*, *synthesizers*, *virtual instruments*):

IN1 And at the same time commonly when one talks about the Meta-Mallette when one knows a little what's inside […], when we talk about **instrument** this is **the small software part** that determines how your sounds and images are gonna react.[xvii]

IN2 So answers could be made by by **instrument** […] so by the **program** when I say **instrument**. So now there is a generic term that I think works pretty well that is **virtual instrument**. And it is more or less accepted, so it starts to be used. We thought a lot about the MM. First we called it **games**, then **synthesizers** and now **virtual instruments**. It seems to me it's the one that works best.[xviii]

In both IN1 and IN2 interviews, these comments included a lot of autonymy markers, e.g. verbal indicators (underlined) of the reflexive work on the terms they use and on the existence (or not) of consensus.

These two first interviews (IN1 and IN2) raised the need to explicitly question the next participants on what they think an instrument is. I therefore started to ask them (at the end of the interview) to provide examples and a definition of the notion of instrument.[10] This revealed that the MI and the MM were sometimes given as examples of instruments (more frequently in the case of the MI). But above all,

[10]These questions were also asked to JIM conference participants in 2009 (see Cance et al. 2013).

defining what an instrument is led users to discuss the instrumental quality of the devices they had been talking about during the whole interview. One of them explained his preference for using *instrumental quality* (or *instrumentality*) rather than *instrument* arguing:

> IN4 I almost prefer using the qualifier than the word because **something can become instrumental**. The notion of instrument, I don't know what it is; the notion of instrumentality maybe more, or instrumental.[xxix]

For this interviewee the substitution of the noun by the adjective (or by another noun built on the adjective) demonstrates the performativity that characterises these instruments:

> The history of music is full of these **things that were not instruments at the beginning** and that have been **twisted by musicians to become instruments**.[xxx]

Therefore this musician developed in his discourse different "conditions of instrumentality" such as expressivity and embodiment:

> […] I don't say that we necessarily need **expressive instruments** to play music but that's what is interesting for me […] because **in this expressivity there is something that goes through the body**, and it's unconscious, uncontrolled,[xxxi]

interactivity and versatility:

> It's precisely this **capability via a device for any tool to serve a singular musical intention in real-time**, **live**. It supposes **interactivity,** a high degree of **versatility** […],[xxxii]

tangibility:

> What interested me in the MI was the fact that it's also a material device, and when I say material I mean **tangible** with real buttons […], **and this resistance is important** because / and for me it's part of what I could call instrumental device or **condition of instrumentality**,[xxxiii]

but also social inscription:

> So what is interesting I think is that there is a **small community around the instrument** and therefore **some possibility of exchange** […], because for me in the **conditions of the emergence of instrumentality** it is not just about / how to say that / individual skills it's about how **you have to work at it collectively** […]. And I think an instrument is not only a certain number of expected characteristics, of interactivity, reactivity and so on. It's its existence as **a shared instrument, a socialised instrument, situated in a society, a time, with a repertoire**.[xxxiv]

This last quote summarises quite well the "conditions of emergence of instrumentality" by emphasising the importance of exchanging and sharing (musical but also discourse) practices and repertoire within a specific context.

5 Naming Strategies in Computer Music Literature Titles: A Stabilisation Under Construction

Having explored how users (as individuals and as members of a specific community of practice) refer to, categorise and define their practices, I shall now look at the terminological practices (here specifically naming strategies) of a larger community. The main idea is to focus on and study how the members of the computer music community communicate with each other about their work.[11]

In order to do so, I refer to and extend a previous analysis of how authors in the computer music community refer to the hardware and software objects they develop and/or work with (Cance et al. 2013[12]). This non-exhaustive analysis examines the titles of the papers presented in international and French conferences (SMC, NIME, ICMC and JIM) during the last seven years (2008–2015). A table in the annex presents all the designations collected.

The tendency observed in 2009 is confirmed: in their conference titles authors mainly use either *instrument(s)* or *interface(s)* modified by a qualifying adjective referring to the artefacts they develop (as already observed in Sect. 4.1), such as *digital (musical) instrument* and *musical interface*. One can also find a few rare examples of *devices* (*musical devices*, *capacitive touchscreen devices*) or *controller(s)* (*musical controllers*, *digital music controller*) and a few designations of a specific instrument qualified by one or various adjectives (*digital piano keyboard*, *virtual piano keyboard*) conferring to it its "digitality". Apart from these few examples, most designations encountered in the titles are structured around *instrument* or *interface* as head noun, mirroring also most of the designations of the MM-MI community.

Interface needs to be qualified by the attributive adjective *musical* to specify the area to which it applies (music), whereas *instrument* already conveys the musical aspect. Therefore many titles combine *interface* and *musical* (*wireless musical interface*, *expressive musical interface*, etc.). On the other hand, *interface* anchors the designations in the technological/computer domain and emphasises the human–computer relationship. This might explain the scarcity of denominations combining *interface* and adjectives such as *virtual* or *digital* in the corpus as well as the frequency of denominations including a combination of *instrument* and the same adjectives (such as in fr: *instrument numérique*, ~ *électronique*, ~ *logiciel*, or in en: *virtual instrument*, etc.).

In other words, according to the domain of reference chosen, emphasis is more on music or on computing: within the musical domain, the noun *instrument* is specified by an adjective referring to computing (and therefore distanced from

[11]Brunner in 2009 already developed a deep and sharp analysis of the "cultural implications embedded in the use and concept of instrument" in the computer music domain from a different perspective.

[12]Published in *La musique et ses instruments* in 2013, this paper was written in 2009 in the context of the CIM09 conference. Therefore the first analysis concerned prior publications.

traditional or *acoustic instrument*), whereas within the computer domain, the noun *interface* is associated with an adjective that confers to it its musicality.

By analysing the other adjectives that qualify *instrument* and *interface* it becomes possible to highlight other particularities of the meaning of these two nouns. When referring to their devices as *instruments*, some of the authors insist on an extra quality of these instruments compared to usual/classical instruments with the use of adjectives such as *augmented*, *ubiquitous* and *configurable*. This pinpoints the *versatility* already described by a part of the 2PIM community.

Interface is sometimes associated with the adjective *tangible*, but also with *malleable* or *flexible* (*tangible acoustic interface*, *flexible interface*). Combined with these adjectives, *interface* refers more to a physical/material interface between the musician and the computer, whereas in noun phrases such as *computer mediated interface* or *interface logicielle* it refers to a software interface between the musician and the computer hardware.

Three other aspects that were discussed by some of the interviewees also emerged from the title analysis:

- Some titles lay emphasis on gesture, using designations such as *gesture controlled virtual instrument*, *gestural multi-touch instrument*, *gestural interface* and *expressive gestural interface*. While the gestural quality is not embedded in the concept of interface, it seems more surprising to find titles combining *gestural* and *instrument*.
- The notion of expressivity that was very important for one of our interviewees (IN4) also appears through the use of the adjective *expressive*, either qualifying *instrument* (*expressive virtual percussive instrument*, *compact expressive instrument*, *instrument for musical expression*) or *interface* (*expressive musical interface*, *new interface for musical expression*[13]).
- Finally, some rare titles include the notion of instrumental quality (*interface de communication instrumentale*, *Instrumenting the Interaction*) emphasised in the interviews and in this volume (see Hardjowirogo ibid.).

6 Concluding Remarks and Perspectives

This comparative analysis of different kinds of discourses on the notion of musical instrument allows us to specify the different semantic values of this concept depending on practices and discourses. Dictionaries define *instrument* in opposition to *tool*, whereas publications in the computer music literature show (a certain) ambivalence between *instrument* and *interface* as already noted by Brunner (2009):

[13]This is also the name of one of the main conferences in the domain: the NIME conference.

> The specific research on input devices for musical expression in fields of computer music ranges in its terminology without any clear coherence. We regard this circumstance as an ambiguity evoked by the performative knowledge produced in research processes […].

The analysis of the descriptions of some digital music devices by their users corroborates Brunner's analysis and demonstrates a concept of instrument with fuzzy boundaries, which is still "under construction". Devices can be considered to be more or less part of the instrument category according to a sort of "family resemblance".

Rather than identifying category membership criteria for the instrument (as an ontological entity with essential properties), this study contributes to specifying the conditions of emergence of instrumentality in the digital era. Once adopted into musical practices, especially collective ones (with a history and cultural and social values), a device can acquire this instrumental quality,[14] as illustrated by one of my interviewees' comments: "One is not born but rather becomes an instrument."[15]

This linguistic perspective enables us to see how this constantly evolving concept of instrumentality is individually and collectively shaped and negotiated. Mainly based on a study that took place in 2008–2009 and that specifically focused on a small community of practices in France, this analysis could be further extended, in order to document, describe and analyse how these practices keep evolving through time, languages and cultures.

Endnotes

- i "Une chose concrète permettant d'agir sur le monde physique." (TLFi)
- ii "Objet fabriqué servant à exécuter quelque chose, à faire une opération." (PR)
- iii "Instrument est plus général et moins concret que outil et désigne des objets plus simples que appareil, machine." (PR)
- iv "Le XVIIIe siècle a été le grand moment du développement des outils et des instruments, si l'on entend par outil l'objet technique qui permet de prolonger et d'armer le corps pour accomplir un geste, et par instrument l'objet technique qui permet de prolonger et d'adapter le corps pour obtenir une meilleure perception ; l'instrument est outil de perception. Certains objets techniques sont à la fois des outils et des instruments, mais on peut les dénommer outils ou instruments selon la prédominance de la fonction active ou de la fonction perceptive." (Simondon 1958: 114)
- v "Objet entièrement construit ou préparé à partir d'un autre objet naturel ou artificiel, conçu pour produire des sons et servir de moyen d'expression au compositeur et à l'interprète."
- vi "L'instrument de musique n'est pas un objet comme les autres, il est un outil à la fois producteur de sons et porteur de sens. Il comporte en effet

[14] Instrumental quality is here preferred to instrumental identity as it bypasses the ontological issue.

[15] This formulation alludes to Simone de Beauvoir's famous "One is not born but rather becomes a woman."

une dimension supplémentaire déterminée par le rôle fonctionnel et symbolique qu'il joue dans la société." (Dournon 1996)

vii "Définir l'instrument de musique n'est pas facile: *quaestio disputata*. Je propose personnellement la définition suivante, inspirée des travaux de Jean-Claude Risset: un instrument de musique est une machine, séparable du corps humain, susceptible d'être réparée morceau par morceau, et permettant de transformer l'énergie produite par le corps de la personne qui en joue en sons considérés comme musicaux par la culture dans laquelle l'instrument est utilisé." (Sève 2011)

viii "Pour définir l'instrument de musique, il vaudrait mieux considérer peut-être l'objet par rapport à son rôle ou à l'usage qui en est fait sans apporter de notion restrictive, telle que la participation de l'homme quant à son élaboration et éviter la ségrégation d'ailleurs magistralement remise en cause de nos jours entre sons musicaux et bruits." (Michaud-Pradeilles 1983: 5)

ix "Pouvons-nous définir le terme d'instrument de musique? Autant peut-être nous demander s'il existera jamais une définition de la musique, qui soit précise et valable en tous les cas, qui réponde également à toutes les époques et à tous les usages de cet art. Le problème des instruments ne touche-t-il pas à celui des limites de la musique? Un objet est sonore ; à quoi reconnaîtrons-nous qu'il est musical? Pour quelles sortes de qualités la musique le mettra-t-elle au rang de ses autres instruments?" (Schaeffner 1994: 9)

x *Pour toi c'est quoi le Méta-Instrument? Qu'est-ce qui le caractérise?*

xi *Pour moi donc le MI et la MM sont des instruments, qui permettent de générer / gérer du son et de l'image en temps réel.*

xii *C'est un peu comme un instrument mais ça marche avec l'ordinateur. Mais sinon pour moi c'est comme un instrument.*

xiii *Je les décris comme comme des outils plus que des instruments pour l'instant, peut-être pas encore ...*

xiv *Les dispositifs MM et MI sont peut-être pas encore assez mûrs pour en faire des / une relation la même relation avec l'utilisateur qu'un instrument avec son son instrumentiste.*

xv *C'est vraiment ça l'idée, d'avoir un dispositif qui soit plus léger [MM] aussi, puisque le MI C'est un dispositif assez lourd. C'est une grosse machinerie avec plusieurs machines, qui calculent l'image, le son en temps réel [...]*

xvi *Toutes ces interfaces sont sont sont très pauvres quand même [...] quand je joue ça ça vibre pas comme une caisse de guitare [...]*

xvii *J'avais envie de d'essayer de créer justement quelque chose avec les échantillons de Mozart et eux ... euh avaient à l'interpréter [...] ce qui est pas une mince affaire en plus avec un instrument un objet un peu nouveau.*

xviii *Y a vraiment un potentiel énorme avec ces ces instrum- / ces interfaces.*

xix *Est-ce que est-ce que tu peux me décrire les différentes / alors c'est pareil je dis interfaces avec des guillemets, je sais pas comment toi tu les appelles /*

mais décrire les les différents outils, instruments, interfaces, dispositifs, qui qui sont utilisables avec la MM?

xx *Comment comment justement, si tu dois décrire, expliciter à quelqu'un qui connait pas du tout ce genre de d'outils, d'in / d'instruments, comment tu décrirais ce genre de de dispositifs, d'instruments, d'outils. Déjà on peut même déjà commencer par discuter sur comment toi tu nommes ce genre de de dispositifs, et comment tu les décrirais?*

xxi *Alors effectivement le terme de dif / de dispositif est suffisamment générique pour pour s'adapter à tous les cas de figure, qui peuvent être très différents. Donc, un instrument de musique, c'est un dispositif, c'est aussi un outil, c'est quelque chose qu'on va travailler, qu'on va / avec lequel on va travailler, qui permet de produire des sons de manière contrôlée.*

xxii *Donc comme moi j'te disais pour moi donc c'est un instrument voilà donc … que j'référence plutôt à un instrument d'musique […] parce qu'y a vraiment un rapport au geste dans cet instrument, donc qui est vraiment proche d'un instrument acoustique du coup, […] et que c'est vraiment l'instrument qui fait corps avec l'instrumentiste.*

xxiii *Du coup la MM contrairement au MI j'dirais moins qu'c'est un instrument de musique mais beaucoup plus un dispositif musical et visuel […].*

xxiv *Non en lui-même je le mettrais pas dans les instruments c'est le MI plus euh ce qu'on appelle l'instrument logiciel […] ça serait plus le couple MI et algorithme d'interprétation […].*

xxv *Je pense oui que ce sont des instruments c'est plus qu'un instrument parce que en fait dedans ils ont des mémoires […] parce que même l'instrument de Serge aussi il a une mémoire.*

xxvi *Un dispositif musical et visuel […] qui va se composer d'un logiciel et d'un nombre d'instruments euh… qui pour le moment sont plutôt des joysticks […] en fait quand j'en parle je vais dire instrument pour joystick dans le sens ou … dans le sens où tu prends ton instrument pour en jouer […] donc tu vas prendre l'objet joystick pour jouer […]*

xxvii *Et en même temps communément quand on parle de la mal / la MM quand on sait un peu ce qu'y a dedans, quand on parle d'instrument, c'est la petite partie logiciel qui va déterminer comment tes sons et tes images vont réagir.*

xxviii *Donc y a effectivement après des réponses qui seraient peut-être à faire par par instrument. […] Donc par programme hein, quand je dis instrument. Alors maintenant y a un terme générique qui marche assez bien qu'est instrument virtuel je trouve. Et qui est accepté à peu près, donc on commence à l'utiliser. On a cherché beaucoup dans la MM, on appelait ça d'abord des jeux, puis des synthétiseurs, et maintenant des instruments virtuels. Il me semble que c'est celui qui fonctionne le mieux.*

xxix *J'ai presque plus envie de d'utiliser le qualificatif que le mot, parce que quelque chose peut devenir instrumental. La notion d'instrument je sais pas ce que c'est, la notion d'instrumentalité peut-être plus, ou d'instrumental.*

xxx *Et bon, l'histoire de la musique est pleine aussi de ces des ces trucs, qui au départ étaient pas des instruments, même a / avant même la synthèse, et qui ont été détournés par des musiciens pour devenir des instruments.*

xxxi *Donc je dis pas que pour faire de la musique il faut nécessairement des instruments expressifs, mais en tout cas c'est ça qui m'intéresse moi quoi. Parce que dans cette expressivité, dans, y a quelque chose qui qui qui passe par le corps, et qui est qui est inconscient, qui est incontrôlé, une espèce d'échappement à la conscience que je trouve nécessaire de conserver.*

xxxii *C'est cette capacité justement, par un dispositif, un outil quelconque, à se mettre au service de de d'un propos musical singulier, en temps réel, en direct. Donc ça suppose de l'interactivité, bien sûr, ça suppose un degré important de de de de versatilité ou de prise en compte en tout cas des des des volontés musicales exprimées, plutôt corporellement quoi. [...]*

xxxiii *Alors moi ce que m'intéressais dans le MI, c'est que c'est un dispositif matériel aussi, quand je dis matériel c'est à dire tangible, c'est à dire avec des vraies touches [...] Et cette résistance est importante parce que / et pour moi fait partie même de ce que j'ai envie d'appeler dispositif instrumental, ou condition d'instrumentalité, c'est aussi cette résistance.*

xxxiv *Donc ce qui est intéressant je pense, c'est le fait que y ait une mini communauté autour de l'instrument, donc possibilité d'un échange. Parce que là aussi dans [...], dans les notions, à mes yeux sont connexes, où les conditions d'émergence de l'instrumentalité, y a y a pas seulement des, comment dire, des compétences individuelles du dispositif. Y a en quoi ça se travaille collectivement quoi [...] Et je pense qu'un instrument c'est pas que un certain nb de caractéristiques attendues, d'interactivité, de réactivité, et tout. C'est son existence en tant qu'instrument de partage, instrument socialisé quoi, en quelque sorte, c'est à dire inscrit dans une société, dans une époque, dans un temps, avec un répertoire.*

Acknowledgments 3DMIN project team and especially S.-I. Hardjowirogo, ANR 2PIM project, Puce Muse, LAM, S. de Laubier, H. Genevois, interviewees and A. Cristinoi for the proofreading.

References

ATILF. (2016). *Trésor de la Langue Française informatisé* (TLFi). Nancy: UMR CNRS-Université Nancy 2. http://atilf.atilf.fr/tlf.htm. Accessed April 16, 2016.

Barsalou, L. W. (2008). Grounded cognition. *Annual Review of Psychology, 59*, 617–645.

Booth, G., & Gurevich, M. (2012). Collaborative composition and socially constructed instruments: Ensemble laptop performance through the lens of ethnography. In *Proceedings of the International Conference on New Interfaces for Musical Expression* (pp. 21–23), Ann Arbor, Michigan, USA.

Bricout, R. (2011). Les interfaces musicales: La question des "instruments aphones". *Methodos, 11*. https://methodos.revues.org/2493. Accessed April 16, 2016.
Brunner, C. (2009). A cultural approach toward the notion of the instrument. In *Proceedings of the International Computer Music Conference (ICMC), Montreal, Canada*. Montreal.
Cadoz, C. (1999). Musique, Geste, Technologie. In H. Genevois & R. de Vivo (Eds.), *Les nouveaux gestes de la musique* (pp. 47–92). Marseille: Editions Parenthèses.
Cance, C., & Dubois, D. (2015). Dire notre expérience du sonore: nomination et référenciation. *Langue française, 188*(4), 15–32.
Cance, C., Genevois, H., & Dubois, D. (2013). What is instrumentality in new digital devices? A contribution from cognitive linguistics & psychology. In M. Castellengo & H. Genevois (Eds.), *La musique et ses instruments* (pp. 283–298). Sampzon: Editions Delatour France.
Clark, H. H. (1996). *Using language*. Cambridge: Cambridge University Press.
Clark, H. H, & Wilkes-Gibbs, D. (1986). Referring as a collaborative process. *Cognition, 22*, 1–39.
Croft, W. (2009). Toward a social cognitive linguistics. In V. Evans & S. Pourcel (Eds.), *New directions in cognitive linguistics* (pp. 395–420). Amsterdam: John Benjamins.
de Laubier, S., & Goudard, V. (2006). Meta-instrument 3: A look over 17 years of practice. In *International Conference on New Interfaces for Musical Expression (NIME06)*, Paris.
de Laubier, S., & Goudard, V. (2007). Puce Muse – La Méta-Mallette. In *Journées d'Informatique Musicale (JIM2007)*, Lyon.
Dournon, G. (1996). *Guide pour la collecte des musiques et instruments traditionnels*. Éd. rév. et augm. Paris: Unesco.
Dubois, D. (Ed.). (2009). *Le Sentir et le Dire: Concepts et méthodes en psychologie et linguistique cognitives*. Paris: Harmattan.
Eckert, P. (2006). Communities of practice. *Encyclopedia of Language and Linguistics, 2*(2006), 683–685.
Hardjowirogo, S. (2016). Instrumentality. On the construction of instrumental identity (this volume).
Hutchins, E. (1995). Cultural cognition. In E. Hutchins (Ed.), *Cognition in the wild* (pp. 353–374). Cambridge: MIT Press.
McKean, E. (2007). *New Oxford American Dictionary*. New York: Oxford University Press.
Michaud-Pradeilles, C. (1983). *L'organologie*. Paris: Presses Universitaires de France.
Mondada, L. (1997). Processus de catégorisation et construction discursive des catégories. In *Catégorisation et cognition : de la perception au discours* (pp. 291–313). Paris: Kimé.
Neufeldt, V., & Guralnik, D. B. (1994). *Webster's New World Dictionary Dictionary of American English*. New York: Prentice Hall.
Rastier, F. (1991). *Sémantique et recherches cognitives*. PUF: Formes Sémiotiques. Paris.
Rey-Debove, J., & Rey, A. (Eds.). (2007). *Le nouveau petit Robert: dictionnaire alphabétique et analogique de la langue française*. Paris: Dictionnaires Le Robert.
Schaeffner, A. (1994). *Origine des instruments de musique: Introduction ethnologique à l'histoire de la musique instrumentale*. Paris: Editions de l'Ecole des Hautes Etudes en Sciences Sociales.
Sève, B. (2011). L'instrument de musique: présentation du numéro. *Methodos, 11*. https://methodos.revues.org/2599. Accessed April 16, 2016.
Simondon, G. (1958). *Du mode d'existence des objets techniques*. Paris: Aubier.
Stowell, D., Plumbley, M. D., & Bryan-Kinns, N. (2008). Discourse analysis evaluation method for expressive musical interfaces. In *Proceedings of NIME08*, June 5–7, 2008, Genova, Italy.
te Molder, H., & Potter, J. (2005). *Conversation and cognition*. Cambridge: Cambridge University Press.
Temmerman, R. (2000). *Towards new ways of terminology description: The sociocognitive approach* (vol. 3). Terminology and lexicography research and practice. Amsterdam: John Benjamins Publishing Company.
Wenger, E. (1998). *Communities of practice: Learning, meaning, and identity*. Cambridge: Cambridge University Press.

From Idiophone to Touchpad. The Technological Development to the Virtual Musical Instrument

Bernd Enders

> *Digitalization transgresses the boundaries of the physically possible, the natural becomes the arbitrary, the arbitrary is being artificially created.*
>
> Jauk (2009, p. 439)

Abstract The history of music can be understood as the increasing digitalization of representation and processing of musical information as notes and sounds. Musical phenomena could be described as a continuous transition from the analog and simple instrument like a wood block to the digital and abstract software instrument like a virtual synthesizer on a touchpad. The development of musical instruments shows an increase in complexity and functionality of handicraft over time, and the music computer forms the last, most comprehensive and most abstract link in a chain of innumerable steps in music and musical technology, starting with the human voice and the invention of drums as a mean of sound and communication and actually marked by digital instruments and virtual sound worlds. Ten developmental stages can be identified with regard to the construction and the usage of musical instruments and multimedia performances. The digital processing of sound information extends the range of artistic presentation of musical processes to unfamiliar, albeit intriguing and expandable dimensions. The aesthetic potential of the artistic approach to musical and multimedia information in the broadest sense proves to be enormous. But it is impossible to predict whether overarching paradigms of music composing and digital culture will emerge or become apparent some day.

Authorized translation of the article "Vom Idiophon zum Touchpad. Die musiktechnologische Entwicklung zum virtuellen Musikinstrument" (in *Musik/Medien/Kunst*, ed. Beate Flath, 55–74. Bielefeld: transcript 2013) from German by Andrea Kampmann (including quotes).

B. Enders (✉)
Forschungsstelle Musik- und Medientechnologie, Institut für Musikwissenschaft und Musikpädagogik, Universität Osnabrück, Osnabrück, Germany
e-mail: bernd.enders@uos.de

T. Bovermann et al. (eds.), *Musical Instruments in the 21st Century*,
DOI 10.1007/978-981-10-2951-6_4

1 Introduction

From both a theoretical and a practical musical point of view, the entire developmental history of music can be understood as the increasing digitalization of the representation and manipulation of musical information and processes. In artistic as well as academic discourse, musical phenomena are being described in ever more abstract terms in terminological and technical respects, basically expressing the transition from the analog and concrete to the digital and abstract.

In this regard, Werner Jauk views the development of music as a process of mediation, "from immediate expressive behavior, its instrumentalization, over gestures, the formalization of communicative expression in presentational and iconic signs to arbitrary codes." (ibid., p. 2) The codes represent the abstract terminology which serves to grasp the instrumentally generated or recorded sound signals as well as the musical notes for the description of musical events (in their various forms) in the digital age of music.

Against this background, the music computer basically forms the last, most comprehensive and most abstract link in a chain of innumerable stages in musical technology, ranging from the human voice as a means of sound and communication to the artistically informed construction of virtual sound worlds. The development of musical instruments, an integral part of technological progress since its earliest beginnings, has always been of state-of-the-art handicraft in each era; therefore, considering the continuous increase in complexity and functionality of handicraft over time, a number of developmental stages can be identified with regard to the construction and, consequently, the usage of musical instruments.

2 The Development of Musical Instruments in Ten Stages from Prehistory to the Digital Age

The discovery of the principles underlying the acousto-mechanical forms of sound production in musical instruments (idiophones, aerophones, membranophones, chordophones) probably occurred parallel to the development of human culture in prehistory and cannot be traced back accurately. As related finds from the Paleolithic Era indicate, humans have made use of simple instruments such as lithophones or bone flutes as well as tools like drum sticks at least as early as 30,000 years ago.

The first stage of instrumentalization, enabling sound production beyond the potentialities of the human body alone through the discovery of the sound tool, was followed by the "mechanization" of instruments, which on the one hand reduced direct body contact with the vibrating body but on the other hand allowed for easier or more efficient operation. The introduction of power-amplifying and precise keyboards, pedals and key mechanisms (controllers), triggering various hammers, valves or levers, to act as intermediaries between the playing human and sound production proper—i.e., as interfaces—already permitted a certain freedom of assigning a triggering action a resulting sound (mapping). In the case of the pipe organ, the technological separation of controller (=console) and generator (=pipe body) is complete: the transmission of data can be achieved by mechanical, pneumatic, electrical or electronic means, even the utilization of an external power source—a criterion of machine systems—already applies here, as the bellows providing the continuous air stream are not operated by the organist himself, in contrast to the pump organ.

The "automatization," or programmability, respectively, of musical processes was first accomplished by means of the pinned barrel invented in 900 A.D., paving the way for mechanical musical instruments such as the medieval carillons, the musical clocks of the late 18th century, the orchestrions and pianolas (now being controlled through punched cards and discs), which enjoyed great popularity up to the invention of the radio and the record player in the early 20th century.

The stage of "electronification" (electromechanical, electro-optical and electronic instruments) was inaugurated (after a few musically negligible experiments) by Cahill's Telharmonium around 1900. For the first time after a millennia-old history of the musical instrument and its continuous improvement, the use of electricity finally marked the invention of new methods for generating sounds. Innovative instruments with unfamiliar sounds and original playing techniques were invented, which were added to the traditional instrumental classification system under the category of electrophones, but did not take root in the classical-romantic orchestra (even though instruments like the Mixtur-Trautonium, the Ondes Martenot or the electric guitar sometimes occur in combination with orchestral instruments, for instance in works by Olivier Messiaen or Harald Genzmer).

With the mid 20th century's electronic music, the separation between composer and virtuoso performer was partially undone again, as a growing automatization of sound processes and/or studio devices adopted from radio (generators, tape recorders, effects units) enabled a direct translation of an artistic idea into sound (Karlheinz Stockhausen) or as the composer oftentimes also functioned as main interpreter of his own works (Oskar Sala).

Michael Harenberg emphasizes the electronic sound world's significance for the history of music: "No other invention of this century had such fundamentally qualitative consequences for the development of music, or changed our understanding of music and its perception so radically as the technical means to produce, record and distribute sound electronically – a milestone of the history of music comparable to the invention of musical notation[1]" (Harenberg 2012a).

In the mid-1960s, analog musical electronics reached a technological peak with Robert Moog's ingenious construction of the voltage-controlled analog music synthesizer, which unified the single devices of the electronic music studio and the electric/electronic musical instruments in a well-balanced modular system. This "modularization" basically distinguished between three sections of an instrument's functionality: (1) generators (oscillators, noise generators), (2) modifiers (filters,

[1]Harenberg (2013). Also cf. Harenberg's elaborations on this issue in Harenberg (2012a, b).

amplifiers, effects units), (3) controllers (keyboard, joystick, sequencer etc.). Although the principle of modularization already applies to earlier experiments, such as the Mixtur-Trautonium by Friedrich Trautwein and Oskar Sala, only transistorization and the development of control voltage led by Robert Moog paved the way for a "trunk-sized" compact studio.

After a series of further developmental stages, the "digitalization" and IC-based miniaturization of music electronics began in the 1970s in the form of the first music computers (Fairlight CMI, Synclavier), whose functionality offered numerous possibilities of modern studio software as well as a menu-driven user interface despite the limitations of contemporary 8-bit technology. With the Fairlight CMI's eye-catcher, the lightpen, the user was able to operate the menu's elements of the underlying software and practically draw any oscillation, envelope function or sequencer notes onto the screen, thus foreshadowing the tapping and swiping of today's tablets and smartphones.

Offering a broad range of synthesis algorithms already integrated into the system as well as enabling the digital storage and control of any sound recorded with a microphone (sound samples), the music computer was hailed as the harbinger of a new musical age and directly appropriated for artistic purposes (e.g. by Austrian composers Hubert Bognermayr and Harald Zuschrader).[2] In other words: both note properties and audio signals were now being processed digitally and connected in the musical process. From a technological point of view, analog note information on a pinned barrel and the oscillations of a record groove now only differed with regard to their coding formats in the musical machine.

Accordingly, all aspects of music production necessary for digital processing were available by the early 1980s: various methods of sound synthesis, the digital storage of sounds, the processing and modification of sounds, and the complete controlling of sound processes, be it manually by means of innovative interfaces or automatically through digital note information (sequencing). In this context, Werner Jauk stresses that the digital sound experience is thus becoming ever more immaterial and abstract: "In generating and playing, digital sound is completely detached from bodiliness. It is not created through an oscillation produced by a vibrating body, it is computed oscillation originating from non-vibrating material, from codes" (Jauk 2009, p. 337).

In 1981, the introduction of the enormously successful MIDI[3]-standard in order to guarantee the compatibility of sound modules with keyboards of different brands finally ushered in the "informatization" of music electronics, as earlier experiments

---

[2]Their *Erdenklang: Computerakustische Klangsinfonie* was premiered in the Brucknerhaus in Linz during the Ars Electronica in 1982.

[3]MIDI (=**M**usical **I**nstrument **D**igital **I**nterface) is a music-specific standard format which not only enjoyed popularity in musical application for a long period, but which furthermore gained a foothold in general computer industry as a standard interface. With regard to its significance, it is therefore comparable to the internationally widespread digital formats PCM for the audio CD introduced in 1981 or the ever more important compressed audio format MP3 used for web transmissions since 1991.

with computer-based composition (Lejaren A. Hiller) in the mid-1950s had not proven feasible due to extensive transcriptions of mainframe-computed data into readable scores (score synthesis).

With regard to information technology, the MIDI system can be viewed as a digital electronic variety or successor of the pinned barrel, which was invented as early as 900 A.D. and served as a mechanical storage of note information in order to trigger sounds with a clearly defined point of onset and duration. Similarly, MIDI signals do not contain sound data, but merely send note information to an electronic or even mechanical sound generator; to be precise, a key number, e.g. the number 60 (here given in decimal format), representing the note c', is transmitted (along with the key velocity for dynamic results). MIDI is furthermore a key-oriented notation code, an originally action-based information system, which (similar to dodecaphony, which is likewise informed by the equal temperament of modern keyboard instruments) irons out harmonic alterations enharmonically and does not discriminate between "c sharp" and "d flat." MIDI documents the hands' musical play on a keyboard and is thus related to action-representing tablatures such as the fingering notation systems for guitarists.

Soon after the introduction of the MIDI standard, MIDI-coded musical data (i.e. note information) were being digitally processed by means of widespread PCs (e.g. C 64, Atari ST), which then opened the door to an entirely new range of possibilities to process musical information so that by the end of the 1980s, computer-based arrangements, sophisticated (harmonically correct) music engraving techniques, creative software for composition as well as artistically and scientifically intriguing analyses were either widely available or least being experimented on in scientific environments (e.g. computer-calculated real-time temperaments such as the Hermode Tuning system (Hermode Tuning Werner Mohrlok e.K. 2015)). Methods of artificial intelligence (AI) suggest future applications transforming the music computer into a creative partner of musical processes, even if these may still be restricted to an experimental-scientific context.

By the early 1990s, the shift of music technology from specifically constructed instruments (hardware synthesizers, electronic organs etc.) to the development of surrogate software components running on basically every regular computer with suitable audio and MIDI periphery (sound card, keyboard etc.) marks the "virtualization" of instruments and music-specific production scenarios. Such a software basis allows for the emulation of multitrack studio tape recorders, mixing consoles, "traditional" effects units such as reverberators or vocoders, as well as the development of new sound processors such as transposers or auto tuning.

Modern studio or MIDI programs achieve a near-complete virtualization of all functions of the music production process; even the hardware predecessors' user interfaces and brand-specific looks are being simulated "realistically" on the screen, although usually only the functions of sound production and modification are subject to virtualization, considering that the interfaces necessary for operation (mouse, musical keyboard, and other controllers) are oftentimes easier to use if mechanical rather than virtual and thus have to remain in the realm of real physical functionality.[4]

Earlier developmental stages of instrumental technology, for example the modularization of an instrument's functions, are being virtually reconstructed. So-called plug-ins, i.e. special software packages, extend a music computer system's functionality by means of suitable interfaces (e.g. Steinberg's VST norm) to include sophisticated studio effects routines or nostalgic imitations of "historic" precursors (e.g. a vintage synthesizer, Hammond organ, Fender Rhodes and many more).

Inspired by professional sample libraries such as the Vienna Symphonic Library (VSL), which impress the user with their ever-growing possibilities of modulation and quality of programmability and have thus found their way into a wide range of motion picture and TV show soundtracks, Harry Lehmann engages in a visionary analysis of a virtual orchestra live in concert, a possibility that could very well be realized by current technology (Lehmann 2012, p. 19ff).

While digital electronic methods are a necessary requirement of highly complex sound synthesis processes like waveshaping or granular synthesis, they also allow for the modeling of unusual instruments which could hardly be physically constructed by traditional means (virtual reality). In this context, physical modeling is held to be a promising method of synthesis, which utilizes mathematical models to simulate the complicated physical processes in vibrating bodies, for instance a vibrating string attached to a resonating body, as accurately as possible and thus ideally comes very close to the acoustic behavior of a mechanical or electro-mechanical instrument or even potentially provides the various forms of modifications involved in the subtle sound variations of any good interpretation. Alternatively, conventional features may be skillfully combined with innovative characteristics to give birth to original musical conceptions which could not have been realized by traditional mechanical and electronic construction principles.

Digitality facilitates the device-based processing of various types of information so that different levels of perception, particularly video and audio, can be perfectly synchronized, paving the way for the enhanced artistic expression and experience of a promising, possibly synaesthetic, multimedia art.

In this process, it is becoming increasingly difficult for terminology to distinguish between the abstract world of computer-based functions and digitally processed information. In one case, the computer system is an entire music studio; in another, a musical instrument; in yet another it becomes a music engraving system, a multimedial playing device, an interactive musical educational program. In the end, the definitions depend on the individual estimation of the features, the expected usage or even strategic marketing, as there are no clear-cut boundaries between one application and another, and many characteristics of both hard- and software may principally be utilized multifunctionally.

---

[4]Cf. Arne Bense's revealing and comprehensive discussion of the complex relationship between real and virtual components of computer-based instruments in Bense (2013).

Against this background, Arne Bense discusses the need for a revised organology, which would have to take the properties of the virtual musical instrument into account as well: "All sections in music production are by now affected by virtualization and are thus subject to various transformation processes – including the musical instrument. It seems to have disappeared between software synthesizers, sound samplers, groove boxes, 'virtual instruments,' digital audio workstations, diverse controllers and musical interfaces. Today, organology is undergoing a phase of reconfiguration – the organology of the computer society has yet to be developed" (Bense 2013, p. 149).

Which part of the interactive graphic digital synthesizer reacTable should be viewed as the actual instrument? Is it the sensational glass top that reacts to manually placeable and movable objects with a coded underside by emitting visual signals that optically simulate the typical structure of a modular synthesizer with oscillators, filters, envelope generators and sequencers, or is it the integrated computer along with the software as the sound machine proper, which however remains hidden from the audience?

The traditional instrumental classification system was challenged as early as 1919, when Léon Theremin's Thereminvox baffled the audience not only by its obscure sound production principle based on vacuum tubes, but moreover with its seemingly magical and surreal playing technique, independent of a visible, tangible interface that would allow the musician to physically engage with the instrument.[5] Even modern performances by usually female theremin virtuosi continue to put today's listening and watching audiences under a similar spell.

Since the digital sound can be created and modified completely independent of playing technique, there are no boundaries to experimenting with innovative forms of interaction between the human and the computer. Fascinating original playing techniques with new controlling devices (mouse, pads, sensors of all kinds) are being artistically put to the test, and newly designed interface technologies specifically geared to the haptics and gestures of the human motor skill, which partly stem from other, non-musical applications, as for instance gesture, voice or game controllers, eye tracking, actuators from robotics and many more, are being adapted for usage in musical electronic devices and instruments.[6]

Digital hardware is becoming ever smaller, more inexpensive and, if combined with refined software, ever more powerful, so that many aspects of the production

[5]In 1930, Karl Gerstberger, a contemporary of Léon Theremin's, attempted to put into words the speechlessness of the listeners, who had probably never seen a thereminvox performance before: "Who would have been immune to the spell of the uncanny, as L. Theremin performed his ether wave music? Was is not as in a tale from the Arabian Nights, when he summoned with a conjurer's hands the 'genie in a bottle' and bade him utter sounds loud and soft, high and low?" (Gerstberger 1930, p. 171).

[6]Cf. Jin Hyun Kim's detailed compilation of the different developmental strands of musical interfaces from historical and systematic perspective, including functional analyses: Kim (2012).

and modification processes mentioned above may already be brought to fruition with sensitive tablets,[7] which are so handy, easy to use and holistically designed that they actually come close to a simple idiophone, e.g. a wood block—at least as long as the battery holds.

Ever since computer systems started interconnecting to form global medial networks, since the immensely significant development and spreading of the internet, which renders the user searching for information nearly completely independent of time and space, new forms of a network-based musical communication emerge which go beyond the mere consumption of music or multimedia. Music production is entering a stage of "globalization" that also encompasses the artistic application of digital music technology. Among these are not only the worldwide exchange of MIDI files or scores, but also the transmission of WAV or MP3 sounds, ringtones,[8] plug-ins, skins and many more.

But what is most intriguing and innovative are those projects that attempt to construct interactive, virtual instruments or sound spaces on the internet, to create collaborative compositions via the web, as for instance Austrian composer Karlheinz Essl has been experimenting on for some years, or even to bring musicians together in virtual space for making music online in real-time, which unfortunately is currently still impractical due to delayed signals. The boundaries between musical instrument and media devices are dissolving, music evolves into "media music" when the platter of a DJ's turntable becomes a creative sound generator, as Rolf Großmann notes in his early discussion of the "medial influence on the construction of musical reality" (Großmann 1997).

As a last developmental step, the "hybridization" of music technology has yet to progress beyond its current experimental stage; its core is the considerable merging of the musical instrument and the human, giving birth to performative man-machine-symbioses, which by digital electronic means basically enhance the possibilities of the human body's interaction with a device, to basically achieve something which had actually been in the humans' power from the beginning, namely the ability to put the unequaled flexibility of the voice as a body-inherent, ubiquitously available instrument to musical use. The performance artist Stelarc, for example, attempts to create artistic symbioses of his own acting body and machine systems (sensors, artificial limbs, robots). Others are experimenting with the transposition of a dance performance into virtual space, synchronizing a virtual avatar with the moving musician's body, or with interactive sound installations, sounding spaces and multimedial creations of 3D virtual reality, which, though not yet commonplace, are nevertheless no longer fantasies of a utopian music culture's science fiction arsenal.

Although first experiments on a playing technique based on the so-called brain controlling, the controlling of electronic instruments by brain waves, are so far

[7]Among the applications available are for instance the "old" Fairlight CMI as well as the "new" reacTable as virtual emulations on current touchpads, albeit with certain limitations.

[8]Frauke Behrendt, for instance, examines the artistic relevance of mobile phone music in Behrendt (2005).

mostly relevant only to current neurological research rather than to musical experience, they do suggest that in the future, making music may be possible without even moving one's hands, as was still required for playing the Theremin (Table 1).

3 New Forms of Composing and Playing in a Digital World of Music

Mainly caused by the impact of the Moog synthesizer, the preference for a keyboard with discrete scale steps for creating electronic music, as well as the universal introduction of the MIDI norm left the subtly shaped, individual (monophonic) tone quality of older music electronics such as the Thereminvox or the Trautonium in the shade. The originally creative mode of experimenting with innovative sound generation and its (theoretically) enormous modification potential was discarded again in musical practice—especially in pop music—in favor of easily accessible sound programs and standardized (usually homophonic) playing techniques. However, additionally installed controllers such as modulation wheels, faders, pedals, pitch bend wheels and wind controllers offered intricate possibilities of modification to the ambitious keyboarder, which for instance are inspiringly appropriated musically by jazz virtuosi such as Jan Hammer.

But the potentialities of digital music technology are not limited to advanced or even completely unconventional playing techniques with various opportunities for articulation, positively demanding to be discovered and put to use for musical interpretation live in concert; furthermore, the ways artistically inclined people are using the mass-produced and inexpensive sound equipment of the computer age questions the traditional understanding of composers and musicians. Since the 1990s, the creative engagement with the rich musical possibilities of the widespread and inexpensive computer systems leads to the emergence of a breed of musician which Karl H. Menzel dubs "PC musician" (Menzel 2005),[9] along with the development of a production space (primarily for pop music) that Werner Jauk circumscribes, with a trace of irony, as the "bedroom studio": "The age of controlling (pre-programmed, barely modifiable) synthesis machines turns into the era of directly working with sound in digital popular culture, now indeed at home, in the bedroom for each and everyone and – in the collective of the bedroom-home-studios interconnected via the WWW – with the understanding of music as open source in the dynamic process of collaboratively making music beyond time and space" (Jauk 2009, p. 315).

---

[9]Cf. the chronological examination of the musical technological development and its musical consequences, ranging from the professional—originally analog—multitrack studio to digital home recording on the PC. Also cf. the current analysis of the reciprocal interaction between music and technology by Wandler (2012).

Effected by digital music electronics, the mass distribution and lowered costs of the technologies necessary for the artistic-compositional production process form the basis of Harry Lehman's criticism of the recent composing avant-garde, who refuse to acknowledge these principally socio-musical changes of music culture "because their value system is still fully anchored in classical music, which only recognizes human subjects, but not social systems." He calls attention to an "absolute exception," composer Georg Katzer, who recently noted laconically: "The computer has democratized the obscure trade of the composer. Today everyone can, even if they are at a musical proficiency level of 'zero,' produce music 'by the boatload,' which, regardless of its artistic value, may moreover be copyrighted" (Katzer 2011, p. 31).

Of course, this does not answer the question of which direction the new music culture, closely intertwined with the availability of digital functionality, will take in the future, which aesthetics will develop against the background of digital music technology, whether art will experience marginalization due to large-scale distribution or whether it will give birth to new creative highlights, whether the creating artist will still exist as an autonomous instance or whether this role in the artistic production process will be replaced by a job-sharing team consisting of programmer, arranger, designer and product manager, and, finally, whether there is still need for the finite, completed work of art, or whether musical phenomena are rather conceived as flexible material for the purpose-built generation of sound environments temporarily designed for specific situations, places and social groups.

Thanks to seemingly inexhaustible possibilities of quickly manipulating the digital audio material, the simple realization, adaptation, transformation, division, variation and recombination of all kinds of sounds, the composition process, in past times considerably more lengthy and wearisome, now undergoes substantial changes. The undo-function available in every piece of modern software and the regular backup files permit the user to simply discard changes, return to the raw material stored in non-destructive form and start from scratch, encouraging the user to experiment with small and big changes without risk. Regardless of the actual planning phase, digital composing and/or arranging thus comes close to an improvising mode in the way the material is handled—moreover, the line to the live act is becoming blurred, as it is principally constantly in flux. The tendency to create, collect and recombine entire series of alternatives, versions and revisions is growing, as Robert Henke vividly describes in his essay "tod durch überfluss" ("death by affluence", Henke 2011), making it necessary to rediscover the "art of leaving out" and to find disciplined compositional strategies.

Already the invention of music notation rendered musical information pictorial, allowed it to be grasped holistically: the temporal rhythmic sequence of musical information previously necessary for sensual experience may be omitted as the eye can subjectively choose any path through the work. Today, the computer basically links these approaches when music is visualized on the screen in the shape of notes or beams and is played synchronously—the computer can even merge them partially when highlighted sound elements are basically halted in an infinite loop (an option for playing an arrangement offered by modern studio programs), or when music is being fragmented, shifted, replicated and recombined in its elements (=-composed) in a patchwork manner: composing in real-time, working on a "living" musical object, as is conceptually supported by innovative music programs (e.g. Ableton Live).

The digital processing of multimedia information extends the range of artistic presentation of musical processes to unfamiliar, albeit intriguing and expandable dimensions. For instance, experiments include dynamic scores such as a flowing visualization of sounding notes reminiscent of a waterfall,[10] the performance of a touchpad musician may be projected as a larger-than-life image onto a large screen, even a completely unskilled drummer can be turned into a virtuoso by means of simulation[11]—a humorously cut sequence of synchronized video bits, each showing a single drumbeat or hit on the cymbals, and many more.

To go even farther, virtual concert performances hold such a power over large audiences that they seem uncannily surreal: the synthetic singer Miku Hatsune—projected onto a life-sized screen on the stage as a quasi-realiter hologram, singing with an artificial voice live in concert (generated by means of the voice plug-in Vocaloid2 by Yamaha), continues to exhilarate a Japanese audience of loyal fans to the point of dancing, clapping and singing along, as these YouTube videos[12] should convince even the most skeptical reader. The relatively high, childlike timbre of the virtual singer is reminiscent of the anime characters extremely popular in Japan, which also inform the visual appearance of Miku Hatsune. "She" is accompanied by a real band, which, positioned around the screen, plays live (in terms of music technology, i.e. synchronously to the playback of the voice track). In combination with the perfectly animated, originally and bizarrely playing "ANIMUSIC" instruments (Animusic 2013), which would never work in the real world, the performance thus achieves a new quality of artificiality, which can go completely without the human musician.

The music computer, the virtual musical instrument as a universal machine for every form of numerically coded sound generation, processing and distribution, has radically transformed the modern music culture in every conceivable area of production and distribution. With its various shapes and modes of application, the music computer forms the climax (at least for the time being) in a chain of successive developmental stages of music technology—from the hand drum to the electronic drum pad, from the bone flute to the MIDI-based wind controller, from the water organ to the sample library with historical organ sounds etc., a development that began in the first days of music culture and which can be systematically divided into ten categories of distinct characteristics and (at least partially) chronological succession.

[10]Already 1995 impressively put into action by Dominique Besson on the multimedia CD-ROM "Les Musicographies," (Besson 1995), which unfortunately only runs on older Macintosh computers.

[11]Cf. Gjeertsen's video "Amateur" (Gjeertsen 2006).

[12]E.g. Hatsune Miku (2011).

In a constantly reciprocal relationship of music culture and technology, the newly discovered possibilities of sound generation by means of instruments have ever been put to artistic use, sparking off further developments, both musical and technical. Thanks to the electronification and digitalization of musical technological instruments and devices, the aesthetic potential of the artistic approach to musical and multimedia information in the broadest sense proves to be enormous.

As of now, it is impossible to predict whether overarching paradigms of music culture will emerge or become apparent some day. Austrian musicologist Werner Jauk describes the newly emerging phenomena of the acoustic and visual arts and medial communication as a distinctive feature of a "digital culture," i.e. a (pop) culture characterized by digital numeric codes, global data exchange, virtual spaces and hedonic aesthetics, which has arisen through the potentialities of the

Table 1 The technical development of the musical instrument (taken from Enders 2005, p. 32f)

1. Instrumentalization	• Externalization of music production outside the body, separation from voice and clapping • Discovery of the drumstick, the whistle, the flute, the string • Development of instruments as sound tools (e.g. fiddle)
2. Mechanization	• Construction of optimized operating devices ("controllers," especially the keyboard as an interface for musical sound controlling) • Construction of special sound generation mechanisms (hammers for striking strings as mechanical amplifiers of finger pressure, plectra for plucking strings, mouthpieces, pitch controlling mechanisms such as keys on a flute) • Transition to the machine by employing external power sources (e.g. pipe organ, electric guitar)
3. Automatization	• Discovery of the pinned barrel (900), programmable instruments • Construction of semi-automatic (e.g. barrel organ) and fully automatic instruments (carillon, music box, orchestrion, player piano), also by means of punched cards and discs, later by electronic means, analog sequencers, MIDI • Recording playing devices (Welte-Mignon reproducing piano, MIDI recording) • Composing devices (composers), mechanical: Componium (1821), electronic: Illiac (1956) • Combinations with animations, controlling through androids
4. Electronification	• Construction of electromechanical instruments (Cahill's Telharmonium, Neo-Bechstein electric grand piano, Hammond organ, electric guitar etc., since 1900, based on vacuum tubes) • Amplification of acoustic (mechanical) instruments through microphone and loudspeaker • Construction of electronic instruments (Thereminvox, Trautonium, Spherophone) • Construction of new non-mechanical interfaces (Thereminvox, ribbon controller) • First multimedia combinations ("color organ," 1725; light organ, sound-on-film technique)

(continued)

Table 1 (continued)

5. Modularization	• Construction of elements for sound generation and modification (prototypically in the Trautonium in the 1930s, since the 1950s as a modular synthesizer system, especially in the Moog synthesizer in 1967) • Largely free combination of sound elements, synthesizer = "sound kit" • Transistorization, leading to miniaturization and a drop in production costs
6. Digitalization	• Digital controllers (punched tape?, polyphonic keyboards), MIDI (since 1981) • Digital sound synthesis (numerically controlled oscillators) • Digital sound modification (realization by means of filters, amplifiers, effects through algorithms) • Digital sound sampling • Multimedia combinations with image and video, leading to the merging of the medially conveyed perceptual channels of video and audio, synthesis of the arts (*Gesamtkunstwerk*) • IC technology, inexpensive mass production
7. Virtualization	• Software-based digital synthesizers (native algorithms, also modularized, e.g. plug-ins) • Simulation and emulation of traditional instruments • Modeling of new (mechanically impossible) instruments, virtual synthesis, physical modeling, granular synthesis • Interactive instruments • Graphic and mouse-oriented user interfaces • Development of new interfaces, e.g. eye tracking, gesture controlling
8. Globalization	• Web-based musical communication, remote interaction with the machine • Global exchange of MIDI files and note information • Transmission of sound, partly compressed, in real-time or as files • Exchange and downloads of virtual instruments, sound designs or skins • Virtual concerts • Transfer of videos (incl. sound)
9. Informatization/ Artificial Intelligence	• Adaptive arrangers (since about 1990) • Musical automata, music-making androids (already since the 18th century) • Creative composing systems • Automatic analysis systems
10. Hybridization	• Brain-controlled multimedia synthesizers (audio and video) • Artistic man-machine-symbioses • Sounding rooms, interactive 3D instruments • Musical (multimedia) experience • Music production and reception without hardware?

digital electronic media and music technology, and whose significance for bodiliness in its interactive engagement with immaterial codes of music and its generation in a virtual environment is being discussed thoroughly (cf. Jauk 2009, among others). For the musical acts and experiences of humans, the digital world of music will continue to engender manifold new perspectives and promising potentialities, which hopefully will not cease to spark artistic highlights in the future.

References

Animusic. (2013). *ANIMUSIC: Computer animated music*. Accessed January 20, 2013. http://www.animusic.com/previews/animusic1.php

Behrendt, F. (2005). *Handymusik: Klangkunst und "mobile devices."* Osnabrück: epOs-Music.

Bense, A. (2013). *Musik und Virtualität: Digitale Virtualität im Kontext computerbasierter Musikproduktion*. Osnabrück: epOs-Music.

Besson, D. (1995). *Les Musicographies*. Échirolles (Isère).

Enders, B. (2005). Mathematik ist Musik für den Verstand, Musik ist Mathematik für die Seele. In B. Enders (Ed.), *Mathematische Musik – musikalische Mathematik* (pp. 9–37). Saarbrücken: PFAU-Verlag.

Gerstberger, K. (1930). Technik und Recht im Bereiche der Musik. In L. Kestenberg (Ed.), *Kunst und Technik* (pp. 157–174). Berlin: Wegweiserverlag.

Gjeertsen, L. (2006). *Amateur*. YouTube. Posted by "Febersyn", November 7, 2006. http://www.youtube.com/watch?v=JzqumbhfxRo

Großmann, R. (1997). Konstruktiv(istisch)e Gedanken zur 'Medienmusik'. In T. Hemker & D. Müllensiefen (Eds.), *Medien – Musik – Mensch: Neue Medien und Musikwissenschaft* (pp. 61–78). Hamburg: von Bockel.

Harenberg, M. (2012a). *Virtuelle Instrumente im akustischen Cyberspace: Zur musikalischen Ästhetik des digitalen Zeitalters*. Bielefeld: transcript.

Harenberg, M. (2012b). *Virtuelle Instrumente: Von der Orgel zum Cyberinstrument*. Accessed January 20, 2013. http://audio.uni-lueneburg.de/texte/harenberg_virtuell.pdf

Hatsune Miku Live Party in Sapporo. (2011). *YouTube*. Posted by "Liexie Noiwa," June 10, 2012. http://www.youtube.com/watch?v=2uOmQmM1mg4

Henke, R. (2011). tod durch überfluss. Strategien zur Komposition elektronischer Musik. *Neue Zeitschrift für Musik, 1*(2011), 24–27.

Hermode Tuning Werner Mohrlok e.K. (2015). *Hermode Tuning: Program Controlled Tuning*. Accessed June 1, 2015. http://www.hermode.com/html/tuning-history_en.html

Jauk, W. (2009). *pop/culture+medien/kunst: Der musikalisierte Alltag der digital culture*. Osnabrück: epOs-Music.

Katzer, G. (2011). À la recherche du son inconnu. *Neue Zeitschrift für Musik, 1*(2011), 28–31.

Kim, J. H. (2012). *Embodiment in interaktiven Musik- und Medienperformances unter besonderer Berücksichtigung medientheoretischer und kognitionswissenschaftlicher Perspektiven*. Osnabrück: epOs-Music.

Lehmann, H. (2012). *Die digitale Revolution der Musik: Eine Musikphilosophie*. Mainz: Schott.

Menzel, K. H. (2005). *PC-Musiker: Der Einsatz computergestützter Recording-Systeme im Amateursektor*. Osnabrück: epOs-Music.

Wandler, H. (2012). *Technologie und Sound in der Pop- und Rockmusik: Entwicklung der Musikelektronik und Auswirkungen auf Klangbild und Klangideal*. Osnabrück: epOs-Music.

Musical Instruments as Assemblage

Paul Théberge

Abstract Traditional analysis and classification of musical instruments is often based on an account of the material characteristics of instruments as physical objects. In this sense, their material basis as a kind of purpose-built technology is the primary focus of concern. This chapter takes the position that musical instruments are better understood in terms of their place in a network of relationships—an "assemblage"—with other objects, practices, institutions and social discourses. Particular attention is applied to the violin, the electric guitar and the phonographic turntable as examples. The assemblage is variable, and the same instrument can be used differently and take on different meanings depending on its place within a particular assemblage; indeed, it is the assemblage that allows us to consider devices like turntables as musical instruments even though they were not designed for such purposes.

1 Introduction: From Musical Instrument to Assemblage

If one were to pose the question, "What is a musical instrument?" the answer would seem obvious: we all can conjure up in our mind's eye the image of a piano, a violin, a trumpet or an electric guitar; or, in our mind's ear, the sound of a wailing saxophone or a drum kit pounding out rhythms. Indeed, the term "musical instrument" is typically used to refer to a relatively narrow range of purpose-built technologies—technologies designed for the production of musically useful sounds; what is, or is not "useful," of course, will vary according to musical culture, genre and context.

The simple status of musical instruments as technical objects, however, is not without its consequences: organology, the science of musical instruments, is in part dedicated to the classification of musical instruments according to their essential

P. Théberge (✉)
Institute for Comparative Studies in Literature, Art and Culture, Carleton University, Ottawa, Canada
e-mail: paul.theberge@carleton.ca

T. Bovermann et al. (eds.), *Musical Instruments in the 21st Century*,
DOI 10.1007/978-981-10-2951-6_5

physical traits and sound producing mechanisms, their individual historical development, and their geographical and cultural distribution. In this sense, modern organology is perhaps best exemplified in the work of Sachs (1940), with it's elaborate classificatory scheme: developed earlier, at the turn of the century with E.M. von Hornbostel, the system attempted to classify all musical instruments, regardless of their cultural origins, in terms of their mechanism of sound production and then further differentiated them according to various physical characteristics. It is perhaps no accident that this descriptive and classificatory system arose as an offshoot of the early period of Comparative Musicology where the gathering of sound recordings and the collecting of musical instruments were key to the field and the colonial context within which it emerged: musical instruments were thus understood as cultural artifacts to be gathered and classified as part of an objective—and objectifying—logic.

The tendency to isolate individual instruments as singular material objects and to categorize them on the basis of their particular physical characteristics is thus not simply an academic pursuit: it is also central to our understanding of musical instruments as both cultural artifacts and as commodities. In the marketplace, musical instruments are valued according to their provenance—a Stradivarius violin or a Fender guitar of a certain vintage is regarded immediately as a sign of quality—as well as their unique characteristics (a particular pattern of pearl inlay or a hand-wired pickup can contribute to an increased evaluation).

But folklorists and ethnomusicologists have also long recognized that common objects and technologies built for other, more mundane purposes can also be put to musical use: for example, the adaptation of household implements, such as spoons and washboards, for use in rhythmic accompaniment has long been a common element of traditional Appalachian and bluegrass music. More recently, as computers have become central to music production in a wide range of avant-garde and popular musics, many musicians and DJs have come to think of their laptops as their primary musical instrument; on a somewhat larger scale, sound engineers frequently refer to the recording studio—comprised of a number of distinct spaces and technologies for the recording, processing and mixing of sounds—as their "musical instrument."

Clearly, these diverse notions of what a musical instrument is, or can be, challenge conventional conceptions of musical instruments as a limited set of purpose-built technologies. To meet this challenge, it is necessary to problematize the conventional characterization of musical instruments as objects and to understand them within the broader contexts in which they operate and take on meanings. To that end, I want to introduce the idea of musical instruments as a kind of "assemblage," a concept that allows one to take instruments into account not only as they are defined by their technical characteristics but also as they are constituted in variable sets of musical practices, genres, institutional settings, social ideologies and discourses.

The concept of "assemblage" was introduced by Deleuze and Guattari (1987, pp. 71–73, 503–505) and has been influential in a number of areas within the humanities and the social sciences. It is intimately related to other concepts used by the two theorists, such as their ideas concerning language and the "machinic" (Wise 1997; Guattari 1993), but an "assemblage," insofar as it can be isolated, is perhaps most succinctly described as "a multiplicity which is made up of many heterogeneous terms and which establishes liaisons, relations between them" (Deleuze and Parnet 1987/2002, p. 69). This multiplicity is extremely variable, historically contingent, and any object, activity, discourse or institution can take part in multiple assemblages. It is the individual components and their interactions that give a particular assemblage its special character: "Thus, the assemblage's only unity is that of a co-functioning" (ibid.).

In relation to music, Georgina Born has made use of the idea of the assemblage to describe the mediating role of technology in musical creativity (2005) and, in a more elaborate fashion, to analyze the links between musical practices and social identity (2011). In the latter work, Born identifies four planes of social mediation, ranging from the micro-social relations of musical performance and practice, to musical audiences and imagined communities, to the more abstract levels of social identity formed by class, race and sexuality, to the institutional forms that govern the production and economies of music (2011, p. 378); the assemblage is constituted by the interactions of all the elements of these four planes of mediation. Born's use of the assemblage in this instance—as a concept that operates at a high level of abstraction and ties together diverse levels of social interaction—is consistent with Deleuze and Guattari's work and with many others who have used the concept in the social sciences.

However, in a more recent work dealing with gender, education and electronic music, Born and Devine (2016) have used the idea of assemblage in what appears to be a more varied and flexible fashion. Throughout their essay they refer to a variety of "technological assemblages" (ibid., pp. 3–4), "material assemblages that incorporate technological objects," "techno-social assemblages" (p. 4), and "music-technological assemblages" (p. 8, 14), with each term underlining subtle nuances in their analysis of the technological mediation of gender. This usage suggests multiple, intersecting assemblages or, perhaps, a complex nesting of assemblages that may be extremely productive in thinking about musical instruments in general. With this in mind, I would like to briefly take up a number of examples, as case studies, to explore how one might think of musical instruments as assemblages.

2 Violins and Fiddles: Identical Instruments, Diverse Assemblages

In some of my earlier work (Théberge 1997), I drew on the ideas of Bourdieu (1984, 1990) in an attempt to understand the dynamics of musical instruments and practices. My aim was, in part, to understand how musical practices essentially redefine the character of an instrument and the sounds it produces. This turn to practice was critical, or at least it seemed to me at the time, as a way of counter-balancing the influence of design in the construction of musical technologies: the idea that "musical instruments are not 'completed' at the stage of design and manufacture, but, rather, they are 'made-over' by musicians in the process of making music" was an important point of departure for my study as a whole (Théberge 1997, p. 160).

In considering the differences between the "violin" and the "fiddle"—two instruments that are physically identical but differentiated by style, genre and practice—I invoked Bourdieu's notion of habitus to describe the structured dispositions, postures, and playing techniques that underlie these differences (ibid., pp. 166–167). While habitus is a useful concept for getting at some of the subtleties of musical practice and, in particular for Bourdieu, their basis in class and other social determinations, I would argue that because the concept depends on processes that are largely unconscious and the product of mimesis, it is not well suited for analyzing the larger range of social, institutional and discursive constructs and interactions that characterize the idea of assemblage.

In this sense, the concept of assemblage can be brought to bear in analyzing what is at stake in naming an instrument a "violin," or a "fiddle," by calling on a wider range of distinctions: not only does the violinist hold the instrument and bow in a different way from the fiddler, but the music they play, their concepts of musical sound, the ensembles and other instrumentalists with whom they play, the venues where they perform and the audiences they address, the educational and economic institutions that offer material support, the discourses that describe and validate their activities, and the social interests that inform those discourses constitute what might be called an assemblage. Considering the violin or the fiddle in this way reveals a network of relationships that is as complex as it is variable.

The idea of the assemblage might also be useful in considering musical instruments as they move from one cultural context to another: for example, the violin has been an important addition to a number of Eastern musical cultures, where it has been taken up in traditional musical ensembles as well as in the production of popular music. In classical Indian music, the instrument is held and played differently from any Western tradition—some instrumentalists sit cross-legged and hold the instrument between the shoulder and toe—and, beyond body posture and technique, the instrument resides within a network of musical relationships that reflect broad musical, social and cultural norms. It is beyond the scope of this short essay to go further with this topic, but it is important to note that such an assemblage would (much like the difference between the Western violin and fiddle) be quite different from that which constitutes the instrument as used within the Bollywood film industry, for example.

3 The Electric Guitar: Technological Assemblage as 'Integrated Circuit'

Modern electric and electronic instruments pose additional challenges to analysis, in part, because they are not singular technologies: for example, unlike a violin, a saxophone or an acoustic guitar, the electric guitar is relatively meaningless without a means of amplification; guitarists also frequently employ a variety of pedals and special effects devices that profoundly alter the sound of their instruments. In a sense, we might consider the guitar-effects-amplifier combination as a kind of "integrated circuit"—or perhaps a rudimentary technical assemblage—a device composed of multiple components, each of which is necessary for the functioning of the device as a whole. At one level, the differences in sound produced by these components are important in defining the stylistic differences between the uses of the electric guitar in a variety of genres, from country, to jazz, to rock and heavy metal music (cf. Waksman 1999). This is not simply a matter of practical or stylistic concern, however: with the multiplication of technical components, there is also the potential for multiple, intersecting (and possibly conflicting) assemblages.

A good example can perhaps be found in a segment of a documentary produced by the BBC, *The Story of the Guitar* (2008), in which guitarist The Edge speaks of his guitar setup for the U2 hit, "Where The Streets Have No Name" (1987). In the song, The Edge employed a digital echo effect that created a rhythmic underpinning to the notes and chords he played on his instrument: because he must coordinate the tempo of his playing precisely with the repeated sounds produced by the echo, The Edge effectively becomes part of the "integrated circuit"—a kind of real-time, musical embodiment of actor-network theory, human and non-human actors working together in a seamless, co-determined process (Latour 2005). But for my purposes here, what is interesting to note is the role of such digital effects devices within rock music culture.

Within rock, there has been a long-standing antipathy towards digital technology: synthesizers and drum machines have been regarded with scepticism, and outright hostility, by many rock musicians and fans. As a result, the use of digital effects devices in rock could potentially be seen as a point of contradiction within the genre: indeed, while The Edge stands in front of huge racks of effects processors in his practice space, where the documentary was filmed, equipment racks of this kind are typically kept off-stage during live rock concerts, far from the sight of fans. In a sense, the musical-discursive assemblage of rock, which emphasizes freedom and authenticity in performance, risks contradiction by a technical-discursive assemblage more typically associated with electronic music, which places a premium on the hi-tech values of precision and automated processes. Interestingly, what might be potentially perceived as an inhuman surrender to the machine-time of the echo is defused by The Edge when he describes playing rhythmically with the device as "almost like playing off a second musician… a kind of conversation."

The Edge's statement in the documentary thus asserts the dominance of the human (and the rock assemblage) over the potentially disruptive effects of an intersecting technical-discursive assemblage associated with digital culture.

4 The Turntable: An Assemblage en Route to a Musical Instrument

By now, the phonograph turntable is widely considered as not simply a playback device—a device designed for the consumption of music—but as a productive device, a musical instrument of the first order: indeed, the inclusion of courses in DJing and turntablism at prestigious educational institutions, such as the Berklee College of Music, in Boston, during the early 2000s marked an important moment in the legitimization of the turntable as a musical-technical assemblage.

Many commentators focus on the DJ or turntablist as the motivating force behind the transformation of the turntable into a musical instrument but an analysis of the assemblage that is the turntable suggests a broader set of social and cultural transformations that must be taken into account. Perlman's (2004) discussion of audiophiles, for example, situates the turntable within a male, domestic culture of technical preoccupations and obsessions that was quite different from the world of DJs. From the 1950s onward, hi-fi enthusiasts understood the turntable as part of a network of specialized audio components that could be fine-tuned into the "perfect" audio reproduction system. The turntable was thus part of an assemblage composed of multiple, specialized technologies, an aesthetics of sound quality based on the idea of 'fidelity,' a press dedicated to the dissemination of esoteric knowledge, a specialized manufacturing and retail infrastructure, and a broad-based understanding of the role of technology in the definition of music and leisure in a suburban, post-war lifestyle.

The turntable that became the centre of a variety of dance-oriented popular musics from the 1970s onward, however, was the focal point of a diverse range of assemblages: for example, Jamaican sound system DJs typically combined turntables with specialized audio mixers, headphones, equalizers, effects, and amplification systems whose chief aesthetic criteria was power and control, not fidelity (Henriques 2011). In hip-hop, DJ techniques of scratching and beat matching became the basis for a broader set of technical processes utilizing digital samplers, drum machines and recording studios, on the one hand, and the valorization of a distinct repertoire of past music made available through specialized networks for record distribution, on the other (Schloss 2004).

If the large-scale social component of these assemblages was based in a street culture that was informed by racial and economic inequalities, the aesthetics and social underpinnings of techno and rave culture, as described by Simon Reynolds (1999), was perhaps the flip side of (or a reaction to) the turntable assemblage inherited from the 1950s leisure culture. Rave DJs initially employed many of the same technologies as sound system and hip-hop DJs—turntables, mixers, headphones and massive amplification systems—however, they quickly added computers, elaborately timed projections and light shows and, more recently, cell phone and messaging technologies, to create a more "immersive" and "interactive" aesthetic environment. While mega-rave events have the air of a massively constructed "integrated circuit" designed for momentary pleasures, each component within the circuit can be analyzed within a larger set of assemblages: for example, Nye (2011) has turned his attention away from records, turntables and amplifiers to the role of DJ headphones in the production of electronic dance music (EDM). Used as a practical device for cueing recordings, Nye links the DJ headset to the dominant use of ear buds in contemporary music listening and to the longer history of headsets in broadcasting and air travel: understood in this way, the DJ headphones become a part of a larger techno-social assemblage linking dance culture to notions of mobility, modernity and travel in the early 21st century.

5 Conclusion: Instrument Design and Assemblage

I would like to end this discussion with a brief comment on the problem of designing new musical instruments in the 21st century, a central theme of the volume in which this essay is found. Instrument builders, interface designers and computer programmers all work and operate within their own techno-social assemblages—assemblages that influence how they think and work within their respective fields and influence the technical design of the devices they build. As Rodgers (2010) has argued, we often see the design process as isolated, but wider social discourses are reflected in engineering practices.

In addition, while instrument makers must, of necessity, focus their attention on the immediate problem of creating new and unique designs for the objects they produce, they must also try to introduce those objects and devices to a world of pre-existing and constantly shifting music-technological assemblages. To do so forces any new device into a precarious and uncertain terrain. This is not simply a matter of knowing the marketplace for one's products nor even a matter of working closely with a select group of musicians: designers of new musical instruments cannot ultimately determine or predict how their inventions will be taken up within musical culture nor how they might be adapted for unintended purposes. The development of the saxophone is a good example of the former problem—it was designed in the 19th century for the world of concert music but only truly became popular when it was taken up by jazz musicians decades later—and the turntable, as described above, an example of the latter.

What the concept of musical instruments as assemblage might offer to the instrument designers project, however, is a larger framework within which one might consider musical instruments not as singular objects, but as components within a network of other instruments, technical devices, social settings and educational, institutional and discursive contexts. Understood in this way, the project is not so much to design objects, but to design relationships.

References

Bourdieu, P. (1984). *Distinction: A social critique of the judgment of taste* (R. Nice, trans.). Cambridge, MA: Harvard University Press.

Bourdieu, P. (1990). *The logic of practice* (R. Nice, trans.). Stanford, CA: Stanford University Press.

Born, G. (2005). On musical mediation: Ontology, technology and creativity. *Twentieth-Century Music, 2*(1), 7–36.

Born, G. (2011). Music and the materialization of identities. *Journal of Material Culture, 16*(4), 376–388.

Born, G., & Devine, K. (2016). Gender, creativity and education in digital musics and sound art. *Contemporary Music Review, 35*(1), 1–20.

Deleuze, G., & Guattari, F. (1987). *A thousand plateaus: Capitalism and schizophrenia (translation and foreword by B. Massumi)*. Minneapolis: University of Minnesota Press.

Deleuze, G., & Parnet, C. (1987/2002). *Dialogues II* (H. Tomlinson & B. Habberjam, trans.). New York: Columbia University Press.

Guattari, F. (1993). Machinic heterogenesis. In V. A. Conley (Ed.), *Rethinking technologies* (pp. 13–27). Minneapolis: University of Minnesota Press.

Henriques, J. (2011). Auditory and technological culture: The fine-tuning of the dancehall sound system 'Set'". *Journal of Sonic Studies, 1*(1), October. http://journal.sonicstudies.org/vol01/nr01/a03

Latour, B. (2005). *Reassembling the social: An introduction to actor-network theory*. Oxford: Oxford University Press.

Nye, S. (2011). Headphone-Headset-Jetset: DJ culture, mobility and science fictions of listening. *Dancecult: Journal of Electronic Dance Music, 3*(1), 64–96.

Sachs, C. (1940). *The history of musical instruments*. New York: W.W. Norton & Co.

Perlman, M. (2004). Golden ears and meter readers: The contest for epistemic authority in audiophilia. *Social Studies of Science, 34*(5), 783–807.

Reynolds, S. (1999). *Generation ecstasy: Into the world of techno and rave culture*. New York: Routledge.

Rodgers, T. (2010). *Pink noises: Women on electronic music and sound*. Durham, NC: Duke University Press.

Schloss, J. G. (2004). *Making beats: The art of sample-based hip-hop*. Middletown, CT: Wesleyan University Press.

The story of the guitar. (2008). Three-part documentary television series, BBC One, London, broadcast 05, 12, 19 October. Segments of the documentary are available online at: http://www.bbc.co.uk/musictv/guitars/

Théberge, P. (1997). *Any sound you can imagine: Making music/consuming technology*. Hanover, NH: Wesleyan University Press/University Press of New England.

U2. (1987). Where the streets have no name. *The Joshua Tree*, Island Records.

Waksman, S. (1999). *Instruments of desire: The electric guitar and the shaping of musical experience*. Cambridge, MA: Harvard University Press.

Wise, J. M. (1997). Living in a Deleuzian World. In *Exploring technology and social space* (pp. 57–82). Thousand Oaks, CA: Sage.

Instrumentality as Distributed, Interpersonal, and Self-Agential: Aesthetic Implications of an Instrumental Assemblage and Its Fortuitous Voice

Deniz Peters

Abstract Philip Alperson, in 'The Instrumentality of Music', extends the commonsense concept of musical instrument to an understanding encompassing the instrument's musical, cultural and conceptual situation. This understanding shifts the focus from a work-based aesthetic to one in which "listeners appreciate the human achievement with specific regard to accomplishment in the context of the demands of the particular instrument involved". With this advanced understanding of instruments and the instrumentality of music in place, I shall discuss a moment of genuine instrumental discovery (as opposed to deliberate design). During an improvisatory extension of the piano's sound board as part of a trio exploration with Bennett Hogg and Sabine Vogel using fishing wire, suspended bansuri flutes, contact microphones, and, vitally, transducers placed inside violins and on the piano's sound board, an unintended feedback loop formed, resulting in an additional voice, curiously turning the trio into a quartet. While the found voice's dynamics and character could be nuanced by varying the dampening of singular piano strings, as well as via the sustain pedal, it could, overall, only be summoned up and influenced in an indirect manner, via an ensemble effort. In analysing the situation of the discovery and in discussing its aesthetic implications, I offer a contribution to Alperson's notion of instrumentality in two respects: performers may together form a single voice, that is, their instrumentality might *join*; and an installation may, under certain conditions, acquire its own instrumental agency and identity, extending the cultural situation to include the natural environment, and the algorithmic.

D. Peters (✉)
Institut 14 Musikästhetik, University of the Arts, Graz, Austria
e-mail: deniz.peters@kug.ac.at

T. Bovermann et al. (eds.), *Musical Instruments in the 21st Century*,
DOI 10.1007/978-981-10-2951-6_6

1 Introduction

If a guitarist grows, files and shapes her fingernails to achieve a particular contact to the string and a particular sound, then is not the fingernail a part of the instrument? And if a saxophone player manages to create a recognisable timbral signature despite the individual characteristics of different instruments she plays, then is not the *embouchure*—the totality of facial, lip, mouth, tongue and throat anatomy and control, including the skill in adjusting the configuration as to produce the characteristic sound—a pivotal part of the instrument? If a particular sonic demand—say, the demand for a more carrying sound—guides a luthier's amendments to the material and design of a violin, then is the guiding intention, which might have sprung from the demands of a social setting in which the music is to be performed, not part of the instrument, implicit as it may be?

Philip Alperson, in his 'The Instrumentality of Music', elucidates how questions such as these have led him to problematise and qualify the "commonsense view of musical instruments", according to which "musical instruments are devices that performers use to make music" (Alperson 2008, p. 38). Alperson enlarges the commonsense concept in two directions: firstly, as to the musicality of instruments, in the sense that (if I read him correctly) music making is not the sheer mechanic activity of producing a sound, but a personally and socially, hence culturally and historically meaningful, intentional activity that turns the instrument into a *musical* one. This activity includes the body in a way that "in some cases, it is difficult to know where the body ends and where the instrument begins", so that "the performer's musical instrument is better understood as an amalgam of material object, the performer's body, and bodily dispositions as habituated by the developments of various musically related skills" (p. 46). Things that might not appear to be instruments at first sight turn out to be instruments after all through their use by people who are not classically thought of as musical performers, but who use them musically: composers, conductors, recital hall acousticians and technicians, and listeners using sound reproduction devices. Alperson shows that it makes sense to include composition software, batons, performance spaces and mobile audio devices (given the musical intentions by those using them) in the category of instruments, very much like "'natural' and 'found' instruments" such as "conch shells, grass reeds, stones […] a typewriter, a steamboat whistle" and so on. All these can become musical instruments when one essential condition is met: "What counts is that an object takes its place in the world of musical practice" (p. 38). Not only is the manipulation of the instrument an intentional activity, however, but also do instruments themselves become what they are through being part of a practice and should thus not be conceived as *separate* entities from that practice. In Alperson's words, "we must understand musical instruments as culturally freighted objects, that is, as objects that arise in the context of the history of musical practice" (p. 46). With this, Alperson arrives at an advanced understanding of musical instruments as "instrumentalities in the context of human affairs" (p. 47). Ontologically, this extends the idea of musical instrument to include its musical, conceptual and cultural *situatedness*.[1] (Remarkably, this means that a piano is *not* an instrument without the immaterial practices which have unfolded and which keep unfolding around it.)

The second direction in which Alperson enlarges the commonsense concept of musical instruments is towards what he calls the "instrumentality of *music* [italics mine]" (p. 46). This is an exquisite effort to view performativity, performance, performer, and performed instrument as intimately intertwined with the musical work. Alperson recognises the performance as an aesthetically appreciable event in terms of its instrumental accomplishment, giving rise to the "*work-in-performance*" (p. 47)—in this understanding, a work is doubly bound in consciousness.[2] Also, Alperson argues that the music's cultural and social agency enters the (thus enhanced) work concept. The idea of music's "full instrumentality" (p. 50) as conceived by Alperson, then, recognises both its aesthetic and "nonaesthetic" (socially instrumental) (p. 49) artistic values. By identifying the enlarged concept of musical instrument as a point of intellectual leverage, Alperson manages to synthesise the reach of music into human life, and the reach of human life into music, into a single, overarching idea.[3]

In what follows I shall consider a case of joint instrumental creation, describing the creative situation and process in and by which a group of three improvisers (including myself) explored the instrumentality of an installatory assemblage. The creative process led to the discovery of a semi-autonomous voice, a voice that in its behaviour and timbral character resembled an Aeolian harp. Reflecting on this voice and on its effect on, and role in, the trio improvisation, as well as on the particular aesthetics amidst which it arose (an aesthetics of interpersonal engagement and intimacy), and building on Alperson's concept, I shall argue towards three points of differentiation, nuancing, and extension: (1) instrumentality can be distributed in the sense that it can be established across various instruments and various players; (2) instrumentality is not limited to the cultural domain but may also include the (natural) environment; (3) a crucial part of hearing the work-in-performance—next to hearing and appreciating the work and the performer's artistic accomplishment—is the appreciation of the *interpersonal* accomplishment within the work-in-performance whenever there is more than one performer. This last point adds a third consciousness to Alperson's pair, in that interpersonal performativity is

[1]Simon Waters, in his 'Touching at a Distance', extends and argues a strikingly similar point, with fascinating insight on the body-sound relation. He summarises: "The constraints and constructs upon which music depends are not only, not even mostly, to be found in the physical object of the instrument, but also in the physiology of this particular body, in the assumptions and embodied knowledge which operate in this particular player, and in the interpenetrations between all of these and the framing acoustic and social environment" (Waters 2013, p. 123).

[2]In the sense that a listener can simultaneously appreciate *both* the work and the performer's instrumental accomplishment.

[3]Alperson's insight opens up a view on all the socially and perceptually fine-grained detail involved in what, in ordinary talk and understanding, has sedimented to the vague and near meaningless expression that this or that practice or object 'blurs the boundaries' between instrument and, say, composition.

present (i.e. expressed) in the work as performed. I suggest that this *interpersonal* aspect of duo, trio, quartet, or any multiplayer performance is a readily overlooked key component of both the musicality of instruments (and their ensemble constellations) and music's (social) instrumentality. In advancing an argument based on what was originally found via a complex creative process, I put forward a case in point towards the view that artistic research may yield conceptual insight.

2 Exploring the Instrumentality of an Installatory Assemblage

In the summer of 2014, during a research residency by Bennett Hogg and Sabine Vogel at the aesthetic lab of the Emotional Improvisation research project based in Graz,[4] we decided to concentrate our trio work on a transfer from an outdoors duo fieldwork situation to an indoors trio setup including a lid-less grand piano. Hogg and Vogel were interested in trying out an instance of transfer—as the question of outdoor fieldwork/indoor performance adaptation is a recurring theme in their own work in Landscape Quartet (a practice of *musicking in and with* nature)[5]—, whereas my specific interest was to observe and better understand which part of the experience would provide the coherence between the two events, thus making the transfer metaphor meaningful. We realised that the shift from windy mountain to piano in a calm room would be a challenge, and apart from transplanting a part of the mountain installation into the project room, we had no preconceived idea of how an experientially meaningful shift would eventuate.

For a number of days, Vogel and Hogg had hiked up to *Schweizeben*, which is a one hour train-ride and three hour walk north of Graz, a city in southern Austria in a region that is situated at the southeastern fringes of the Alps. The site at *Schweizeben* was at about 1000 m altitude, sufficiently removed from the surrounding valleys to be outside the general reach of sounds from human activity. The weather on those days was often windy, with an imminent thunderstorm, and it was the wind's agency as well as the quasi-social relation between trees that became the

[4]The project which I direct is called Emotional Improvisation: Musical, Interactive, and Intermedial (Austrian Science Fund FWF/AR188, 2014–2019) and hosted by the University of the Arts Graz.

[5]In a nutshell, the practice continuously developed by Landscape Quartet (Bennett Hogg, Sabine Vogel, Stefan Östersjö and Matt Sansom) resists standard soundscape art techniques and approaches that represent nature in sound, instead playing in and with the natural environment, with a particular emphasis on connecting with it, and on giving it a voice. Hogg reasons that he chooses "to work *in* 'Nature', taking up an ecosystemic approach to the working process as a way to resist the ecomimetic tendencies of producing work *with* 'Nature'; not trying to 'render' the natural through the cultural" (Hogg 2013, pp. 264–265). Vogel (2015, p. 332) describes Landscape Quartet's practice of "reconfiguring" an outdoors work indoors.

Fig. 1 Sabine Vogel at *Schweizeben* playing a bansuri flute while listening to the Aeolian installation via headphones

primary focus of attention within Vogel's and Hogg's improvisatory experimentation (Fig. 1).[6]

The instrumental installation on the mountain consisted of three bansuri-flutes hanging off a nylon fishing wire spun between two trees. Hogg and Vogel tied a hydrophone to the wire on which the bansuri flutes were suspended, and the duo explored ways of improvising with the live-recording of the wind playing the fishing wire and the bansuris themselves in a number of varied setups including a blown violin. That the wind could actually play the suspended bansuris was unexpected, and it was just as unexpected that the bansuris would resonate with the fishing wire, amplifying its aeolian harp effect.[7] This experience led Hogg and

[6]In 'Tuning-in' (Vogel 2015, esp. pp. 327–329), Sabine Vogel describes her practice of "tuning-in" whereby she—via "slow walking, listening, feeling, smelling, watching, observing, meditating, being still and aware" (Vogel 2015, 327)—develops a sense of the environment, and experiences the reciprocity of its presence and her own presence in it.

[7]Vogel (2015, p. 328) mentions her ongoing practice of placing her bansuris so that the wind can play them. In earlier pieces she had stuck flutes into the soil (for a work on hills at Allenheads) and hung them off a beam (for a work at Klagshamns Udde), but, according to personal conversation, she had never before suspended them off a fishing line that itself functioned as an Aeolian harp, nor had the bansuri resonated with a string so that they could be heard without microphones and electronic amplification.

Vogel to mount contact microphones directly onto the bansuri flutes in the subsequent indoors setting.

Vogel and Hogg brought back recordings of aeolian-induced sonic passages, and of a passage of interlocking cowbell sounds from a group of cattle that on one occasion grazed close to the site. Back in the project room we were joking that now I, at the piano, would have to be the wind and the cows. And we went ahead and tied the fishing wire to a piano string. The resonance did carry, but so softly that we ended up amplifying the sound—adapting an installatory element used in the outdoors experience. For this Vogel and Hogg tied a contact microphone to a suspended bansuri flute, sending the signal to a transducer, which is a small vibration speaker that instead of a membrane uses any surface it is attached to as resonator.

Rather than putting the transducer into the piano directly, we wanted to include Hogg's instruments in this growing resonance chain, and so he attached it to one of his violins. Hogg wrote a software patch to send and modulate the sound from a small microphone mounted on the other violin via his laptop to a transducer placed in one of the circular holes of the piano's iron frame, i.e. directly on the piano's wooden soundboard, a soundboard made from spruce.

We improvised a number of pieces, trying out different installatory tweaks, introducing further transducers into the piano to play back the aeolian and cowbell recordings, and a further flute, violin and microphone. The exploratory process was driven by the question whether we could somehow enter the same or at least a comparative experience to that of the environmental fieldwork (which was marked by an overall sense of connectedness with natural agency). After every alteration of the instrumental assemblage we improvised a piece or a set of pieces, subsequently analysing changes to the level of interconnection and ensemble playability the alteration had effected (Fig. 2).

And then it happened. Not only could we hear that the piano string's sound was being extended and modulated through the resonance chain going via the flute and past the first violin; the sound obscurely ended up feeding back through the piano, in such a way that I could feel its vibrations in the keys and in my fingertips.

A feedback-like voice arose, a voice that could move between a number of pitches, amongst other factors depending on the pitches played on the piano. I discovered I could finely influence the dynamic of that voice via the pedal, and found that nearly every millimeter of pedal movement made a difference.

Further on I found that even the raising or lowering of a single key's damper changed the spectral character of the voice. Rather than systematically dissecting the limits and behaviour of that voice in a mechanistic and isolated fashion, we kept on improvising with it, so as to not allow for our constructive intentionalities to override the experience of discovery. We understood that we were dealing with an unfamiliar other, and we kept the shared imaginative connection we had immediately made between the natural agencies on Schweizeben and that voice's agency intact. Intently, we tried to explore the newfound instrumentality and increase our

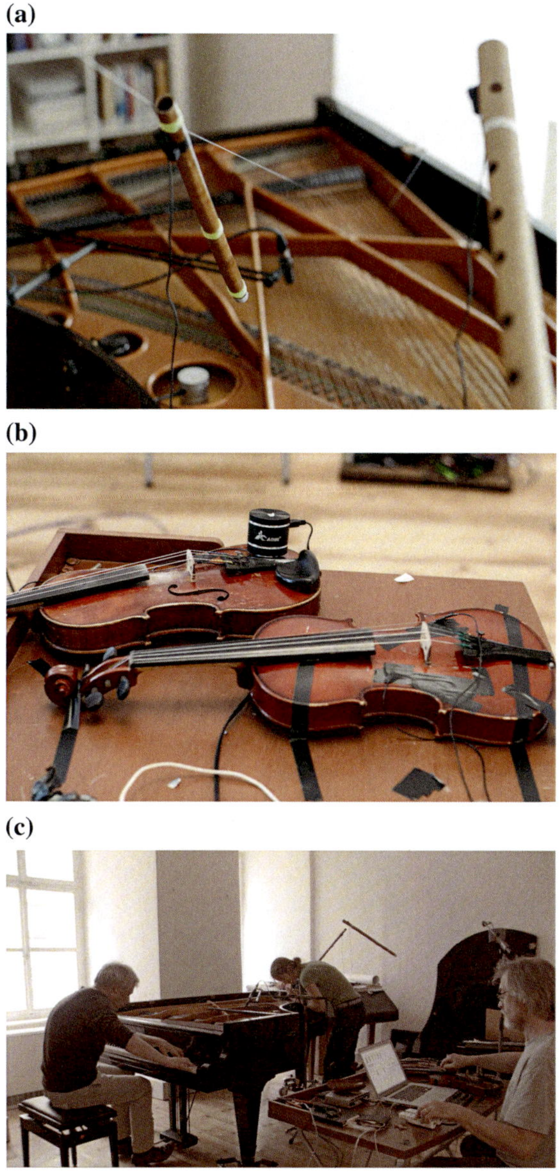

Fig. 2 Images of the installation in the project's aesthetic lab. **a** Detail of suspended bansuri flutes and contact microphones, plus transducers inside the piano's frame; **b** Hogg's violins, one with transducer, the other with small microphone; **c** performance situation (Peters, Vogel, Hogg)

interpersonal presence in the resonating circle that had formed. We discovered that, for example, if I slowly lifted the pedal to decrease the resonance's energy, Hogg could steer against it, using a controller to increase the microphone gain. At this point we had reached a setup in which Hogg and myself felt that we were able to hear and 'feel' each other's nuanced presence in the overall sonic development very clearly, with Hogg describing the experience as that of a "circuit"; Vogel however uttered that she, playing the flute and not the bansuris, felt "excluded". It was only after some further adaptation—such as making the bansuris accessible to Vogel by removing surrounding objects to extend the space around the far end of the piano—that both Vogel and Hogg could be heard to give sonic impulses that helped raise and sustain, or dissolve the voice. The instrumentality had spread to the entire trio. Yet on top of this, the voice retained a behaviour of its own. We noticed our trio had become a quartet.

The fourth, semi-autonomous voice suggests that, next to the separate instruments, the interconnectedness of the instruments creates a new instrument—the one producing that very fourth voice. What, then, *is* the instrument in the given scenario? Following what Alperson calls the 'commonsense understanding' of musical instruments—i.e., the notion that "musical instruments are devices that performers use to make music" (p. 38)—the present instrumental complex could be read either as an installatory extension of the piano; or, just as well, as an installatory combination of piano, strings, flutes, violins, microphones and transducers to be performed by three players. But the instrumental assemblage further invites a closer analysis in terms of the advanced understanding for which Alperson argues, in which "the musical instrument is better understood as an amalgam of material object, the performer's body, and bodily dispositions as habituated by the developments of various musically related skills" (p. 46). This certainly seems to hold for each performer individually in the given case. But, beyond the individual level, it seems that all three performers together work on establishing a *shared* instrumentality. The striving for this finds its expression in Vogel's remark that she, at first, felt excluded from "the circle"; her insistence in becoming part of the circle motivated a phase of collaborative creative experimentation.

Is there sufficient reason to think that a shared instrumentality was in fact gained? First, a reminder: the bodily extension of an instrument—like in the case of Alperson's embouchure example—is not simply a given, but an achievement, as it may depend on *skilful* conduct; only upon establishing skilful bodily extension is an instrument in Alperson's enlarged sense created (see his first category of enlarged understanding, ibid.). Now, whenever an instrument is played by multiple performers, and when, *also*, its bodily extension is multiple, then a compound sound or even single sound as in the present example might become the result of a joint intentionality. In the present case physical vibrations are intersected by electronic circuitry and digital algorithmic modulation; yet resonances and feedback loops within the instrumental assemblage can and do form. Whenever they do form, they do so as a consequence of shared decisions and actions, actions that afford a careful balancing—a balancing of instrumental bodily extensions *and* intentionalities. As Alperson points out, in improvisatory music (and, one could add, notwithstanding notated elements) the spontaneity of musical decisionmaking is foregrounded. In the present case, three improvisers engage their intentionalities (recalling Alperson's second category of enlarged understanding, ibid.) at any given moment or for any duration to create the assemblage's compound instrumentality—what could be called its instrumental identity. The identity deviates from those historically formed ones of the individual instrumental components; for example the piano—despite its very characteristic Bösendorfer sound—stops sounding like a 'classical' piano as the new instrumentality's practice unfolds, modulating a set of contributing historic practices (this time recalling Alperson's third category of enlarged understanding, ibid.). Plus, the assemblage's resonance and feedback behaviour would produce sounds that are not direct consequences of instrumental actions, but arise from the assemblage's internal dynamic, giving the impression that the sounds are caused by an additional agency—like in the case of the wind at Schweizeben. The presence of this additional agency and the joint decisionmaking it affords from the players are also an important part of the emerging practice that comes with the newly developing instrumentality. On these grounds it seems justified, I think, to say that what otherwise would merely be a material assemblage has indeed become an instrument in Alperson's richer understanding.[8]

3 Distribution, Natural Environment, Interpersonal Accomplishment: Three Extensions to Alperson's Concept of Instrumentality

Shared instrumentality in the sense of the example discussed in the preceding section—involving an assemblage of multiple instruments and combining the intentionalities of more than one performer—is covered only to some extent by Alperson's concept.

In his diagram illustrating his extended understanding of musical instruments, Alperson includes, next to the performer's instrument, the instruments of composer, conductor, and audience (Fig. 3). Alperson counts baton and orchestra as the conductor's musical instrument (ibid.), and thus, indirectly, considers collectivity. The baton is, clearly, an instrument only in combination with the orchestra, as it does not make a sound itself (other than perhaps a nearly inaudible swishing noise as it cuts through the air, or a tapping noise with which some conductors call for order), but rather mediates the conductor's movements, which are directed towards sound production. To call the orchestra an instrument, however, seems accurate only insofar as there are situations when the individuality of its members is attenuated within (or abstracted away into) a collective sound producing body. The members of the orchestra are, in such cases, subjugating a part of their work-directed intentionality to that of the conductor. But a closer look reveals that amongst each other, as a collective or compound body, a shared, sound-directed and work-directed

[8]An appreciation of the factors that went into the constitution and ongoing restitution and unfolding of the new instrumentality during listening makes a difference. If a listener did not notice the existence of an intentional dynamic between the performers directed at establishing and maintaining a shared instrumentality, then much of the identity of the performance—the substance of its intricate interpersonal dimension—would be missed.

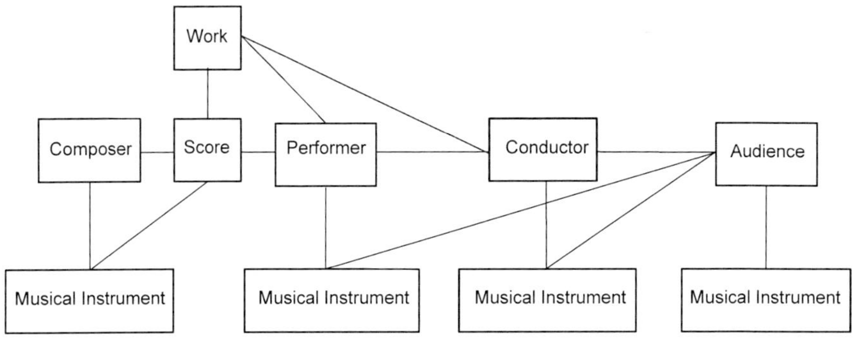

Fig. 3 Alperson's diagram of his third version of the 'standard presentation situation' (p. 45)

intentionality can also build: orchestras have their own characteristic 'sounds' and interpretive practices *with* which the conductors work (more or less successfully). The issue becomes even more apparent when considering that small ensembles without conductors—say, string quartets—still arrive at becoming 'one instrument' via a more fluid negotiation of sound- and work-directed intentionality between its individual members. The issue, then, is that through group consciousness or joint intentionality (like in the case of non-conducted ensembles, but even in the case of conducted orchestras), instrumentality becomes distributed between more than one person and more than one instrument. This more complex case is not explicitly featured by Alperson in his pertinent list of various intentionalities (p. 46), in all likelihood for space reasons. Put differently, while an ensemble is a combination of individual performers, each using her instrument in the discussed widened understanding, sophisticated ensembles form a compound intentionality, without which the sounds made individually would not merge into a cohesive overall sound and interpretation. Like in Alperson's observation of the orchestral case, their instruments join into a single instrument with *distributed* instrumentality.

The suggestion of distributed instrumentality invites yet another closer look. If individual instruments (each made up of material objects, bodies, and skills) combine to form a larger unit, and if intentionalities combine also, then this distribution needs to be taken into account in the idea of instrumentality. However, the combining of intentionalities, as well as the reaching of stretches of joint intentionality or states of group consciousness, are not only extraordinary achievements, but also a question of a choice made for a particular aesthetic position. When in the preceding section I noted and argued that "the instrumentality had spread to the entire trio", this shows that instrumentality can shift between its individual (monadic) and distributed (shared) forms. Performers—for instance in the case of a disjunct improvisatory setting in which three subjectivities independently contribute to a sonic situation—start out with three separate instrumentalities. Now, depending on the *aesthetics* of the improvisation, these instrumentalities might remain separate if performers choose to restrict their interpersonal activity to negotiating *individual* sonic territories for example (each strand of activity here being musical in itself).

That is, an audience (and the performers themselves) would here be in the presence of three individual minds representing their individuality. But if performers choose to increase the interdependence of their decision-making, if they choose to (and competently) create *shared* sonic gestures for instance, then, in *these* instances, their intentionalities align, and instrumental intersubjectivity is reached. (In the example with the fourth voice arising from joint action, the presence of this fourth voice is direct evidence of such collective decision-making successfully in place.) An ongoing passage will then sound as if it were created by a single being—while in fact being created by a *compound* being, via a balancing of individual and group consciousness.[9] Shared instrumentality is, therefore, fragile or even precarious,[10] a state to be reached rather than being ordinarily given, since a musical process—improvised or composed and interpreted—might shift and drift between keenly negotiated subjective co-presence, stretches of achieved intersubjectivity, and arbitrary sound making.

A second conclusion that can be drawn from considering the production and reception aesthetics of the discussed trio example is that, given suitable instrumental design, natural agency—in the form of environmental agency—can enter the musical setting and its instrumentality as another contributor. With 'natural agency' I do not only mean that, say water may enter into the assemblage by way of the three modes identified by Stefan Helmreich ("evoking", "invoking", "soaking", see Helmreich (2012), 'Underwater Music: Tuning Composition to the Sounds of Science', p. 153), or that animals and plants may enter it. I also mean any other dynamics an (instrumental) environment may have on its own accord or upon stimulation. This includes self playing media, circuits, algorithmically driven instruments and, more generally, media-instruments[11] whenever they are capable of autonomic variation and of (at least a small level of) interaction. The alterity of all these potentially agential entities (in their agency going beyond the composer's instruments already identified by Alperson, p. 45) enriches the image of the instrumentality of music.[12]

And a third extension of the understanding offered by Alperson concerns the point made about the way in which, in appreciating the work-in-performance, "we might speak of a double consciousness of the performance of the work and the performance in the work". As was seen in the present discussion of the achievement

---

[9]cf. Sherrie Tucker's intriguing account of the experience of rubber band group improvisation: "At first I experienced choices. [...] Then, we all became a body/organism and I stopped thinking in terms of what my own impulses were [...] and tuned in to what the organism was doing. [...] It wasn't a passive experience to fold into the organism, but a different way of being. [...] There was a shift in consciousness. Maybe it was the reorientation from self-consciousness to group-consciousness?" (Hahn 2016, p. 158).

[10]I am grateful to Férdia Stone-Davis drawing my attention to this in personal conversation (2016).

[11]A term fittingly coined by Sarah-Indriyati Hardjowirogo in her 'Medien-Musikinstrumente' (forthcoming).

[12]Simon Waters's idea of "treating performance as a complex dynamical system in which the feedback loops and interpenetrations between performer, instrument, and environment are fully recognised" (Waters 2013, p. 122) anticipates my point.

of shared instrumentality and joint intentionality, the performance 'in the work' is *in itself* twofold: not only can a listener appreciate *technical* accomplishment and virtuosity; she can also appreciate the performers' *interpersonal* accomplishment and virtuosity. The diagram (ibid.) could thus be expanded to include an arrow that is directed from performer back to performer, representing both the interpersonal distribution of instrumentality between performers within ensembles of any size, and the interpersonal sophistication a listener can appreciate.

This last point clearly offers a remarkable opportunity for an extended unpacking in many intriguing aspects, ranging from the vivid attainment (or falling short of) and display (or lack) of interpersonal phenomena such as trust, intimacy, responsibility, through to the sounding presence or absence of a Buberian I-You relationship (Buber 1923) finding its expression in music. I can, within the present context, only highlight this potential. But I think that I have covered sufficient ground to show that not only is an advanced consideration of instruments and instrumentality, with Alperson, instrumental to a more refined and subtle understanding of *music*; it is also high time to consider the crucial interpersonal dimension of musicking and creativity more carefully:[13] its culture, ethics, aesthetics, and future potential as a site of profound human encounter, societal and individual value, and growth.

References

Alperson, P. (2008). The instrumentality of music. *The Journal of Aesthetics and Art Criticism, 66* (1), 37–51.

Born, G. (Forthcoming). After relational aesthetics: Improvised musics, the social, and (Re) theorising the aesthetic. In G. Born, E. Lewis, & W. Straw (Eds.), *Improvisation and social aesthetics*. Duke University Press.

Buber, M. (1923). *Ich und Du*. Leipzig: Insel Verlag.

Hahn, T., et al. (2016). Banding encounters: Embodied practices in improvisation. In G. Siddall & E. Waterman (Eds.), *Negotiated moments: Improvisation, sound, and subjectivity* (pp. 147–167). Durham, NC: Duke University Press.

Hardjowirogo, S.-I. (Forthcoming). Medien-Musikinstrumente. In R. Großmann & S.-I. Hardjowirogo (Eds.), *Musik und Medien*. Laaber: Laaber.

Helmreich, S. (2012). Underwater music: Tuning composition to the sounds of science. In T. Pinch & K. Bijsterveld (Eds.), *The Oxford handbook of sound studies* (pp. 151–175). Oxford: Oxford University Press.

Hogg, B. (2013). When violins were trees: Resistance, memory, and performance in the preparatory experiments for *landscape quartet*, a contemporary environmental sound art project. *Contemporary Music Review, 32*(2–3), 249–273.

Vogel, S. (2015). Tuning-in. *Contemporary Music Review, 34*(4), 327–334.

Waters, S. (2013). Touching at a distance: Resistance, tactility, proxemics and the development of a hybrid virtual/physical performance system. *Contemporary Music Review, 32*(2–3), 119–134.

[13]Georgina Born's consideration of the microsocialities of performance and her current contribution to a social aesthetics of music (Born, forthcoming, esp. pp. 56–59) is a brilliant and rare example of this.

Interactivity of Digital Musical Instruments: Implications of Classifying Musical Instruments on Basic Music Research

Jin Hyun Kim and Uwe Seifert

Abstract The introduction of the computer as musical instrument and the development of interactive musical instruments have led to completely new purposes and questions for music research; as a result, it no longer seems adequate to rely on the traditional classification of musical instruments, which is based on the purpose of instrument design and presentation of instruments in public or private exhibition. Based on insights from the philosophy of science, this paper suggests pursuing another purpose of and approach to instrument classification appropriate for basic music research. We argue that (digital) computing systems, to some extent, have the potential to act as autonomous and artificial social agents. This argument is based on the conceptualization of machines as (abstract) automata. In addition, we exploit concepts from dynamic systems theory in a metaphorical manner to find a more appropriate point of view to develop new research questions. Discussing interactivity, for which embodiment and situatedness are prerequisites, we suggest taking interactivity, agency, and autonomy into account to develop an appropriate classification system of musical instruments and at the same time to rethink the traditional concept of musical instrument. Whether a musical instrument can be defined as broader than a device that has the function of generating sounds, i.e. whether it can be viewed as an embodied, situated or even social agent, remains a challenging question for basic music research. To discuss this question, not only sound generating actions, but also other musically meaningful actions that involve agency should be taken for granted.

J.H. Kim (✉)
Institute for Musicology and Media Science, Humboldt-Universität zu Berlin,
Unter den Linden 6, 10099 Berlin, Germany
e-mail: jin.hyun.kim@hu-berlin.de

U. Seifert
Universität zu Köln, Musikwissenschaftliches Institut,
Albertus-Magnus Platz, 50923 Cologne, Germany
e-mail: u.seifert@uni-koeln.de

T. Bovermann et al. (eds.), *Musical Instruments in the 21st Century*,
DOI 10.1007/978-981-10-2951-6_7

1 The Hornbostel-Sachs System and Musical Instrument Classification

A musical instrument is defined, in general, as a sound-generating tool used explicitly for making music. In traditional musicological studies on the classification of musical instruments, the system developed by von Hornbostel and Sachs (1914) is representative: physical-mechanical devices manufactured to generate sounds are foregrounded, with focus on the mechanism of a sound generator, including its material and the ways of vibrating it. Musical instruments that generate sound by means of electrical energy were subsequently included into the Hornbostel-Sachs (HS) classification system in 1940 as electrophones. Hugh Davies' system of classifying electrophones into electroacoustic, electromechanical and electronic musical instruments (Davies 2000) provides a further subset of the HS system, extending and at the same time sustaining the HS system. However, with the invention of digital computers and computer programs describing the generation of sounds and using these computers as musical instruments (Mathews 1963)—for the sake of brevity, referred to here as 'the computer-as-a-musical-instrument'—these classification systems may no longer be adequate.

In the context of computer music, the term "instrument" refers to a program describing an algorithm that realizes or performs a musical event (Dodge and Jerse 1985). The principles of algorithmic sound generation are based on information processing, which is, following Norbert Wiener, to some extent not material or physical: "Information is information, not matter or energy" (Wiener 1961, p. 132). Information and not energy is conceived of as constitutive for musical events and essential for a model of musical sounds. Musical events that are generated by means of a mathematical construction are algorithmically computed. To what extent can 'the computer-as-a-musical-instrument' be included into the HS system? Since algorithmically generated sounds can be heard by transforming numbers into electrical energy vibrating the loudspeaker's mechanism, one may try to classify the computer as an electrophone, taking into account the resonating mechanism of loudspeakers. This attempt to latch onto a single aspect of making algorithmically generated sounds audible, however, would lead to a modification of the HS system's criterion. In the first HS system (von Hornbostel and Sachs 1914), in which electrophones were not included, the original criterion for classifying musical instruments, i.e. the material of the sound generator, was already slightly modified and extended by the class aerophones. Aerophones that allow for sound generation by air are classified by how the sound generator is vibrated rather than by the material of the sound generator whereas for idiophones, membranophones, and chordophones the material that is vibrated to produce sound, such as a part of the instrument itself, e.g. a membrane or string(s), plays a substantial role and in turn becomes—in terms of traditional logic—a principium divisionis. For this reason, the logical consistency of the HS system was put into question (Simon 2004). A further modification of the classification criterion applied to the computer concerning its resonating mechanism would render the HS system definitely improper since it would bring together both the properties of the sound generator and those of the resonator in terms of inclusive disjunction—in other words: a musical instrument could be classified taking both the properties of the sound generator and those of the resonator into account. As a result, some instruments that would be assigned to one class when focusing on the properties of the sound generator might also belong to another class regarding the properties of the resonator and an instrument could be even in both classes; according to the adequacy conditions for classification this would prove to be incorrect.

2 Scientific Concept Formation: Classification and Taxonomy

2.1 Classification and Taxonomy I: The Logical Concept

To elaborate, a classification system, i.e. a taxonomy, is based on classifications of a domain into classes (sets). Logically, there are two adequacy conditions for a tenable classification. Firstly, exclusiveness: the classes in a classification are mutually exclusive, i.e. members of one class cannot belong to another class. Secondly, completeness: a classification must be complete, i.e. there is no member of the domain that is not a member of a class. A classification system or taxonomy, then, forms a hierarchy of classifications. As a correct classification system consists of correct classifications at all its levels, classification systems, as classifications, can be correct or incorrect. The correctness of the system depends on correct classifications and their fulfillment of the adequacy conditions. So, from a purely logical point of view, the concept of classification system is well-defined. Moreover, classification is the most fundamental scientific concept (Hempel 1952; Stegmüller 1974, pp. 19–27). Logical standards of scientific classification should be respected and maintained in empirical research. However, the question of whether logical standards are sufficient for empirical research is posed: How is the logical structure of classification related to empirical data? What is the empirical content of a classification system? If we are developing a correct classification system, there is no guarantee that it will be scientifically fruitful and exhibit empirical content.

2.2 *Classification and Taxonomy II: The Problem of Empirical Adequacy*

In the empirical sciences, a scientifically fruitful classification system allows for gaining new insights and formulating new empirical laws. Exclusiveness and completeness are logical adequacy conditions of correct classifications. A purely logical correct classification system forms a necessary condition for fruitful empirical research—but it is not sufficient. The question to be discussed in detail is (Stegmüller 1974, p. 21): Given a classification C of a domain D in k classes (sets) S, how do we know that C satisfies both adequacy conditions?

Two answers are possible: (1) Both conditions are satisfied because they are a logical consequence of the definition of the k classes S. A logical analysis, then, is sufficient to show that C is a correct classification of D. But in such cases we are (only) dealing with logical truths. (2) For the empirical sciences, the more interesting and important cases deal with empirical truths. In such cases it is logically possible that extensions of (empirical) concepts, i.e. the classes S, do not encompass the whole domain D, i.e. the completeness condition is violated. It might also be logically possible that extensions of concepts intersect, i.e. the exclusiveness condition is not respected. In empirical interpretations, adequacy conditions might indicate scientific laws or—as in biology—are scientific laws concerning that domain. In such a case, a classification respecting the exclusiveness criterion expresses an empirical fact that one, and only one, combination of properties can be ascribed to an animal.

In asserting an empirical classification of empirical phenomena, two ideas are interwoven and assumed. Firstly, an assertion is a summary of empirical facts, meaning that until now no other entity has appeared. Secondly, there is a hypothetical assumption involved that the assertion will hold even in the light of future discoveries. A revision of a classification may be dependent on to what extent it allows for new or broader scientific insights. According to Wolfgang Stegmüller (Stegmüller 1974, p. 27), the relation between the logical structure and empirical content of qualitative or classificatory concept forms, i.e. of classification, can be summarized as follows: Classification systems are based on conventions because the manner of classifying a domain in classes is a scientist's free choice. But conventions that are guided by considerations concerning simplicity and fruitfulness of the system are not the only relevant thing for building classifications or classification systems; empirical facts are also necessary to ensure that both adequacy conditions have been satisfied. Moreover, hypothetical assumptions are necessary to support the claim that both adequacy conditions will be satisfied for the future. So, even at the most fundamental level of scientific concept formation, i.e. classification or classificatory concepts, there is an interplay between concept formation, systematic observation, empirical-hypothetical generalization, empirical confirmation and intuitive considerations of simplicity and fruitfulness.

We may conclude that, from a logical point of view, there is no doubt about what constitutes a correct classification system. Moreover, classificatory concepts are the most basic scientific concepts, and any claim that a correct scientific classification system could violate both adequacy conditions leads to pre- or unscientific thinking and concepts with scientifically less information content. Furthermore, the philosophy of science shows that scientific purpose and potential for new insights are of main importance in setting up a classification system.

3 The 'Computer-as-a-Musical Instrument', Interactivity and Automata: A Necessity of a New Approach to Musical Instrument Classification?

Considering the literature on musical instrument classification (Kartomi 1990; Montagu 2007; Simon 1992), in conjunction with readings from the philosophy of science (Hempel 1952; Stegmüller 1974), it is doubtful that—as some scholars propose—giving up the logical structure of correct classification systems helps tackle problems involved in the HS system. Furthermore, to us, it is doubtful that any approaches extending the HS classification system without explicitly discussing the purpose of a classification system can be fruitful at all. Coming back to the question of whether it makes sense to classify the 'computer-as-a-musical-instrument', we may also ask, is it necessary to revise the classification of tools for making music—which is based on their mechanisms of generating sound—when taking the conception of the 'computer-as-a-musical-instrument' into account? Two important changes took place within the last century which are difficult for the HS system to cope with, and need to be taken into account: (1) The need to study (social) interaction and interactive systems, i.e. interactivity; and (2) the need to study the autonomy and agency of cognitive artificial systems participating in artistic creation and performance. Point 2 is necessary because, with the invention of digital computing machines, the conception of a machine or automaton has changed dramatically. A (digital) machine or automaton is no longer a purely mechanical system, such as a clock.[1]

From a philosophical or epistemological point of view (Nelson 1989), the main difference between the classical conception of an automaton or machine as a mechanical system and the logical conception of an (abstract) automaton emerging in the 1950s consists in the importance of energy input and external control for the former. The latter conception takes into account signal input or information as input, i.e. the (physical) signal is a carrier of information. So, it is viewed as an information processing system. Furthermore, internal states and control underlying

[1]However, this was not recognized by Rebecca Wolf, who traces the history of the concepts of the automaton, machine, and clock (Wolf 2014).

a system's observable input-output behavior, i.e. rules guiding behavior (Nelson 1989), are taken into consideration. Internal control can be thought of as represented by rules, which describe state transitions and output based on current input and the state of a system, i.e. an (abstract) automaton with state transition function and output function. In this context, it is important to distinguish between embodied and free rules concerning the physical realization of an automaton. Embodied rules form the physical bases of a system's basic potential capacities, the processor, and enable systems to carry out free rules, i.e. algorithms. Free rules can be thought of as programs that can be added to a system by a programmer or by learning, which are then carried out by the system's embodied rules, i.e. its processor. In sum, this abstract conception of an automaton captures the idea of an information processing system with internal states and the system's input-output behavior that is determined by two functions expressed through a system of rules: an automaton is a rule system and as such a logical or mathematical object.

From an engineering point of view a rule system is a specification of a physical machine's behavior. According to Minsky (1967) a particular machine is typically identified by its functioning or behavior rather than by its material construction. Abstract automata, as already indicated, capture a machine's functioning or behavior in form of an abstract description, i.e. its input-output behavior in terms of rules. So, an abstract automaton is "a specification of how a physical object ought to work" (Minsky 1967, p. 5) or function. A machine, then, may be viewed as a physical realization of an abstract automaton: Digital computing machines are realizations of abstract automata. So, in general, an automaton is an abstract specification of a machine. The digital computer is a realization of an automaton and not a mechanical system like a clock, because its behavior is rule-governed and based on information. The concept of an (abstract) automaton is at the core of the theory of automata. In general, today's automata theory provides—with Turing-computability and the Chomsky hierarchy—a reference point and precise scientific conceptualization for the intuitive and pre-scientific term "automaton." Therefore, the 'computer-as-a-musical-instrument' may be viewed as the realization of an automaton. Moreover, automata theory provides the theoretical framework for cognitive science, which assumes as working hypothesis that cognition is some kind computation, by giving an explication of the intuitive notion of computation.

Since the digital computer is a realization of an (abstract) automaton, in particular a finite transducer, a substantial difference between classical mechanical systems and automata as information processing systems must be taken into account to understand the extent to which a 'computer-as-a-musical-instrument' differs from traditional physical-mechanical musical instruments. Using a (digital) computer, i.e. an information processing system and (abstract) automaton, as musical instrument involves embodied and free rules, i.e. algorithms. The question of how the conception of a musical instrument is changed therefore deserves thorough discussion.

4 Interactivity: Towards a New Approach to Musical Instrument Classification

4.1 Interactivity I: Interactive Versus Algorithmic Computation

In the course of the development of digital musical instruments, there is a development from a disembodied automaton to an embodied and situated one. The 'computer-as-a-musical-instrument' is based on algorithms realizing computational functions transforming input (i.e. arguments) into output (i.e. values). According to the American computer scientist Peter Wegner, in the first computer programs developed in the 1950s and 1960s, algorithms, whose descriptions are programs, realize computational functions. In this metamathematical paradigm of computation, the input is completely defined before the start of computation and the output provides a solution to a general problem at hand (Wegner 1997; Wegner and Goldin 2003). The procedure for calculating function values and the arguments of the input domain are specified in advance and cannot be changed during the execution of a program. Such algorithms of computation following identically reproducible rules ensure the identical recurrence of output. Therefore, computation of output values from their inputs by such an algorithm taking place in a closed system, which is not affected by factors of the environment external to itself, is not conceived of as an interactive system (Wegner 1997). The first idea of 'the computer-as-a-musical-instrument', based on its capacity to algorithmically generate sounds, was realized within the dominant paradigm of computation in the 1950s and 1960s, which can be characterized as non-interactive.

4.2 Interactivity II: Interactive Music Systems

Since the 1970s, 'the computer-as-a-musical-instrument' has played an important role in the context of live electronic music, which consists of the interplay between sounds generated by traditional musical instruments and electronically generated and manipulated sounds. For this purpose, so-called interactive music systems have been developed to capture data from a live performance and transform them into parameters for algorithmic manipulation of sounds. In early interactive music systems such as score-driven interactive music systems used, for instance, in Pierre Boulez's piece … *explosante-fixe* … (1973, 1974), Philippe Manoury's *Jupiter* (1987) and *Pluton* (1988), the principle of a knowledge-based system developed by a traditional approach to artificial intelligence served as a basis for interactive live performances. The score-following technique allows the computer system to monitor musical events coming from a live performance and to compare them with the score, which serves as a kind of represented knowledge of the computer system, so as to process computer-generated sound parts. Such kinds of interactive music systems using knowledge-based processing have a hierarchical structure of interaction processes—from the sensing, to processing, to the response stage (Rowe 1993). A knowledge-based process of interpreting information coming from the sensing stage, taking place in the processing stage, is separated from the sensing stage as well as the response stage. In other words, an exchange between internal and external processes does not take place during the processing stage. Output events of machines as a response to input events are determined in this isolated stage and realized by top-down organization. Hence, knowledge-based interactive music systems are conceived of as decoupled from the environment and therefore not interactive.

In a further stage of interactive music systems, not only sound events of a live performance, but also bodily actions and information provided by different kind of sensors are taken as input data. This kind of interactive music system therefore includes sensor systems to detect bodily actions and algorithmic mapping from parameters of bodily actions as parameters for sound synthesis and manipulation. If sensors are attached to an object that can be used gesturally, this object can serve as a controller for sound synthesis, comparable to a control unit of a traditional musical instrument, such as a piano key. This kind of controller is called a gesture controller. Recently, research on gesture controllers has flourished in the context of interactive music, and given rise to the "new digital musical instruments" (Miranda and Wanderley 2006). Accordingly, new digital musical instruments differ from those instruments that are also digital but do not contain a gesture controller. In our view, the term "interactive musical instruments" is more appropriate for referring to such instruments than "new digital musical instruments" since the latter does not specify what kind of aspects are added to digital musical instruments; several interface designers and composers already define or characterize their instruments as "interactive musical instruments."

4.3 Interactivity III: Interaction and Feedback in Research on Interactive Musical Instruments

In investigating interactive musical instruments, Thoben (2014) distinguishes three components: an interface, a mapping, and a sound-generating component, taking into consideration that an interface provides haptic and visual feedback and a sound-generating component provides acoustic feedback. Although he uses the term "interact" in relation to the (human) performer and the interface, as he writes "The performer interacts with a musical interface […]",[2] his attention is paid to a component of 'interactive' musical instruments that remains passive in the sense that it does not exhibit agency and is not able to initiate actions.

[2]Translation of Thoben (2014, p. 433, Fig. 1) by the authors: "Der Interpret interagiert mit einem musikalischen Interface, […]."

Let us elaborate this point: When "interaction" is used in the social sciences and the humanities, it is assumed that communicating systems involved in information exchange exhibit intentions, intentionality and make decisions to some extent. Traditionally, in the social sciences and humanities these capacities have been ascribed only to humans. Hence, in such conceptual framework only humans can interact: the terms "interact" and "interaction" can only be correctly ascribed to humans; machines and animals are excluded. Whether this is still appropriate today is another question and is under discussion but—in our opinion—might be reasonably doubted. In computer science and research on human-machine interaction [sic!] "interaction" is often used without any further specification in a loose and metaphorical manner rather than as a precisely defined term according to a consistent scientific terminology. In general, "interactive" or "interaction" mainly refers to technological capacities that the computational artifact offers the user, e.g. use of an interpreter or compiler.

Thoben's article uses the terms "interact" and "feedback." According to our reading the connotation of the term "interaction" as used by Thoben is almost synonymous with that of "feedback." For Thoben the interface provides haptic and visual feedback and the user "interacts" with the interface. He also speaks of auditory feedback provided by the computer system following some algorithmic processing. So, for Thoben "feedback" seems a relevant term to analyze the artistic human-machine relationship, often called "human-machine interaction." But even if one thinks of this relationship in terms of control theory, "feedback" would refer to the human as a system exhibiting closed loop control. This is just one part of that relationship: the human being receives feedback—not the computer system or the interface. In general, nothing is said about the artifact's capabilities. Therefore, the appropriate description of that relationship mentioned by Thoben must be—instead of the "performer's interact[ing] with a musical interface […]"—that the performer acts upon the interface and, as a consequence, the interface reacts by sending signals to the information processing unit: The interface exhibits no agency. Furthermore, "interaction", as described by Thoben, is only ascribed to the relationship between a performer and an interface. But that is just one part of the whole artificial computational system involved in that relationship between human and machine. If we take the whole information processing system into account, e.g. our current digital computer system consisting of hardware (with interfaces) and software or a laptop on "wheels" with sensors, i.e. a robot, we are dealing with computationally bounded agents. In such a case it is justified under some circumstances to talk about (social) interaction between a human and an artificial agent because some of these systems exhibit more than reactive behavior and also carry out decisions. To some extent their behavior is likely to be best described as being based on "intentions." This point of view leads to new research questions for music research as it raises general epistemological questions such as what the "cognitive-emotional" and behavioral boundaries of such artificial computational systems are. A particular question to be investigated for music research is what their boundaries in artistic contexts are. Moreover, as more autonomous but computationally bounded agents enter the arts, music, and education in addition to (empirical) aesthetics, normative issues as well as ethical concerns will become increasingly important and relevant for society in general and in music research in particular. Questions to be addressed are who the creator of an artwork or piece of music is; who will be responsible if in an art installation some person gets hurt because the computational agent involved is capable of learning and decision-making during interaction. The general question is to what extent a computationally bounded agent will be able to follow our current normative theories and whether these theories are adequate. Also questions concerning machine ethics come to mind.[3]

To sum up: Thoben's article indicates scientific thinking of the natural sciences and engineering. Even if one restricts oneself to an engineering approach to human-machine relationship in artistic contexts, relevant notions need to be clarified. Therefore, as Thoben uses "feedback" and "interaction" as synonyms, at least the term "feedback" should have been clarified in a scientific manner, e.g. as in control theory, and its application to analyze the relationship of human and machine in the social context of artist actions should have been discussed as well. In addition, he restricts his description to a human "interacting" with interfaces neglecting the whole artificial system as part of that "interaction." Accordingly, Thoben focuses on the classification of interfaces and interface designs. This might be appropriate from a purely technological point of view, but as a result, his classification, which is conceptually based on the traditional HS system and an instrument designer's perspective for "new interfaces for musical expression", cannot provide insight into the new quality of interactive musical instruments and fails to develop new research questions.

4.4 Interactivity IV: Interactive Musical Instruments

It is obvious that using 'the computer-as-a-musical-instrument' is based on a mechanism which is not comparable to the traditional conception of a physical-mechanical system. The material of the sound generator or the ways of vibrating it are not essential for the characterization of a device, which is governed by free rules that are carried out by embodied rules. Taking the aspect of interactivity into account, the following questions should be thoroughly addressed.

[3]One is reminded of the science-fiction author Isaac Asimov's well-known "three laws of robotics" indicating the importance of an ethics for machines which is now becoming a real social necessity about 60 years ago. A machine ethics (Anderson and Anderson 2011) or android epistemology (Ford et al. 2006) is not only urgently needed for military, educational, or social applications, but also for the arts.

1. Is the term "interactive musical instruments" appropriate for a class within a new classification system for musical instruments? What new and broader scientific insights can be provided by such a classification system?
2. Does interactivity need to be integrated into criteria for a new system classifying musical instruments? What would then be the purpose of such a classification system?

The term "interactive musical instruments" is a candidate to use in place of "new digital musical instruments" when emphasizing the theoretical concept of "interactivity." Taking into particular account the increasing number of sociological and psychological studies on musical interaction as well as research in media science or sound studies on interactive art in connection with "autonomous artifacts," i.e. interactive systems, interactivity might be integrated into criteria for a new system classifying musical instruments; such a classification system of musical instruments follows scientific purposes of basic research in musicology other than those for presenting instruments or designing instruments (Kartomi 1990; Simon 1992). These considerations lead to the claim that it is necessary to pursue another approach to classification of musical instruments for music research.

4.5 Interactivity V: Embodiment, Situatedness, and Agency

Taking up Peter Wegner's statements, a system of computation, which Wegner characterizes as closed regarding its environment, i.e. there is no information exchange between the computational system and its environment during a computation,[4] proves to be non-interactive. To exhibit interactivity, a system should be able to change the procedure for calculating function values and the arguments of the input domain during computation, not working on a completely defined input before the start of computation. In addition, a system that allows for interaction with its environment during the execution of a program is embodied and situated in an environment, and equipped with sensors capable of sensing environmental events as well as actuators to act on its environment (Kim 2012; Schmidt 2010). Embodied and situated systems can therefore be conceived of as prerequisites for interactivity in practice.

Accordingly, it can be assumed that embodied and situated interactive musical instruments allow interactivity between artists, artifacts, and musical events. In such cases, interactivity can be conceived of as a core property of those interactive musical instruments. As a result, a new dimension of instrumentality comes into play, given that a property of a musical instrument is its 'instrumentality.' What implications does this have for the conceptualization of a musical instrument? Traditionally, a device that has the function of generating sounds is considered a musical instrument. This implies that a musical instrument is a reactive apparatus

[4]This means that the system does not interrupt to ask for new information during computation.

that a human being intentionally acts upon. Instrumentalities accordingly comprise those properties of an object which is acted upon to generate sounds, such as material resistance and tactile feedback, which are related to motor actions used to generate and control musical sounds. Those, however, are reactive properties. By contrast, an embodied and situated interactive musical instrument such as a musical robot—that not only simulates and extends the mechanism of sound generation, but also exhibits some kind of agency to 'interact' with humans—acts as an autonomous (and social) agent that is not only reactive, but could initiate other agents' musical actions.[5] So, interactive artifacts can be conceptualized as embodied and situated agents. If a new classification system integrating interactive musical instruments is developed, its purpose might be to investigate the interactivity, autonomy and agency underlying each musical practice that can be characterized as (inter)action fulfilled by a certain goal. Musical agency can be considered in terms of the initiation of musical actions that influence other musical (albeit receptive) actions.

4.6 *Interactivity VI: Dynamic Systems and Autonomous Agents*

Dynamic systems theory might provide some useful metaphors to generate first ideas for dealing with such complex concepts. In dynamic systems theory an (embodied) agent and its environment are conceived of as being coupled in constant interaction, forming dynamical systems. State variables of the dynamical system agent A and motor outputs M have effects on the environment E and sensory inputs S have effects on the agent A. Via S and M, both systems (A and E) are tightly coupled. In general, although coupled to another system, a dynamical system cannot completely specify the trajectory of another dynamical system (Beer 1996). Therefore, the interaction of both agent and environment is best viewed as "mutual sources of perturbations, with each system continuously influencing the other's potential for subsequent interaction" (Beer 1996, p. 182). The main idea is that "an agent's behavior does not simply arise from within the agent itself, but rather through its interaction with its environment" (Beer 1996, p. 183). According to Beer (1996, p. 183), one important lesson can be drawn from dynamic systems theory: "[W]e must learn to think of an agent as containing only a latent potential to engage in appropriate patterns of interaction. It is only when coupled with a suitable environment that this potential is actually realized through the agent's behavior in that environment." This raises two questions: (1) Are there degrees of interactivity for different kinds of artificial agents? (2) To what extent do these different kinds of interactivity influence the potential for artistic and musical action when interacting with human or other artificial agents?

[5]For more on agents as dynamical systems and robotics in music research, see Schmidt (2010).

The first question can be dealt with in terms of an agent's autonomy,[6] referring to a proposal for a classification of autonomous agents by Stan Franklin and Art Graesser (Franklin 1997; Franklin and Graesser 1997). They define an autonomous agent as "a system situated within and part of an environment that senses that environment and acts on it, over time, in pursuit of its own agenda and so as to effect what it senses in the future" (Franklin and Graesser 1997, p. 25)[7] and give several examples of different agents. Franklin and Graesser provide a rough classification of autonomous agents in analogy to biological taxonomy, in which autonomous agents are divided into biological, robotic, and computational agents. The property of being a natural kind[8] provides the criterion for this classification at the highest level because "[e]very culture and even very young children readily distinguish between animate organisms, artifacts and abstract concepts" (Franklin and Graesser 1997, p. 30).

Biological agents subsume cognitive agents; only humans and some higher animals are currently classified as cognitive agents. Computational agents are further divided into software and artificial life agents. Software agents consist of task-specific agents, entertainment agents, and computer viruses. A further classification via matrix organization, i.e. feature collections, is then proposed.[9] Such a classification (albeit preliminary) in conjunction with new terms such as 'agent', 'interactivity', and 'agency' might be taken as a reasonable starting point for empirical research in connection with computational modeling and cognitive robotics (D'Mello and Franklin 2011; Seifert and Kim 2008) guiding research on interactive art and music.[10] As a consequence, research taking into account such a classification assesses each system's latent potential as well as the latent potential of a whole coupled system and its relevant state variables. It might also connect naturally to computational investigations of complex systems in the social and biological sciences (Axelrod and Tesfatsion 2006). In cognitive science and artificial intelligence some authors (e.g. Beer 1996, 2014; Port 2003) view dynamical systems and agent-based approaches to cognition as the most promising route to computational cognitive modeling and the explication of embodiment and (social) interaction. So, there is a strong relation to current music research on embodiment

[6]'Autonomy' related to artificial systems is discussed in Vernon (2014).

[7]For a detailed comparison of different agent concepts see Schmidt (2010, pp. 35–44).

[8]In different areas of research on classification or categorization such as developmental psychology or cognitive anthropology natural kinds are distinguished from classification systems that are merely conventional. Natural kinds exist independently of our classificatory activity and are not merely conventional (Kornblith 1999). For example, classifying the world into animate and inanimate objects might be a natural kind whereas classifying the world into different kinds of musical instruments is a conventionally culture-dependent categorization.

[9]Note that matrix organization enables the intersection of sets.

[10]For more about agents as dynamical systems and robotics in music research, see Schmidt (2010).

(Leman and Maes 2014), rhythm (Large and Kolen 1994), joint action (Keller 2008) and alignment (Bharucha et al. 2012). Moreover, it introduces a more 'ecologically' valid point of view for investigating internal processes if embodiment and interactivity are studied empirically in connection with computational cognitive modeling in new media art environments (D'Mello and Franklin 2011; Seifert and Kim 2008; Verschure and Manzolli 2013).

5 Concluding Remarks

As our reflections concerning scientific concept formation and interactivity showed, there is the question of whether a new classification system of musical instruments, integrating interactive musical instruments including autonomous agents, leads in turn to a rethinking of the concept of a musical instrument, since conceptualizing a musical instrument as an apparatus does not cope with interactive musical instruments under development, i.e. taking its interactivity, autonomy and agency into account in the sense that it can exhibit musically meaningful, interactive behavior and initiate musical actions. The traditional concept of a musical instrument involves a minimal condition of making music, namely sound generation. Musical actions emerging during interaction, however, are not only sound-generating actions, but also other musically meaningful actions, such as actions allowing people to rhythmically adjust to each other and to be musically together, complementing or becoming attuned to each other.

As a result, further conditions of making music, which are related to non-verbally meaningful interaction that allows for shared affect and social bonding, can be taken into account. Whether and how this rethinking could in turn lead to a new definition of a musical instrument remains a challenging question for basic music research. Where musical actions are desired, there is a development of tools by which these actions can be executed effectively. We are at present confronted with a 'tool' that is capable of initiating musical actions which we may entrain and become attuned to. How can we define and integrate it into a classification system also directed towards the traditional tools called 'musical instruments'? If the term 'musical instrument' should not be abandoned, musicologists must widen the extensions of instrumentality regarding musical events that cannot be reduced to sound events, but are rather understood by non-verbally sounding events that make sense to their participants during their interaction, and are described by "musical" in the context of each cultural community. This attempt might only succeed within the scope of basic music research, which allows music scholars to pursue an explication of the concept of music.

References

Anderson, M., & Anderson, S. L. (Eds.). (2011). *Machine ethics*. Cambridge: Cambridge University Press.

Axelrod, R., & Tesfatsion, L. (2006). Appendix A: A guide for newcomers to agent-based modeling in the social sciences. In L. Tesfatsion & K. Judd (Eds.), *Handbook of computational economics, Vol. 2: Agent-based computational economics* (pp. 1647–1659). Amsterdam: Elsevier.

Beer, R. D. (1996). A dynamical systems perspective on agent-environment interaction. In P. E. Agre & S. J. Rosenschein (Eds.), *Computational theories of interaction and agency* (pp. 173–215). Cambridge, MA: MIT Press.

Beer, R. D. (2014). Dynamical systems and embedded cognition. In K. Frankish & W. M. Ramsey (Eds.), *The Cambridge handbook of artificial intelligence* (pp. 128–150). Camdridge: Cambridge University Press.

Bharucha, J., Curtis, M., & Paroo, K. (2012). Musical communication as alignment of brain states. In P. M. Rebuschat, M. Rohrmeier, J. A. Hawkins, & I. Cross (Eds.), *Language and music as cognitive systems* (pp. 139–155). Oxford: Oxford University Press.

D'Mello, S., & Franklin, S. (2011). Computational modeling/cognitive robotics complements functional modeling/experimental psychology. *New ideas in Psychology, 29*, 2017–2027.

Davies, H. (2000). Electronic instruments: Classifications and mechanisms. In H. J. Braun (Ed.), *'I sing the body electric': Music and technology in the 20th century* (pp. 43–58). Hofheim am Taunus: Wolke Verlag.

Dodge, C., & Jerse, T. A. (1985). *Computer music: Synthesis, composition, and performance*. New York: Schirmer.

Ford, K. M., Glymour, C., & Hayes, P. J. (Eds.). (2006). *Thinking about android epistemology*. Menlo Park: AAAI Press.

Franklin, S. (1997). Autonomous agents as embodied AI. *Cybernetic and Systems, 28*(6), 499–520.

Franklin, S., & Graesser, A. (1997). Is it an agent, or just a program? A taxonomy of autonomous agents. In J. Müller, M. J. Wooldridge, & N. R. Jennings (Eds.), *Intelligent agents III Agent theories, architectures, and languages—ECAI'96 Workshop (ATAL)*, Budapest, Hungary, August 12–13, 1996 [=Lecture notes in computer science, Vol. 1193] (pp. 21–35). Berlin: Springer.

Hempel, C. G. (1952). *Fundamentals of concept formation in empirical science*. Chicago: The University of Chicago Press.

Kartomi, M. J. (1990). *On concepts and classifications of musical instruments*. Chicago: The University of Chicago Press.

Keller, P. E. (2008). Joint action in music performance. In F. Morganti, A. Carassa, & G. Riva (Eds.), *Enacting intersubjectivity: A cognitive and social perspective on the study of interaction* (pp. 205–211). Amsterdam: IOS Press.

Kim, J. H. (2012). *Embodiment in interaktiven Musik- und Medienperformances – unter besonderer Berücksichtigung medientheoretischer und kognitionswissenschaftlicher Perspektiven*. Osnabrück: epOs.

Kornblith, H. (1999). Natural kinds. In R. A. Wilson & F. C. Keil (Eds.), *The MIT encyclopedia of the cognitive sciences* (pp. 588–589). Cambridge, MA: MIT Press.

Large, E. W., & Kolen, J. F. (1994). Resonance and the perception of musical meter. *Connection Science, 6*(2–3), 177–208.

Leman, M., & Maes, P.-J. (2014). Music perception and embodied music cognition. In L. Shapiro (Ed.), *The Routlegde handbook of embodied cognition* (pp. 81–89). London: Routledge.

Mathews, M. (1963). The digital computer as a musical instrument. *Science, 142*(3592), 553–557.

Minsky, M. (1967). *Computation: Finite and infinite machines*. London: Prentice-Hall.

Miranda, E. R., & Wanderley, M. M. (2006). *New digital musical instruments: Control and interaction beyond the keyboard*. Middleton: AR Publications.

Montagu, J. (2007). *Origins and development of musical instruments*. Lanham, Maryland: Scarecrow Press.
Nelson, R. J. (1989). *The logic of mind* (2nd ed.). Dordrecht: Kluwer Academic Publisher.
Port, R. F. (2003). Dynamical systems hypothesis in cognitive science. In L. Nadel (Ed.), *Encyclopedia of cognitive science* (pp. 1027–1032). Chichester, West Sussex: Wiley.
Rowe, R. (1993). *Interactive music systems: Machine listening and composing*. Cambridge, MA: MIT Press.
Schmidt, L. (2010). *Embodied cognitive science of music: Modeling experience and behavior in musical context* (Doctoral dissertation). Universität zu Köln. http://kups.ub.uni-koeln.de/5342/. Accessed February 20, 2016.
Seifert, U., & Kim, J. H. (2008). Towards a conceptual framework and an empirical methodology in research on artistic human-computer and human-robot interaction. In I. Pavlidis (Ed.), *Human-computer interaction* (pp. 177–194). Vienna: In-Tech.
Simon, P. (1992). Die Hornbostel/Sachs'sche Systematik und ihre Logik. *Instrumentenbau-Zeitschrift—Musik International, 46*(7–8), 64–66.
Simon, P. (2004). *Die Hornbostel/Sachs'sche Systematik der Musikinstrumente: Merkmalarten und Merkmale. Eine Analyse mit zwei Felderdiagrammen*. Mönchengladbach: Verlag Peter Simon.
Stegmüller, W. (1974). *Probleme und Resultate der Wissenschaftstheorie und Analytischen Philosophie, Band II: Theorie und Erfahrung. Erster Halbband: Begriffsformen, Wissenschaftssprache, empirische Signifikanz und theoretische Begriffe*. Verbesserter Neudruck. Berlin: Springer.
Thoben, W. (2014). Elektronische Musikinstrumente: Interfaces und Controller. In S. Weinzierl (Ed.), *Akustische Grundlagen der Musik* [=Handbuch der Systematischen Musikwissenschaft, Bd. 5; ed. by H. de la Motte-Haber] (pp. 433–445). Laaber: Laaber.
Vernon, D. (2014). *Artificial cognitive systems: A primer*. Cambridge, MA: MIT Press.
Verschure, P. F. M. J., & Manzolli, J. (2013). Computational modeling of music and mind. In M. A. Arbib (Ed.), *Language, music, and the brain: A mysterious relationship* (pp. 393–414). Cambridge, MA: The MIT Press.
von Hornbostel, E. M., & Sachs, C. (1914). Systematik der Musikinstrumente: Ein Versuch. *Zeitschrift für Ethnologie, 4–5*, 553–590.
Wegner, P. (1997). Why interaction is more powerful than algorithms. *Communications of the ACM, 40*(5), 80–91.
Wegner, P., & Goldin, D. (2003). Computation beyond Turing machines. *Communications of the ACM, 46*(4), 100–102.
Wiener, N. (1961). *Cybernetics, or control and communication in animal and machine* (2nd ed.). Cambridge, MA: MIT Press.
Wolf, R. (2014). Musikautomaten. In S. Weinzierl (Ed.), *Akustische Grundlagen der Musik* [=Handbuch der Systematischen Musikwissenschaft, Bd. 5; ed. by H. de la Motte-Haber] (pp. 409–431). Laaber: Laaber.

Part II
Design Make Create

Movement Meets Material—An Improvisational Approach to Design

Johanna Schindler and Amelie Hinrichsen

Abstract How can we integrate an embodied musician-instrument relation to musical instrument design? To answer this question, we have proposed a design process where musical instrument prototypes are developed taking a specific improvisation practice from contemporary dance. Over the course of four improvisation sessions, we invited an acoustic musician, an experimental electronic musician and a contemporary dancer to develop a solo performance with given material. Their improvisations inspired the design of three instrument mock-ups, which integrated movement, material and sound. After four subsequent improvisation sessions the process resulted in two refined instrument prototypes. Using improvisation as a performance setting, our developmental process revealed that for live set-ups the instrument benefits from a reliable system, which allows the musician to perform in a spontaneous and flexible manner. To further engage the musician with the instrument, the sound synthesis process should reflect genuine material sound qualities of the object. Emphasizing its identity as an instrument, we refer to this as *material authenticity*, a notion, which raises questions on the relationship between material, digitality and sound.

J. Schindler (✉)
Zeppelin University, Am Seemoser Horn 20, 88045 Friedrichshafen, Germany
e-mail: johanna.schindler@zu.de

A. Hinrichsen
University of the Arts Berlin, Grunewaldstrasse 2-5, 10823 Berlin, Germany
e-mail: a.hinrichsen@udk-berlin.de

T. Bovermann et al. (eds.), *Musical Instruments in the 21st Century*,
DOI 10.1007/978-981-10-2951-6_8

1 Introduction

Within the community of electronic experimental music, it is nothing new that discussions have circled around bodily gestures and expression.[1] A closer look at this community reveals that live set-ups are often established according to very specific and individual needs. For example, when Jeff Carey expresses the wish, "to have a physical grip on my sound,"[2] he is searching for a corporeal and immediate response within the set-up or instrument he performs with. This "physical grip" supports bodily communication with his instrument and the audience during the performance as well as addressing his muscular memory during the composition and rehearsal processes of new pieces. For Carey, this also implies the disappearance of notebooks from stage. Without such an external controller interface the audience is "much more present" to Carey and "it comes to a nervous feedback".[3] Going one step further, Pierre Alexandre Tremblay's work aims at "the body and the interface [...] becom[ing] one"[4] through continuous practice. He considers his instrument as an extension of his body. These statements reveal three constituent elements for live performances: the musician, the instrument, and the audience. Within this triad, the notion of corporeality is crucial to music making and, in our case, for instrument design. It alludes to the phenomenology of perception, which we need to take into account. Referring to Jean-Luc Nancy's statement, "the body is the openness" (2007, p. 105), we can say that on stage, it is through the sensible body that the musician enters the physical world in the form of her instrument and its sounds.

Against this background, we approached the following research question: How can we design a digital musical instrument, which integrates expressivity in sound and movement from the beginning? Miranda and Wanderley (2006) define five major decisions within such a design process: a selection of gestures "to control the system" (p. 4), a number of sensors to measure these gestures, algorithms for the sound synthesis process, a mapping and the "feedback modalities" in addition to the sonic output the digital musical instrument produces (ibid.). In more specific regard to developing a new piece, electroacoustic composer Young (2015) claims that, "the most serious problem in composing electroacoustic music is not the seemingly limitless range of available sounds, but the problem of how to direct sound materials through the inevitable labyrinth of possibilities that the composer faces to achieve aesthetically satisfying results" (p. 159). To contrast these infinite

---

[1]Since 2001, the international conference on New Interfaces of Musical Expression (NIME) brings together researchers and musicians who discuss the design of new musical interfaces.

[2]Jeff Carey, Interview, April 3, 2014.

[3]JC, Interview.

[4]Pierre Alexandre Tremblay, Interview, November 11, 2014.

possibilities digital sound synthesis opens up, the design process aimed at the following: The instrument should incorporate a distinct sonic and material-visual identity as well as possibilities for the performer to get in touch with the generated sound and the audience. Therefore, we combined direct acoustic sound signals with pre-defined options of digital sound modification and mapped them to the musical instrument. We intentionally designed a system opening up both intuitive and predetermined options of use and interaction with the object. A set of limitations challenges the musician to work with or around a given structure and stimulates creativity at the same time. Additionally, we avoided the presence of a computer in order to keep the musicians focus on the instrument itself while playing.

Given the expertise of contemporary dance to relate bodily movement, space and sound, we looked to this discipline for inspiration on the design process of the instrument. Improvisation is a well-established method in the performing arts to enhance spontaneous and immediate play and interaction on the one hand, and to develop new pieces—or in our case: a new digital musical instrument—on the other. In contemporary dance, improvisation in particular has become relevant as a means to liberate the dancer from hierarchical structures imposed by teachers and choreographies and to develop an individual style (see also Carter 2000, p. 181). We therefore utilized an improvisation strategy, which contemporary choreographers Nik Haffner and Christina Cuipke[5] developed, in the design process. This strategy can be described as follows: Haffner and Cuipke use short improvisation cycles to come up with new choreographies, during which they have a dialogue without words. One of them improvises a short sequence of movements for about one minute. The other watches, reflects and answers with a sequence relating to what she just saw, but focusing on certain elements that seemed interesting to her. After taking three to five turns, they end up with a short sequence of movements. The one watching is allowed to take notes, but the dancers should not talk to each other during the whole process. It is a playful method for which observation, identification, response and surprise are crucial elements. The material—that is, the choreography—develops over time and almost by itself. It is through the interplay between several people that elements emerge.

In our specific research setting, we asked two musicians (one with an electronic the other with an acoustic background) as well as a contemporary dancer to improvise with each other during four sessions. The strategy described provided a formal structure which continuously demanded the participants and the designer to make choices. In this manner, the participants' individual inputs evoked more concrete ideas regarding sound generation as well as motion patterns and playability each time, which fed into the instrument we aimed to design.

[5]Nik Haffner is the Artistic Director of HZT (Hochschulübergreifendes Zentrum Tanz) Berlin. Christina Cuipke holds an MA in Choreography from Amsterdam School of the Arts. During a workshop at HZT Berlin in 2014, Nik Haffner introduced the method to one of the authors.

1.1 Improvisation in Performance and Design

Whilst in the performing arts, improvisation has been frequently discussed in regard to concrete practices (see e.g. Bormann et al. 2010; Brehm and Kampfe 1997; Nakano and Okada 2012) or in the form of manuals on improvisation in Jazz, acoustic music, digital sound generation and experimentation in music (see e.g. Crispin and Gilmore 2014; Keep 2009), improvisation in the field of design is regarded as a tool to support design work on various levels (Gerber 2007, p. 1069). In the 1980s, designers started to approach human-computer-interaction (HCI) and interface development (cf. Mareis 2014, p. 133). In doing so, they did not only focus on the interfaces' functionality; rather, they considered and designed them to enable agency (ibid., p. 135). To gather inspiration on potential human-machine-interactions designers developed human-centered methodologies such as "informances"[6] (Burns et al. 1994, p. 119f) and "experience prototyping"[7] (Buchenau and Suri 2000). Loke and Robertson state that design research, which aims to understand human activity and movement within interactive systems, "is [...] described, with a particular focus on enactment and physical role-playing [as well as] methods for working with the moving body and felt, kinaesthetic experience" (Loke and Robertson 2009, p. 395). Gerber even suggests actively applying Johnstone's rules of theatrical improvisation (1989) to brainstorming sessions "to support collaboration, spontaneity, learning through failure, and storytelling" (Gerber 2007, p. 1072). Consequently, we can observe collaborative, embodied approaches based on movement and improvisation especially in HCI design-research (e.g. Hummels et al. 2007; Klooster and Overbeeke 2005; Larssen et al. 2007; Loke and Robertson 2013; Moen 2005). Furthermore, recent studies within the domain of musical instrument design focus on the sound-movement relation and implement improvisation in the development process. For example, Sylleros et al. (2014) developed several sounds, and asked musicians to think of gestures, which they would use to produce these sounds and then integrated individually adapted gestures into a refined prototype (p. 94). Using a different approach, Donnarumma et al. (2013) mapped two neuronal signals resulting from muscle contraction and limb movement to an instrument, and asked novice performers to improvise with it. The clear connection between the bodily signals and a limited number of sound signals facilitated non-expert musicians to perform with the instrument. (pp. 130ff) Again related to bodily signals, Lee and Yeo (2012) mapped several dancers' respiration to sonic parameters, which were then naturally connected to their performance movements. In the subsequent interviews, the

[6]Informances can be described as improvised role-plays with the use of props in front of an audience of peers or clients. Designers use this method to "build an increased empathy for the people that [have been] identified as the users" (Burns et al. 1994, p. 120).

[7]"[A]n Experience Prototype is any kind of representation, in any medium, that is designed to understand, explore or communicate what it might be like to engage with the product, space or system we are designing." (Buchenau and Suri 2000, p. 2).

dancers stated that they perceived the resulting sounds as generally supportive for their dance performance (p. 4).

The instrument design studies above demonstrate how both intuitive, associative actions as well as spontaneity are centrally important in "mak[ing] (something) from whatever is available" (Oxford Dictionary 2015). This traditional definition of improvisation hints to the emergence of elements during the improvisatory action. However, Sennett (2009) rejects the idea of spontaneity, "if 'spontaneous' represents a mindless occurrence" (p. 236f). With explicit reference to Jazz music, he claims instead that improvisers act within a strict framework of "rules of economy" and via impulses that demand deliberate decision-making, because "otherwise they lose focus; [...] the harmonic reversals are disciplined by what came before. Above all, the jazz musician has to select elements for his or her own instrument [that] someone playing a different instrument can respond to" (ibid.). Sennett's example underlines the development of existing elements as well as the emergence of entirely new parts over the course of improvised actions. In this context, Redgate (2015), who is famous for redesigning the classic oboe to play experimental contemporary music,[8] states that he uses improvisation to "push boundaries and to look for new and original ideas and novel solutions to problems" (p. 213). While the current literature seems to agree on the aspect of emergence during improvisation, several authors would disagree with Redgate on the suggestion that it "leads to solutions". Bormann et al. (2010) claim that the freedom to give up rules during improvised activities sometimes results in discarding elements rather than developing them further (p. 12). It is this specific productive action, which reveals problems and malfunctioning aspects in the first place (ibid.). Also the authors quoted above (Donnarumma et al. 2013; Gerber 2007; Lee and Yeo 2012; Sylleros et al. 2014) concluded that their studies revealed further necessary refinements for example regarding the integration of corporeal aspects into the prototypes they had developed, instead of directly resulting in final instruments.

Improvisation has become a frequent tool in various design fields to integrate corporeality into the developed objects. Yet, in instrument development processes, improvisation is involved at a late stage only, namely to address questions of direct sound-gesture-mapping.

1.2 The Role of the Environment

Elaborating on the specific combination of improvisation and contemporary dance Brehm and Kampfe (1997) state that improvisers take up impulses from their surroundings and transform them into something "meaningful" and "satisfying" (p. 15), thereby altering the environment. Such an ongoing back and forth between improviser and their surroundings renders the elements emerging during

[8]See also Redgate (2016).

improvisation fragile and transient (see also Douglas and Coessens 2012, p. 183). At the same time, it allows the improvisers to choose and focus on the elements they consider most promising. One could say that improvising consists in a constant mutual exploration and negotiation of possibilities between the improviser and the environment. Within the field of electroacoustic research this complex dynamic system between performer, instrument and environment is referred to as "performance ecosystem" (Waters 2007, p. 2).

Gibson's theory of affordance (1986) approaches the material and usable components of the environment from an ecological perspective, that is, he looks at what the environment offers, or "affords," to fulfill an animal's needs (pp. 127f). Each animal or human might perceive affordances differently and make use of them or not, but an affordance is stable and "always there to be perceived" (p. 139). In regard to our specific setting the environment consisting of the space, the technological setup, material, instrument mock-ups, the generated improvised sounds and movements, the improvisation structure, the participants, their own and others' bodies and body memory as well as disciplinary backgrounds. All of the above afforded distinct behavior in relation to the given task of improvising a solo and each of these affordances influenced the participants' judgments on whether they considered the others' and their own actions as meaningful or not. Consequently, the affordances of the material(ity), for example, which the performing participants, designer and ethnographer interpreted individually, influenced the resulting prototypes to a strong degree.

1.3 Methodological Setup

Over the course of nine months we, that is, a designer (in the following referred to as D) and an ethnographer (E), convened for four improvisation sessions for which we invited a contemporary dancer and choreographer (CD), a musician with a background in acoustic music and contemporary composition (AM), and a musician with a background in media art and electronic music (EM). Following the conceptual approach developed by the designer (Amelie Hinrichsen), we asked the participants to improvise one-minute solos (which we refer to as a *quick round*) according to the improvisation method explained above. At the end of the session, the participants' experiences were discussed regarding surprising moments and the potential of the material they improvised with (see Sect. 2.1). In addition to recording the sessions on video and feedback discussions, D and E closely observed the three participants, took notes, pictures and—in the case of the ethnographer—more detailed Fieldnotes as Emerson, Fretz and Shaw describe (2011, pp. 21ff).

To analyze the first session, D applied the improvisation technique described above to her own practice: She watched the video recordings of each *quick round* and took fifteen minutes to improvise with the same materials the participants had been using (see Sect. 2.1). These first immediate responses resulted into three *material constellations* (see Fig. 1a–c).

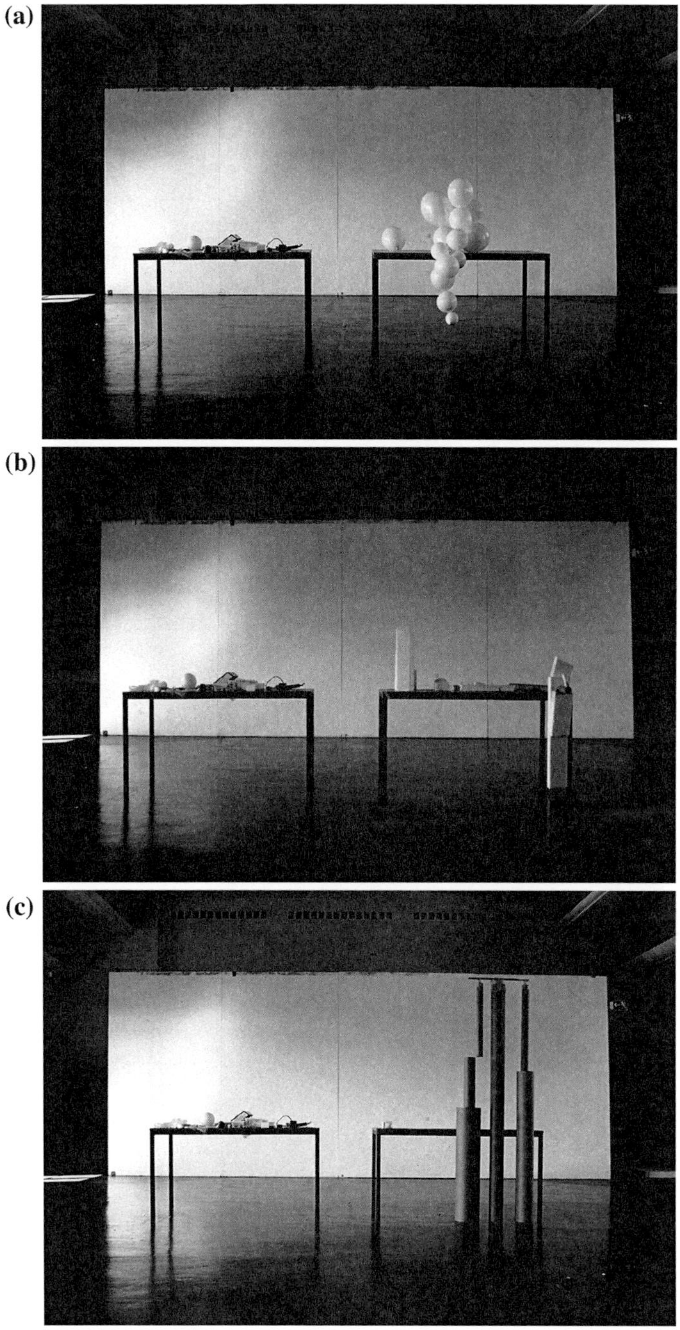

Fig. 1 Material Costellations

Afterwards she combined video stills, photographs, notes, keywords and sketches within three corresponding mood boards to identify and group similarities and differences in the participants' actions regarding movement patterns and sonic output. Similar to the participants, she transferred and translated these 'codes' she had identified in the participants' solos in an intuitive manner, i.e. immediately and guided by personal preferences. D then developed three *instrument mock-ups* based on impressions of the first session (see Sect. 2.2). The use of these mock-ups during Session #2 (see Sect. 2.3) guided the decision-making process for more refined *improvisational prototypes* (see Sect. 2.4). These were played in the third session, technically refined once more and tested in a final, fourth improvisation session.

Parallel to this, E wrote detailed descriptions immediately after each session to capture her impressions. On the basis of these texts, the pictures and video-recordings, she carried out a qualitative analysis. That is, she coded the material both according to pre-determined categories ("improvisation", "materiality", "corporeality", "sound"), and in an open manner to examine further aspects, which would influence an live setup of instruments, but which we had not considered as important beforehand (Emerson et al. 2011, p. 171ff).

Following these individual analysis procedures, D and E exchanged their results once before and after each improvisation session. In this manner, D could integrate E's observations into the designed mock-ups and later prototypes. Additionally, we discussed the structure of each improvisation sessions according to our own perception and in consideration of the participants' feedback on time cues, technological setup etc.

Along with the descriptions of the mock-ups and the prototypes, the "design (ed)" and the ethnographic perspective are combined in the following descriptions. The first two of four improvisation sessions are depicted in detail, because they were most relevant for the resulting design of the final prototypes. After each description we will show how the results of our analyses led to these instrument mock-ups and later prototypes. In addition, the technical refinements for Sessions #3 and #4 are briefly outlined (Sect. 2). In Sect. 3, the influence of improvisation on the design of the instrument is discussed as well as how materiality and corporeality are mirrored in the improvisation process. This is followed by concluding remarks on the limitations and opportunities of the chosen approach, as well as comments on its implications for the broader notion of a digital musical instrument.

2 Analysis of the Improvised Design Approach

2.1 Session #1: Material Exploration

In the first session we asked the participants to improvise with three different physical materials. We had supplied materials that complemented each other and offered diverse qualities in their sound, size and materiality. We chose transparent film, cardboard tubes, white balloons and polystyrene foam (see Fig. 2).[9] Transparent film, familiar from the context of domestic painting work, was chosen for its subtle qualities of extreme lightness and sensitivity to airflow. The tubes' simple shape and resonant body offer diverse possibilities for interaction without being too direct in terms of symbolic content or connotation. Balloons in contrast are rather concrete and complete in their appearance. They convey a distinct symbolic content and allude for example to birthday parties and funfairs. Finally, closed-cell extruded polystyrene foam combines both soundproofing and sound generation qualities. Its structural characteristics contrast significantly to those of balloons and transparent film.

Both the participants' different backgrounds and the range of material to improvise with were selected to ignite a transdisciplinary dialogue evoking ideas for the design of the instrument we wanted to develop. In addition, the musicians were encouraged to bring some of the tools they usually work with (see Figs. 3 and 4). The dancer's tool—his body—was present and did not require any further equipment.

We asked the participants to subsequently improvise a short solo taking turns three times (*quick round*, as stated above). A different participant initiated each round so that three rounds of improvisation took place; he or she could choose one of the provided materials. Afterwards, the participants performed an improvisation of five minutes as a trio—still using the same kind of material, which had been selected in the beginning. In a final collective improvisation, the participants could choose any of the given materials and combine them; no external time cues were given. A square of approximately 2 m^2 was marked on the floor as performance space. Stepping inside the square indicated the start of the improvisation, stepping outside its end (see Fig. 5).

2.2 *Analysis of Session #1: Instrument Mock-Ups*

2.2.1 Balloons

Observations:

During Session #1, it was observed that there was an overall characteristic of contrast in how the participants dealt with the balloons. Moments of careful handling and soft sounds, even silence, were juxtaposed with the application of a lot of force, with one balloon even bursting. The acoustic sounds produced ranged from loud squeaking to soft hissing. One key moment was the following: AM slowly

[9]We purposefully offered four different kinds of material for three rounds of improvisation only so that one material would remain unused in the end of Session 1.

Fig. 2 Selection of materials provided for Session #1

Fig. 3 Tools for acoustic modification brought along by AM

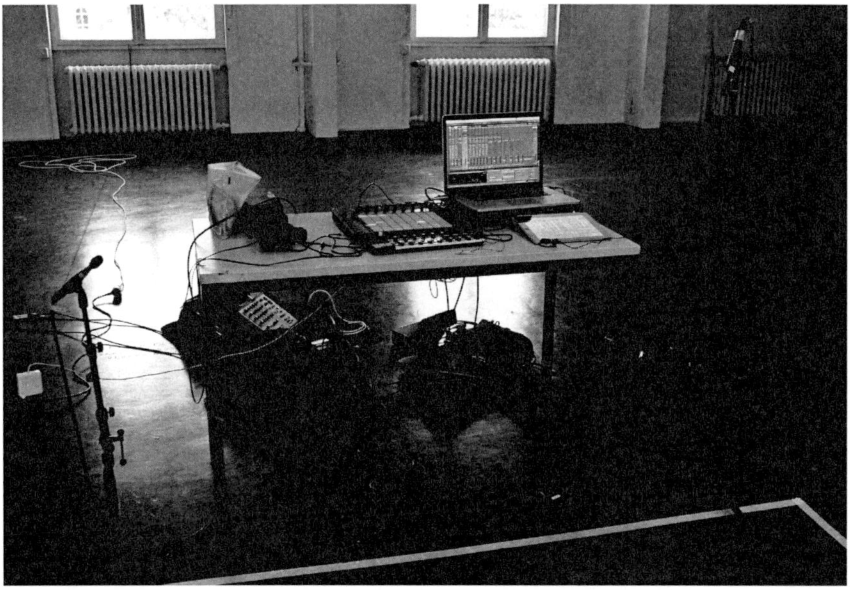

Fig. 4 Tools for digital modification brought along by EM

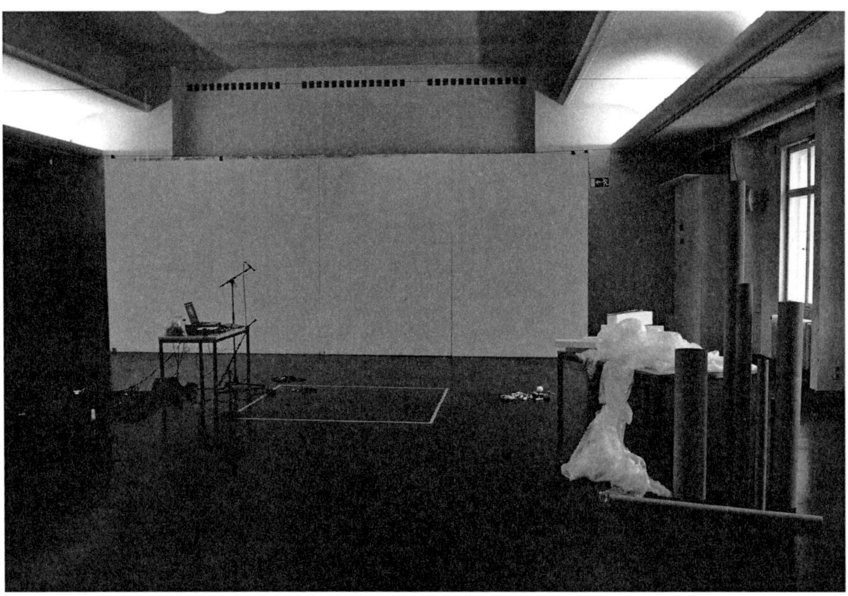

Fig. 5 Performance space

deflated one balloon, producing a shrill squeaking sound. EM imitated and digitally modified this playing mode. With the help of reverb, pitch shifting and grain delay, he produced several layers of long-lasting sonic textures, squeaking more softly than the original sound, which seemed to be filling the entire space. A further key moment can be described as follows: CD used a lot of force pushing the balloons hard on the ground, even performing push-ups on them, producing trembling in his arms (see Fig. 6a, b). Both AM and CD held and squeezed the balloons repeatedly, which EM picked up to alter the texture of his sounds through squeezing and rubbing a balloon with the thumb (see Fig. 6c, d). Furthermore, AM used the bouncy quality of the balloons to play rhythmical patterns (see Fig. 6e).

Designer's response:

To account for the intense use of body weight and close contact especially by CD, the *instrument mock-up* consisted of three balloons, which were all attachable to the body with adjustable straps. One balloon was equipped with a piezo element, the second with a condenser vibration pick-up.[10] Incoming audio signals could be modified with some reverb, frequency shifts and rhythmical delays during Session #2. A headset-microphone was installed in the third balloon, resulting in low frequency sounds when the balloon was held and squeezed, imitating the sonic qualities played by EM.

2.2.2 Polystyrene Foam

Observations:

During two moments in the use of the polystyrene foam, D noticed a space-structuring aspect. Firstly, when CD was holding one piece of polystyrene foam between his teeth and pausing in various positions (see Fig. 7a) and secondly, when AM was arranging several pieces of polystyrene foam in an upright position within the performance space (see Fig. 7b). Acoustically, loud sounds like cracking and crashing produced by stabbing and scratching (see Fig. 7c) were dominant, even though they were contrasted with very soft sounds like crackling resulting from stroking and tapping on the polystyrenes foam. A remarkable moment for D was EM's answer on AM's first run. After AM had first slowly and softly broken apart small pieces of polystyrene foam and let them fall to the ground (see Fig. 7d), EM attached a contact microphone to a piece of polystyrene foam and modified the direct output. Instead of the expected crackling of a dry amplification, we heard subtly pitched sounds. In parallel to this, the amplified microphone output with rhythmical delays and reverb remained in the audible background. The most dominant motif for D regarding movement was one of rotation, a windmill-like

[10]The same kind of miking was used by EM during Session 1.

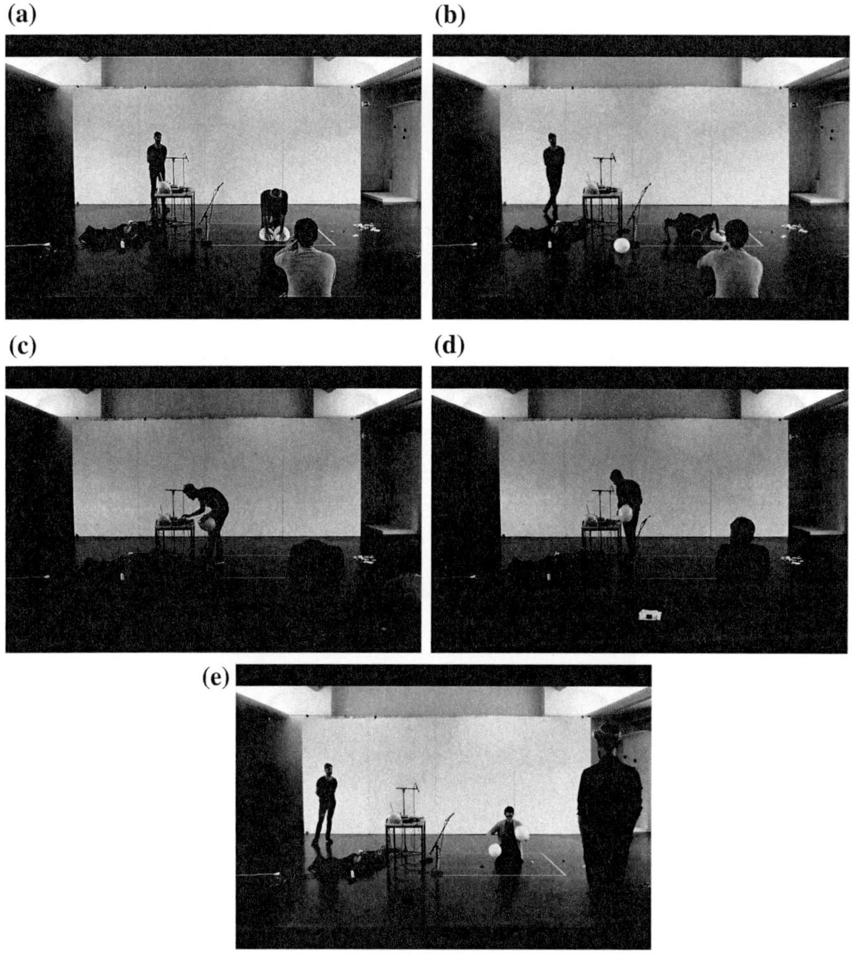

Fig. 6 Key moments identified during Session #1 improvising with balloons

movement performed by CD, which was varied during the collective improvisation (see Fig. 7e).

Designer's response:

The mock-up resulting from these key moments was a V-shaped paddle with a wooden handle, clearly and playfully suggesting its use for shoveling air and wind generation. By providing a second smaller paddle, the possibility to create a spatial structure as described above was taken into consideration. Using it to play and stroke the bigger paddle resulted in constellations of geometrical shapes, structuring

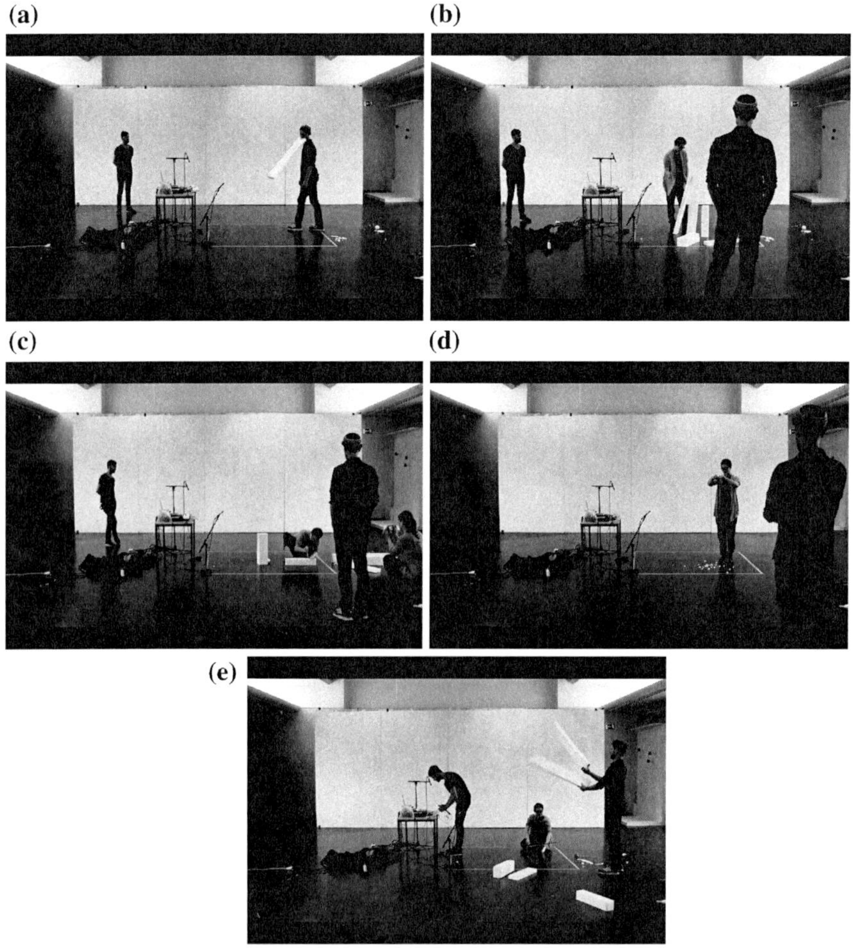

Fig. 7 Key moments identified during Session #1 improvising with polysterene foam

both the improviser's action and the performance space. In terms of sound generation, contact microphones (two piezo elements and one condenser vibration pick-up) were attached to the bigger piece of polystyrene foam to enhance percussive playing. Once more, reverb and filters enabled the modification of incoming audio-signals. The designer applied a digitally modeled marimba corpus to the signal picked up from the top piezo element, and decided to process the other two captured signals each with a resonator network as well as one with a reverb and the other with a "Ping-Pong" delay. The high-pitched, bell-like sound of the top piezo imitated EM's first sonic response to AM's "falling snow". The other signals resembled distorted, guitar-like sounds, which implemented a reference to AM's attempt to break and destroy the material.

2.2.3 Tubes

Observations:

Key moments of the tube setup consisted, for example, of CD putting all the tubes inside one another (Fig. 8a) and AM taking them apart afterwards. Then she arranged them on the floor and used them for rhythmic playing (Fig. 8b). We observed playing modes such as percussive playing and blowing into the tubes, as introduced by AM (see Fig. 8c) as well as swinging the tubes like a pendulum, which CD brought up (see Fig. 8d). The amplified sound quality of EM's first run provided the guiding motif for D's translation of the quick round. Multi-delays with time shifts provided irregular rhythmical patterns of low-frequency, wood-like percussive sounds with a high level of tension and energy.

Designer's response:

D's immediate associations to the observations were related to floating on water, the rocking of a wooden boat, sailing, barrels hitting against each other, (natural) forces and adventure (see Fig. 8e–g). Hence, for the visual representation, D decided to stay within these moods and themes. The resulting mock-up was a telescope-like object consisting of three inwardly collapsible tubes. Varying the artifact in size by pushing and pulling this telescopic construction would enable tonal changes depending on how far each element is pulled out. The designer equipped the mock-up with a piezo element in the middle to enhance the percussive playing techniques observed at repeated occasions. Metallic and wooden filters modified its output. Also, she attached a headset microphone to amplify further activities such as blowing and swinging. Consequently, the mock-up can be interpreted as a telescope but can additionally turn into a foghorn or a quarterstaff while blowing, hitting or swinging it.

2.2.4 Resulting Mock-Ups

After Session #1, we considered the resulting three *instrument mock-ups* as representations of three possible instrumental worlds, each suggesting different modes of interaction in respect to sonic modification. To open up the possibility of fading each channel in or out and looping the main output, each setup was equipped with a compact USB control surface. The mock-ups differed in the physical material generating sound, in their construction as well as in their *degree of distance* to the body. Particularly, the balloons were meant to provide the closest relation to the body by being attachable to different limbs. The tubes featured a second degree of distance in matters of corporeal relation, since they needed to be held by the performer who could directly touch them and play them with her hands. In contrast to direct manual touch and as a third degree of distance, a second paddle was provided for playing the polystyrene foam mock-up (Fig. 9).

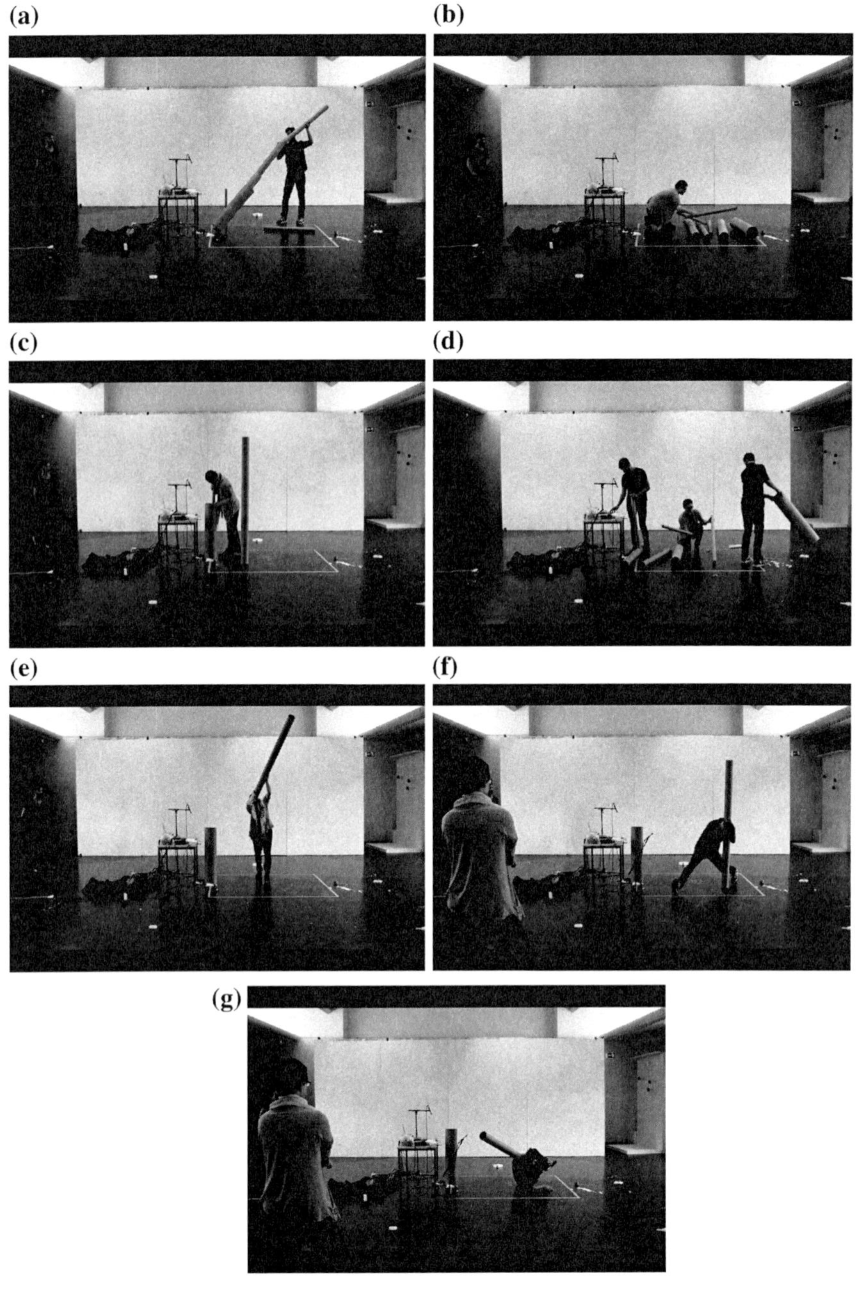

Fig. 8 Key moments identified during Session #1 improvising with tubes

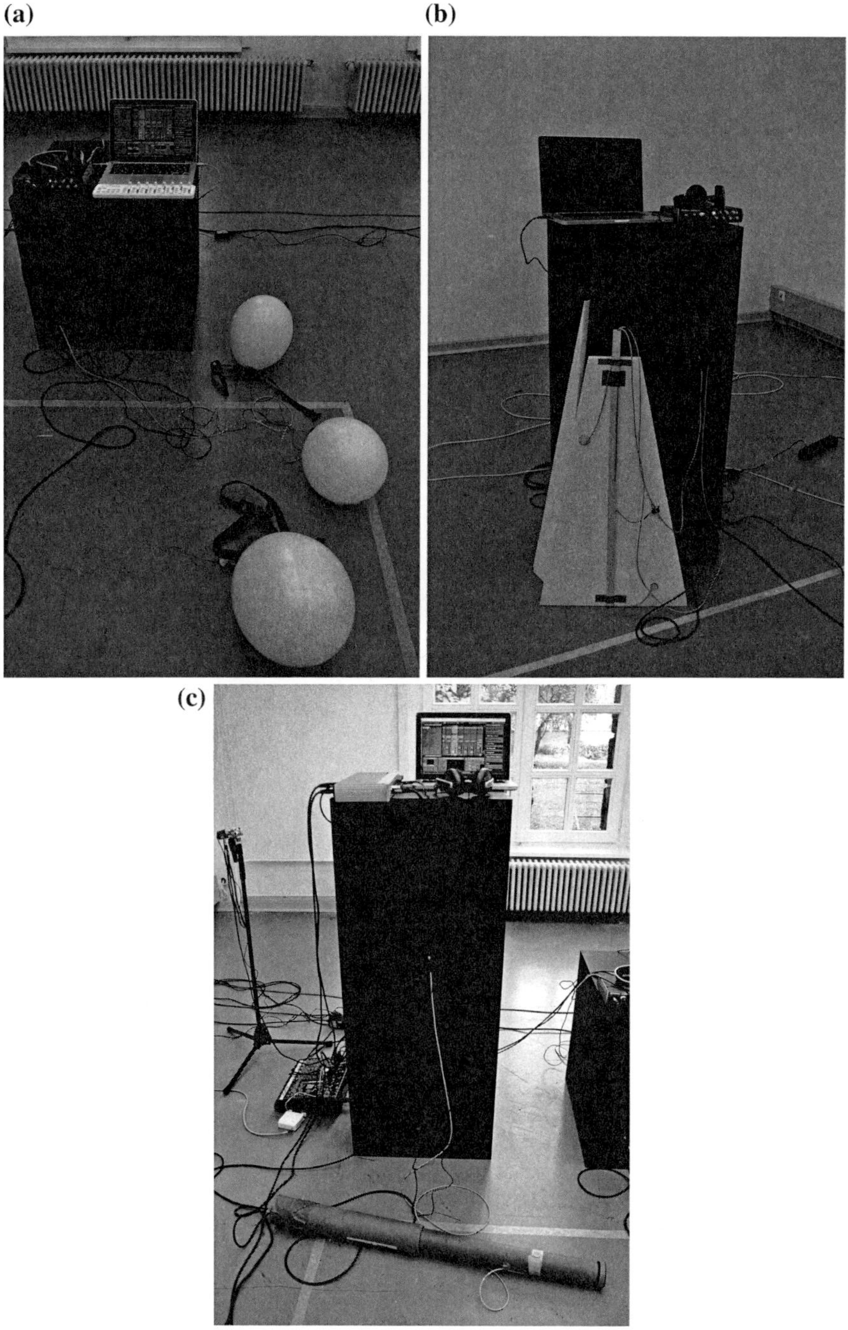

Fig. 9 Instrument Mock-ups for Session #2

2.3 Session #2: Exploring Instrument Mock-Ups

For the second session, the improvisational structure was maintained according to the previous session, as were the participants and their order of activity.[11] Instead of the “raw” material of Session #1, this time the three *instrument mock-ups* were used. (1) Three air-filled balloons, equipped with straps and belts to individually fasten them to the body of the performer, (2) two triangular polystyrene boards (one bigger than the other) with the larger one equipped with a handle and the other left entirely blank, and (3) various telescoped cardboard tubes with variable total length. Each instrument was installed on a black pedestal accompanied by a notebook (see Fig. 10).

2.4 Analysis of Session #2: Instrument Prototypes

The designer translated the insights gained from the observations and feedback during the second session into refined prototypes. This process was complemented by an individual exchange between D and one participant per instrument. CD was interviewed for the refinement of the balloon mock-up since he had chosen it in Session #1, and EM informed the further development of the tube mock-up.[12]

The following paragraphs reflect upon the use of the mock-ups with specific attention to their improvisational, material and corporeal qualities. We combined our observations and the participants’ immediate feedback with material extracted from interviews carried out with EM and CD following Session #2. At the end of each paragraph, implications for the design of the *improvisational prototypes* are outlined.

Improvisation:

In contrast to the first session in which materials were destroyed, participants considered the *instrument mock-ups* to be fragile, hence handled them very carefully. During the feedback time, EM pointed out that, especially in an improvisational setting, he required his instrument to be both physically robust and also with regards to its software. He needed to be able to rely on “a system that works”.[13] CD reported that improvising was similar to exploring. According to him, instruments for improvisation need to “challenge your understanding”[14] so that they contribute

[11]In Sessions 3 and 4, we proceeded in the same manner.

[12]The acoustic musician had to cancel her participation after the second session for personal reasons; so two prototypes were further developed. Following Session 3, D again consulted EM and CD regarding the technical adjustments of these prototypes.

[13]EM, Interview, December 29, 2015.

[14]CD, Interview, December 23, 2015.

Fig. 10 Performance space for Session #2

to the "journey of discovery".[15] AM and EM also stated that an improvisational setting required more sonic variation than was offered by the *instrument mock-ups*.

Materiality:

Compared to the dynamic, nuanced and immediate acoustic interaction with the material itself (in Session #1), the participants considered the digital level static and uniform. During the later interview, EM added that in his musical practice he preferred the access to a large range of frequencies.[16] In this context, he referred to a concept, which we call *material authenticity*, an idea, which is congruent with our aim of giving the prototype a recognizable identity. In his opinion, the instruments' material identity, which generates distinct acoustic signals, should be maintained in the production of sound. He considered the recognition of genuine material elements crucial, especially against the background of the infinite number of sonic qualities we are able to produce through digital sound synthesis.

Corporeality:

In Session #2, the participants were more present and aware of each other during collective improvisations than in Session #1; they improvised in a more inclusive

[15]ibid.

[16]EM Interview.

manner and played together (see Fig. 11b, c). We were surprised, however, that the participants varied the intended *degrees of distance* (see Sect. 2.2). After all, the mock-ups had been designed with the aim to be used respective to bodily relations we had observed during Session #1. The balloon mock-up, for example, was used in an unexpected manner: the participants only rarely strapped the balloons directly to their bodies. Instead, they used the strap clips to produce sounds (see Fig. 11a) or to fasten the balloons to each other (see Fig. 11d).

In one of AM's solos, we observed her turn towards the controller to change the amount of output channels for the balloons (see Fig. 11a). This produced a break in her performative, bodily presence: moving towards the controller and arranging settings interrupted her playing; musician, instrument and controller were separated. In this context, AM's preference of the tube mock-up is worth mentioning: "the didgeridoo created a relation to the body or system of the instrument." The artifact she performed with had a clear connotation, which led her to identify it as an independent entity. While the bodily contact between her and the instrument was obvious, she focused on the corporeality of the instrument, rather than on the immediate relation between her body and the tube mock-up.

2.4.1 Balloons

Observation and Conversations:

The manner in which the balloon-material constellation was designed is called a *compound situation*: several things of the same kind are arranged as a whole. A comparable image reappeared when CD held all three balloons very close to each other (see Fig. 11f). After having discussed with CD whether it was more interesting to deal with one or several objects, D decided to aim for a *compound of one situation*, that is, one object that consists of many parts like a modeling balloon. Furthermore, CD stated that he generally experienced lightness as a limitation: "I lift the balloon up and I'm not getting any weight information. […] It's not affecting my balance. […] I like playing with things with a bit more weight because then you can relate to them more, you can alter your balance, you can counterbalance, moving it around affects you."[17] In addition, CD indicated a preference for something roughly the size of another human body, because his hands were not as "intelligent" as his body, which he referred to as his main and specific tool.[18]

Designer's response:

To have a greater physical effect, the designer added weight to the prototype by filling an inflatable structure with other kinds of material. In the end, she stayed

[17]CD Interview.

[18]CD Interview.

(a) (b) (c) (d) (e) (f)

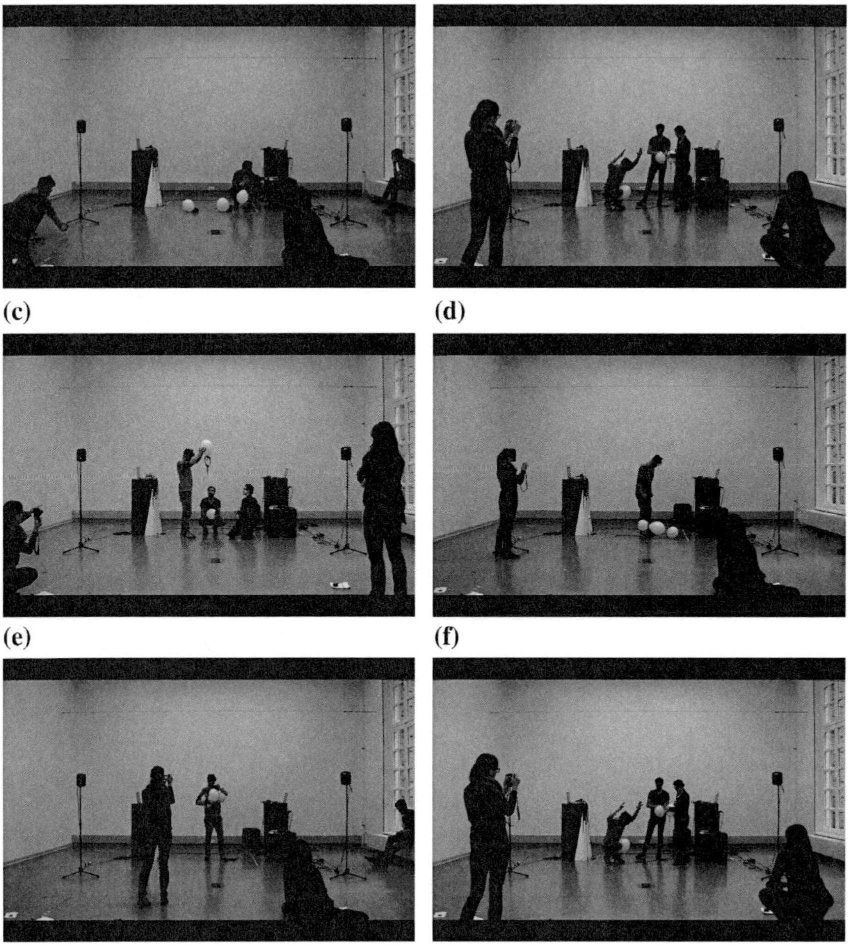

Fig. 11 Key moments identified during Session #2 improvising with balloons

with small polystyrene foam balls for their sonic and material behavior: the balls are electrostatically loaded and move towards or away from hands touching the inflatables from the outside. In this manner, D provided a new haptic experience coupled with a further layer of acoustic quality. Additionally, in reference to CD's general comment on corporeality, the prototype was adjusted to an average body size, made out of five elements connected by joints. During the interview, CD also remembered the moment when he pressed the balloon with the headset microphone between his knees as a satisfying object-sound-movement relation (see Fig. 11g). Consequently, the designer kept this element in the prototype for the following sessions.

2.4.2 Tubes

Observation and Conversation:

AM carried out an interesting movement pattern with the tube mock-up: she pulled the tubes out and pushed them back in by changing her body position (see Fig. 12a, b). To do so, she supported one end of the tubes with the tip of her right foot and held the top part with one hand. Bending her torso forward and straightening her back produced a telescopic shrinking and enlargement of the tube. EM stated that the transition from acoustic to digital sounds had to work smoothly and should be done in a "smart" way. This does not necessarily mean that switches and controllers should be on the instrument itself. Yet, for the specific case of the tube setup, he would prefer an inclusive solution in order to increase the instrument's mobility.

Designer's response:

Taking this into account, D decided to include a light-sensor on the bottom and an LED on the top. In this way, the extraction of the tubes could be mapped to the sound synthesis process. She further applied two pressure sensors to the top part and two potentiometers on the ends of the top and middle tubes. Also she implemented a broader range of frequencies in the sound synthesis, since EM considered this an important attribute of a good setup. The observation of AM's practice of telescoping the tubes described above was translated into an additional hand-strap on the top part of the tube prototype.

2.5 Further Technical Refinements: Improvisational Prototypes

Following Session #3, the improvisational prototypes were readjusted once more according to our observations and the participants' feedback.

While the tube set-up technically stayed the same and the construction was only slightly modified—a button was added on the top and a foot-strap on the bottom, the balloon setup was refined in more detail. D equipped it with a control unit containing two potentiometers, two buttons, an x-y-joystick, a microphone, and a hand strap. It enabled the performer to record loops, change the delay time and the feedback, as well as filters. Furthermore, D added two bend sensors on the object itself to directly map the objects' curvature to sonic aspects. The acrylic glass joints

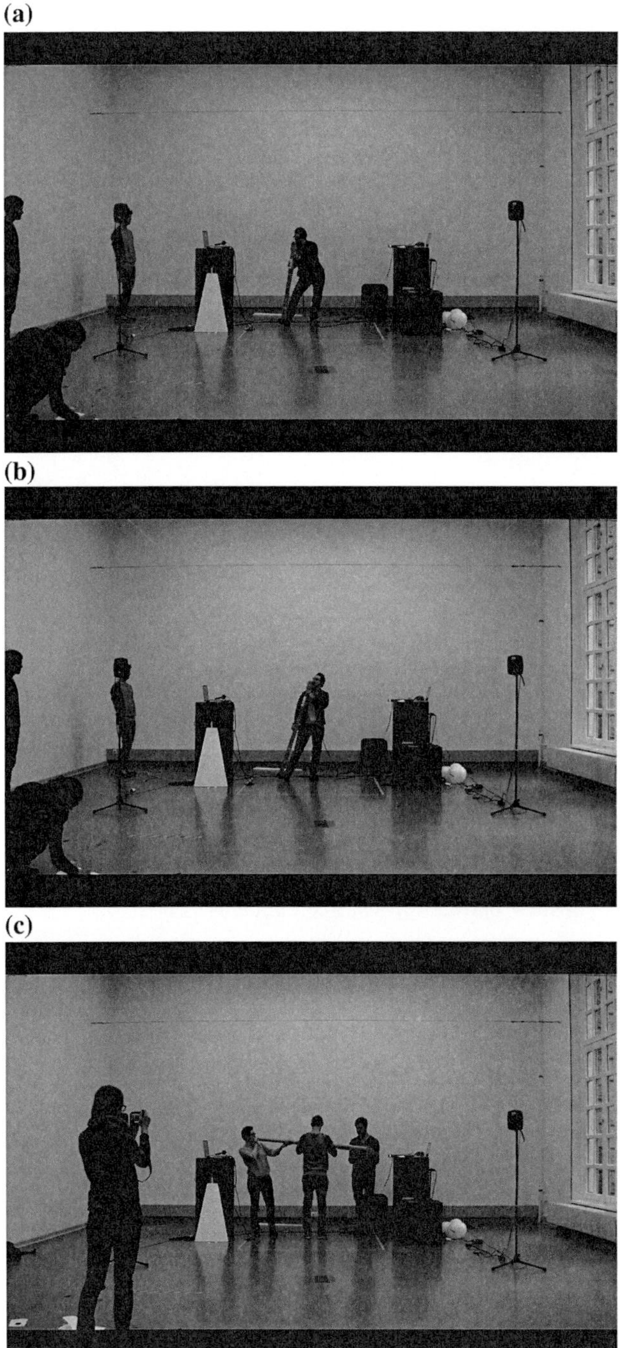

(a) (b) (c)

Fig. 12 Key moments identified during Session #2 improvising with tubes

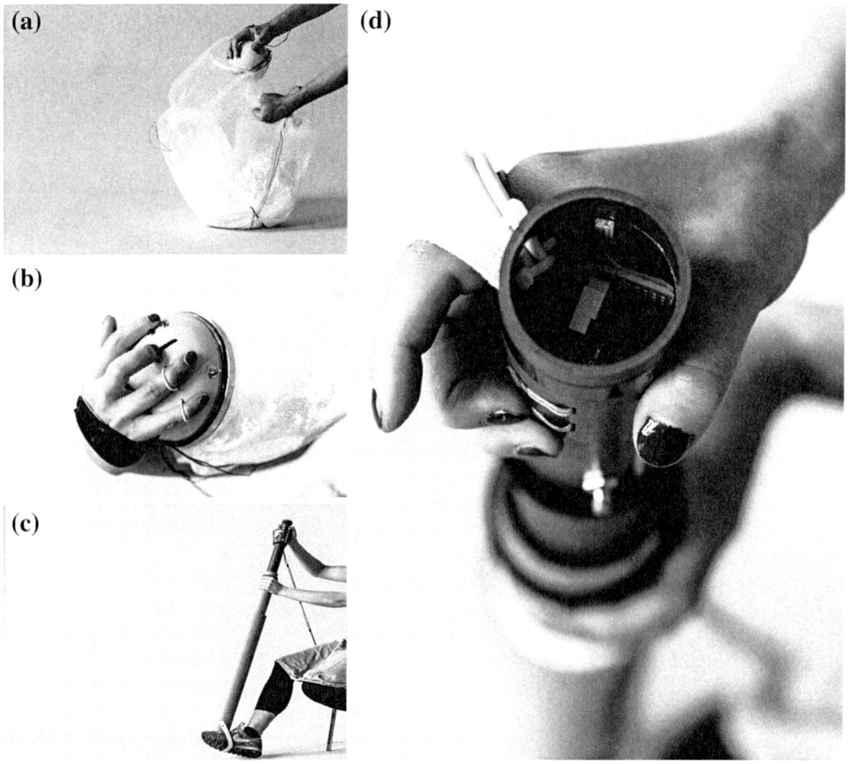

Fig. 13 AM produced rhythmic figures in nearly each round

were replaced with rubber bands to turn the setup back into a compound of one as well as to enable airflow between the single elements (Fig. 13).

3 Discussion

Our observations and results described above are discussed in line with the triad improvisation-materiality-corporeality. However, since we perceived these last two aspects as closely related during the improvisation sessions, we will discuss them jointly (Sect. 3.2). The closing paragraph serves as a reflection upon our improvisational design approach and discusses the idea of material authenticity, a term the designer coined during the development process. It merits special attention since it opens up further questions in regard to the design of digital musical instruments and therefore offers a connection for future research in this field.

3.1 Improvisation

In the setting presented, the interactive improviser-environment-relation described in the introduction to this article was a crucial part of improvisation; it consisted of a constant mutual exploration and negotiation of bodily, sonic, and material possibilities. For example, the irreversible deformation and even destruction of the material provided altered the participants' 'environment' significantly, in that they effectively changed the material in its sonic as well as haptic qualities. Furthermore, sounds played by one participant left an impression on the other participants, hence influencing their actions and playing style.

A persistence of individual playing styles was noticeable across the sessions, even though several playing modes disappeared from Session #1 to Session #2. Instead of developing ideas introduced by others, the participants preferred to engage around individual ideas, sometimes carrying on even across the sessions. For example, after Session #1, CD stated that he had needed "some time to calm down", because in the beginning he had aimed at "a perfect minute [of improvisation] every time". Therefore, it seemed on the one hand that he picked up on his own stylistic elements to fulfill his high self-expectations, and—in congruence with Brehm and Kampfe (1997)—on the other hand, to produce something meaningful.

In a group of improvisers such as this, one would usually expect elements to emerge through the participants' interaction. But as stated above, it was only rarely that the participants collaboratively developed elements. One could assume, however, that the improvisation setup within which we asked the participants to develop their solos constituted an overload of demands. Being confronted with a strict structure and timeframe, rich material qualities as well as a transdisciplinary group, the participants had to follow the "rules of economy" (Sennett 2009, p. 236) in order to orientate themselves. Despite this possible overload of demands, the participants appreciated being challenged in order to come up with ideas—just as Sennett expresses in the following quotation.

> Getting better at using tools comes to us, in part, when the tools challenge us, and this challenge often occurs just because the tools are not fit-for-purpose. […] the challenge can be met by adapting the form of a tool, or improvising with it as it is, using it in ways it was not meant for. However we come to use it, the very incompleteness of the tool has taught us something. (Sennett 2009, p. 194)

With the explicit reference to tools, that is, in our case to material, mock-ups, and refined prototypes, the quotation makes a good transition to further reflections upon materiality and corporeality.

3.2 Materiality and Corporeality

Let us take one more look at the challenges mentioned above. For example, after Session #2, CD stated that, "the limitations made you think". Here, limitations clearly refer to the modified material. In Session #1, the material provided was relatively indirect in its connotations and possibilities of use. In comparison, the designed and objectified mock-ups in the second session offered much clearer affordances, that is, functions and usability in Gibson's sense of the term (1986, p. 127). We already showed that the participants perceived the constitution of the materials and treated them according to the ideas the material quality provoked—or as Gibson says, "The object offers what it does because it is what it is" (ibid., p. 139). Behavioral patterns distinct for each participant reappeared with their interactions with each material: AM produced rhythmic figures in nearly each round, CD constrained himself through using the material as a body extension inhibiting his movements, and EM repeatedly combined several pieces of one material group to explore their sound qualities. This reveals that the material affordances and the participants' embodied practices influenced both their improvisation and the instrument design to an equally high degree.

Furthermore, we can conclude that corporeality exists on three levels. Firstly, on the aforementioned individual level of embodied practices, which the participants used for their solos. Secondly, on the level of artist-instrument-relation. It was most obvious in the direct, very engaged and close bodily contact between the participants and the raw material during Session #1, which seamlessly transformed into the careful handling of the objects during Session #2. Finally, on the level of bodily presence of the participants. AM stated e.g. that EM was not accessible during collective improvisations, which—again—changed in Session #2, where the collective improvisations consisted of more inclusive exchange.

Interestingly, the participants perceived one affordance similarly: in Session #2, the moment of destruction disappeared completely. Instead, participants all treated the mock-ups very carefully, which can be linked to the mock-ups' more objectified shape and material identity. This points to the conclusion that as soon as there is a manufactured object to handle, it should be robust and stable so that the musician can use it in a free, intuitive and spontaneous manner, without being afraid to break anything.

3.3 Limitations and Outlook

The presented design process can be characterized as transdisciplinary and iterative. Transdisciplinary, because the process involved four different disciplines: Contemporary dance, acoustic music, electronic music, and product design. It was iterative, because the participants involved reconvened several times. In each session, all participants were working towards and discussing aspects of digital musical instruments and live set-ups. The process became more directed each time. It departed from material exploration and instrument mock-ups to playing with refined improvisational prototypes. The design and functionality of these prototypes emerged step by step, which was also manifested in the behavior of the participants: their activities underwent a change from individually exploring material possibilities to playing together 'on' one object. Similarly, different performance practices informed the subsequent discussions. The participants stated varying needs according to whether they performed in a predetermined setup with a composed piece, or whether they were supposed to be improvising. Furthermore, we showed that the participants' exploratory manner of dealing with the proposed material as well as their reaction to the perceived sounds and movements influenced the design process to a high degree.

In conclusion, the chosen improvisation practice opened up a broad range of aspects that can impact and enrich the design process. At the same time, it revealed the interdependent relation between (1) individual practices, experiences, and demands, (2) a transdisciplinary group constellation, (3) spatial settings, (4) implemented technologies, and (5) provided materials. It is due to this complex network of aspects that we are unable to clearly state which of them predominantly influenced the participants' improvisations and, consequently, the developed prototypes.

We showed that improvisation in a performative context can significantly shape the design of a musical instrument. It revealed insights into the interplay of materiality and corporeality—aspects we consider to be crucial for an instrument design process. Furthermore, spontaneity and flexibility are important factors to consider in an improvisational setting. An instrument played during improvisation therefore needs to be a stable and reliable system providing the performer with immediacy and variability at the same time. In regard to the chosen improvisational approach it is noteworthy that CD's bodily engagement decreased significantly while playing with the improvisational prototypes during the final session. From Session #1 to Session #2 we noted the disappearance of several playing modes. We explained this with the emergence of form. We consciously designed a digital instrument usable in a performative, improvisational live set-up. It was never intended to become an artifact for dance performances. Yet, the applied method stemmed from contemporary dance and both CD's movements and input strongly informed the developed instruments. It is noteworthy to perceive such an important disciplinary foundation to slowly disappear in the concretization process of the instrument.

One more aspect connected to the overall design process merits further attention: Digital sound synthesis opens up infinite possibilities of sound generation and manipulation. We think that in this labyrinth of sonic possibilities, materiality can act as a *leitmotiv* for an electronic instrument's audio-visual character, which will consequently form its 'identity'. In our sessions, the instruments' materials informed their emerging identities in that they provided genuine and recognizable sound signals and playing patterns. We thus speak of an instrument's *material authenticity*. When for an electronic live set-up, an artifact is combined with digital sound generation, we consider it important to integrate this authenticity. When connecting these two spheres, mechanic-analogue and digital sound generation, in one hybrid instrument, we should take care that they 'speak the same language'. In this manner, the performative and musical flow between the musician and the instrument can be strengthened and mutually enhanced.

Playing with refined prototypes during Session #3 already raised the participants' expectations towards the instruments. For our final session, we had achieved the aim of two digital musical instruments including a limited set of sound synthesis based on their material components. These were cardboard for the tube instrument and transparent film filled with tiny polystyrene foam balls derived from the balloon mock-up. Material authenticity marked the sounds produced both during the quick and the collective improvisation rounds: Filters and delay, for example, modified the participants' blowing into the cardboard tube, but the sound of in- and exhaling remained audible. Also, the trickling polystyrene foam balls were not only visible. Their falling onto the microphone produced a quite loud electronic ticking sound. Their original, 'natural' sound, however, was barely perceivable; an aspect to which the participants expressed criticism during our feedback discussions at the end of the session.

We would therefore like to finalize our contribution with the following questions, simultaneously paving the way for further research on material authenticity of future digital musical instruments: Which relation between material qualities and sensorial sensitivity is most suitable to enable both rough material treatment and subtle sonic results? How can the resulting instrument system offer enough room for material and sonic exploration without inhibiting the musician through its complexity, and without boring him through being too obviously authentic? And last but not least: Can we actually define an instrument's or a digital sound's authenticity, or is this subject of subjectivity, that is, individual perception?

Acknowledgments We would like to thank the three artists, Lea Danzeisen, Joshua Rutter and Tobias Purfürst, for their voluntary participation in the four improvisation sessions. Only with the help of their artistic input and feedback we were able to arrive at the above-mentioned insights and results. Also we would like to thank Till Bovermann for taking part in Session #4 with technical support in sound programming.

This paper is based upon work of 3DMIN, an interdisciplinary research project of the University of the Arts Berlin and the Technical University Berlin supported by the Einstein Foundation Berlin.

References

Bormann, H.-F., Brandstetter, G., & Matzke, A. (2010). Improvisieren: eine Eröffnung. In H.-F. Bormann, G. Brandstetter, & A. Matzke (Eds.), *Improvisieren. Paradoxien des Unvorhersehbaren. Kunst—Medien—Praxis* (pp. 7–19). Bielefeld: transcript.

Brehm, M. A., & Kampfe, C. M. (1997). Creative dance improvisation: Fostering creative expression, group cooperation, and multiple intelligences. In *China-U.S. Conference on Education. Collected Papers.* http://files.eric.ed.gov/fulltext/ED425401.pdf. Accessed October 14, 2015.

Buchenau, M., & Suri, J. F. (2000). Experience prototyping. In *DIS '00: Proceedings of the Third Conference on Designing Interactive Systems.* ACM, New York, NY, USA, pp. 424–433. https://www.ideo.com/images/uploads/news/pdfs/FultonSuriBuchenau-Experience_PrototypingACM_8-00.pdf. Accessed June 2, 2016.

Burns, C., Dishman, E., Verplank, B., & Lassiter, B. (1994). Actors, hair-dos and videotape: Informance design; using performance techniques in multi-disciplinary, observation based design. In *Proceeding CHI'94 Conference Companion on Human Factors in Computing Systems*, pp. 119–120. Boston, MA. https://designforservice.files.wordpress.com/2010/02/actors-hairdos-videotape.pdf. Accessed June 2, 2016.

Carter, C. L. (2000). Improvisation in dance. *The Journal of Aesthetics and Art Criticism, 58*(2), 181–190.

Crispin, D., & Gilmore, B. (Eds.). (2014). *Artistic experimentation in music. An anthology*. Leuven: Leuven University Press.

Donnarumma, M., Caramiaux, B., & Tanaka, A. (2013). Muscular interactions. Combining EMG and MMG sensing for musical practice. In *Proceedings of the 13th International Conference on New Interfaces for Musical Expression*. http://research.gold.ac.uk/10634/1/marco-donnarumma_muscular-interactions-combining-emg-mmg-music-practice_NIME2013.pdf. Accessed May 25, 2016.

Douglas, A., & Coessens, K. (2012). Experiential knowledge and improvisation: Variations on movement, motion, emotion. *Art, Design & Communication in Higher Education, 10*(2), 179–198.

Emerson, R., Fretz, R., & Shaw, L. (2011). *Writing ethnographic fieldnotes* (2nd ed.). Chicago: University of Chicago Press.

Gerber, E. (2007). Improvisation principles and techniques for design. In *CHI '07 Proceedings of the SIGCHI Conference on Human Factors in Computing Systems*, pp. 1069–1072. https://dschool.stanford.edu/sandbox/groups/k12/wiki/73bac/attachments/4ffd6/Gerber_Design_Improv.pdf?sessionID=b5013b50d4dae144a4c570db24b9933747c64e87. Accessed May 27, 2016.

Gibson, J. J. (1986). *The ecological approach to visual perception*. New York: Taylor & Francis.

Hummels, C., Overbeeke, K. C., & Klooster, S. (2007). Move to get moved: A search for methods, tools and knowledge to design for expressive and rich movement-based interaction. *Personal Ubiquitous Computing, 11*(8), 677–680. http://link.springer.com/article/10.1007%2Fs00779-006-0135-y#/page-1. Accessed May 31, 2016.

Keep, A. (2009). Instrumentalizing: Approaches to improvising with sounding objects in experimental music. In J. Saunders (Ed.), *The Ashgate research companion to experimental music* (pp. 113–130). Surray/Burlington: Ashgate.

Klooster, S., & Overbeeke, C. J. (2005). Designing products as an integral part of choreography of interaction: The product's form as an integral part of movement. In *Proceedings 1st International Workshop on Design and Semantics of Form and Movement (DeSForM '05)*, pp. 23–35. https://pure.tue.nl/ws/files/1913459/Metis197549.pdf. Accessed June 1, 2016.

Lee, J.-S., & Yeo, W. S. (2012). Real-time modification of music with dancer's respiration pattern. In *Proceedings of the 12th International Conference on New Interfaces for Musical Expression*. http://www.nime.org/proceedings/2012/nime2012_309.pdf. Accessed May 25, 2016.

Larssen, A. T., Robertson, T., & Edwards, J. (2007). Experiential bodily knowing as a design (sens)-ability in interaction design. In *Proceedings of the 3rd European Conference on Design and Semantics of Form and Movement*. https://opus.lib.uts.edu.au/bitstream/10453/7029/1/2007000387.pdf. Accessed May 27, 2016.

Loke, L., & Robertson, T. (2009). Design representations of moving bodies for interactive, motion-sensing spaces. *International Journal of Human-Computer Studies, 67*.

Loke, L., & Robertson, T. (2013). Moving and making strange: An embodied approach to movement-based interaction design. In *ACM Transactions on Computer-Human Interaction (TOCHI)—Special Issue on the Theory and Practice of Embodied Interaction in HCI and Interaction Design* 20, no. 1.

Johnstone, K. (1989). *Impro: Improvisation and the theatre*. London: Methuen Publishing.

Moen, J. (2005). Towards people based movement interaction and kinaesthetic interaction experiences. In *Proceedings of the 4th Decennial Conference on Critical Computing: Between Sense and Sensibility*. ACM.

Mareis, Claudia. (2014). *Theorien des Designs zur Einführung*. Hamburg: Junius.
Miranda, E. R., & Wanderley, M. M. (2006). *New digital musical instruments: Control and interaction beyond the keyboard*. The Computer Music and Digital Audio Series, Vol. 21. Middleton: A-R Editions.
Nakano, Y., & Okada, T. (2012). Process of improvisational contemporary dance. In *34th Annual Meeting of the Cognitive Science Society*. https://mindmodeling.org/cogsci2012/papers/0362/index.html
Nancy, J.-L. (2007). *Corpus* (2nd ed.). Zürich/Berlin: Diaphanes.
Oxford Dictionary. (2015). http://www.oxforddictionaries.com/definition/english/improvise#improvise__2. Accessed October 5, 2015.
Redgate, C. (2015). Creating new music for a redesigned instrument. In M. Dogantan-Dack (Ed.), *Artistic practice as research in music: Theory, criticism, practice* (pp. 203–217). Surrey/Burlington: Ashgate.
Redgate, C. (2016). Personal website. http://www.christopherredgate.co.uk. Accessed May 24, 2016.
Sennett, R. (2009). *The craftsman* (3rd ed.). London: Penguin Books.
Sylleros, A., de la Cuadra, P., & Cádiz, R. (2014). Designing a musical instrument: Enlivening theory through practice-based research. *Design Issues, 30*(2), 83–96.
Waters, S. (2007). Performance ecosystems: Ecological approaches to musical interaction. In *Proceedings of Electroacoustic Music Studies Network Conference 2007*. De Montfort/Leicester. http://www.ems-network.org/IMG/pdf_WatersEMS07.pdf
Young, J. (2015). Imaginary workscapes: Creative practice and research through electroacoustic composition. In M. Dogantan-Dack (Ed.), *Artistic practice as research in music: Theory, criticism, practice* (pp. 149–166). Surrey/Burlington: Ashgate.

Instrumentality, Time and Perseverance

Giuseppe Torre and Kristina Andersen

Abstract In this article we discuss how the act of perceiving a digital object as a musical instrument can be considered as directly proportional to the amount (and quality) of time invested in its development and refinement to suit individual needs rather than generic ones. In that regard, the purpose-free approach to the design of generic controllers contrasts with a view of personalised tools developed and continuously redefined by the artist to fulfil artistic and musical needs. In doing so, the time invested relates to the artist/designer's perseverance in a never-ending process of subjectification of the digital instrument identity. The discussion provided in the article is supported by a case study on one of the pioneers and developers of digital musical instruments: Michael Waisvisz (1949–2008) and his work on The Hands (first exhibited in 1984—last performance dated 2008). We argue that this almost 30-year long and engaged process of development and experimentation can be seen as a model, through which we can allow other musical devices to evolve from controllers of digital musical matter to instruments that may provide integrated and embodied possibilities for musical expression.

1 Introduction

Over the past three decades, an ever-increasing number of researchers have studied, designed and performed with a highly diverse set of digital tools better known as digital musical instruments (DMI). This effort has been largely driven by the desire to develop new tools that enable a performer to connect with the vast realm of sounds allowed by digital technologies.

G. Torre (✉)
Department of Computer Science and Information Systems, Digital Media and Arts Research Centre, University of Limerick, Limerick, Ireland
e-mail: giuseppe.torre@ul.ie

K. Andersen
STEIM Institute, Amsterdam, The Netherlands
e-mail: kristina@tinything.com

T. Bovermann et al. (eds.), *Musical Instruments in the 21st Century*,
DOI 10.1007/978-981-10-2951-6_9

To some extent, the use of the term 'instrument' in the acronym DMI reveals an intention to draw similarities with traditional instruments (e.g. a guitar) and as such to more closely relate to the concepts of expressivity and musical purpose. Arguably, there seems to be a general discomfort amongst both experts and audience in attributing the status of "instrument" to the vast majority of these novel digital tools that, often, are better described as controllers.

While the words instrument and controller are not synonyms, they are both used almost interchangeably in both academic writings and conversations. Arguably, this may to some extent be due to the fact that the word interface appears in the title of one of the most prestigious conferences in the field concerned here: the New Interfaces for Musical Expression conference (NIME). The idea underpinning most of the research presented in its proceedings is, indeed, to develop objects that would facilitate the performer to interface with the vast realm of musical sounds. Yet, the word interface connects better with an interaction design culture than an artistic one and in fact, NIME was born as a workshop at an interaction design conference in 2001 (the ACM Conference on Human Factors in Computing Systems—CHI).[1] In this context, the performer is often seen predominantly as a user, and the instrument is an object interfacing and mediating the interaction between a human and the computer (i.e. the digital realm). At the same time, the NIME community needed to find its way away from CHI and, in an effort to establish its focus, researchers started highlighting the differences between the controlling elements (i.e. a controller) and the expressive mechanisms (i.e. a musical instrument) in a DMI (Dobrian and Koppelman 2006). This distinction delineates two distinct types of focus, the artistic and musical purpose of an instrument and the technical and interactive properties of a controller.

1.1 Heritages

Cultural heritages plays a key role in allowing a digital musical object to be defined as a musical instrument in that it helps forming judgments according to widely accepted parameters used for traditional ones such as affordances, virtuosity, touch, expressivity and, most importantly, a music literature. Instead, the idiosyncratic approach that characterises the development of DMIs has impeded the formation of a tradition shared by both audience and performers. In turn, this has forced the discussion on the proper nomenclature of these devices into a lexical issue in which a never-ending need for defining attributes such as the one of 'digital' are invoked to highlight peculiar interaction and modus operandi paradigms (e.g. the digital paradigm).

[1]Incidentally, the interchangeable and confusing way in which both terms of *instrument* and *controller* are used is evident from the very first paper published in the first proceedings of this: Principles for Designing Computer Music Controllers by Perry Cook and dated 2001.

The paper by Cance et al. (2009) presents a good picture of the linguistic conundrum that the definition of a DMI has created. Yet, Cance et al. also conclude that "the instrumentality of these new devices (DMIs), as well as of "classical" instruments, does not result from their intrinsic properties only. It is constructed through musical play, interactions between musicians and the design and development of the instruments."

Similarly, Cook used the word 'remutualizing' to suggest that the development of new digital instruments should follow the traditional workflow in which the "design used to be the result of mutual evolution of performer and craft, and that, with care, designers can reintroduce this symbiosis in our modern electronic instruments" (Cook 2004). We agree with Cook's suggestions in that this would re-centre time as the dominant factor to the making of an instrument. Time is intended as the unavoidable temporal length required to continuously refine an object. In doing so, time becomes related to the willpower and dedication of the developer. Time invested can be then considered to be proportional to the developer's perseverance.

1.2 Three Phases of Development

In light of the considerations above, we retrace the life of an instrument in three phases: an initial experimental phase in which a performer's needs and crafting solutions are explored. A second phase, standardisation phase, in which the best crafting solution is accepted by (and suited for) an audience who also contribute to the making of a musical literature dedicated to the new instrument. A third phase referred to as the customisation phase. Here the instrument is tailored to individual needs although still retaining the main characteristic of the standard one (for example the addition of an extra sensor, the introduction of a mapping technique or a different colour for the main body and so on).

With the three phases above in mind, it can be seen how the life of most DMIs, as presented in the pertinent literature (e.g. NIME) ends in the experimental phase. The DMI is built on the basis of the needs or some technical issue that the researcher (usually the performer, too) wants to investigate. The end of this preliminary research marks also the end of the DMI's life. This, of course, does not allow for the creation of a dedicated instrument literature or for the subsequent customisation phase. The life of the device is too short to fully bloom into an instrument that is also recognisable as such by an external audience. The DMI remains a DMI with all the problematics, misunderstandings, lexical and epistemological issues associated with it.

Only a few DMIs have followed the three phases of development. Using Jordà's words, it is also important to note that "many new instruments are being invented" but "too little striking music is being made with them" (Jordà 2004). The list of professional musicians that have used extensively DMIs to the point of becoming virtuosos is small and includes performers like Jon Rose and his series of deconstructed violins and Laetitia Sonami and her Lady's Glove. One other prominent example is Michel Waisvisz (1949–2008) and his instrument The Hands. In the remaining part of this article we report and discuss excerpts taken from a previously unpublished interview with Michel Waisvisz from April 2008, conducted by Kristina Andersen at STEIM, and apply our three-phases hypothesis to the development of The Hands. In this interview Waisvisz looks back at his work on The Hands, outlining the motivations and story behind their development. All indented quotes in the following are taken from this interview.

2 The Story of the Hands

Michel Waisvisz' work on The Hands started at the STEIM Lab in 1984, and it engaged him for over a quarter of a century, during which time The Hands were his main performance instrument. The basic system comprised of a pair of data gloves, each of which is made of a small keyboard on the player's hands, accelerometers, pressure sensors and an ultrasound distance sensor. Over the 25-years development three versions of The Hands were designed and built: version 1 from 1984, version 2 from 1990 and version 3 from 2000 (Torre et al. 2016) (Figs. 1, 2 and 3).

Throughout this process Waisvisz remained the designer and lead on the technical and artistic design while a large number of individuals contributed to the building and coding, amongst those: Wim Rijnsburger, Hans Venmans, Peter Cost, Bert Bongers, Frank Baldé, Tom DeMeijer, Maurits Rubinstein, Jorgen Brinkman and David Bristow.

Each version presented a different number of sensors and design. The sensor technology in use, however, remained almost unchanged, except for the addition of a clip microphone on the left data glove starting from version 2. The Hands connected to a box secured around the back of the performer, hosting a microcomputer that converted the incoming analogue data from the data gloves into MIDI messages and sending them to the host computer for further sound processing and manipulation (ibid.).

2.1 *The Experimental Phase*

The experimental phase is generally characterised by the exploration of the performer's musical need. Waisvisz had arrived at STEIM in 1973 with a musical desire to "touch" (Waisvisz's quote) electronics, and by the mid-seventies, with the help of Peter Beyls, Nico Bes and Johan den Biggelaar, he created the 'Crackle Synth' and its offspring, the 'Crackle Box'. He motivated the physical approach in the design of this electronic music instruments by stating that:

> Machines are precise with numbers, but the human hand is more precise with musical time.

Fig. 1 Waisvisz and The Hands version 1 (photo from STEIM Archive)

Fig. 2 Waisvisz and The Hands version 2 (photo from STEIM Archive)

Fig. 3 Waisvisz and The Hands version 3 (photo from STEIM Archive)

This project was seminal for the subsequent development of The Hands. However, it took ten years and a breakthrough in industry standards for it to eventually bloom.

> In '84 the DX7 turned up, the Yamaha DX7, which was the first synth that was widely spread that had MIDI, and so when they explained to me what MIDI actually did… I took that of course in my way, and I thought, I could make a crackle box and have that translated into MIDI. And then, you know, I could just have two Crackle-boxes, one on each hand, or something like that and then I realised we needed keys, 'cause MIDI was very much key-oriented.

With the design based on such a simple premise and the keys being central to the MIDI concept, the first prototype was quickly built.

> So with Johan de Biggelar in three month's time we basically build this little plate with a number of keys and like a sonar sensor and a pressure sensor and a potentiometer, and that was it. And with that, after four months I did this concert in de Concertgebouw… and I used it only in a part of the piece, because it wasn't finished…

In 1984 Waisvisz performed with The Hands for the first time in the Concertgebouw in Amsterdam. This early version was, however, far from being considered a success in the opinion of its developer/performer:

> To be honest, I was a bit disappointed with The Hands, because it was so complex in the beginning. You had to realise so many things, to keep track of what is switched on, and that if you move to another octave, you should switch them off, and indeed how to have them move to another octave? 'Cause we used these quick keys or switches to move to other octaves, so with a tiny keyboard of 12 keys, you had an 8-octave instrument. So I kind of dropped it through the summer of '84… and you know, I wasn't sure that it was such a nice instrument, I had such problems playing it, and it needed so much work….

The instrument was nearly impossible to play, and it was immediately clear that it would take a lot of time investment to not only make it playable, but, maybe more importantly, learn to play it. The concept had been demoed and now the question was if it would be worthwhile to continue working on it.

> ...so we left it for a while,... kind of threw The Hands away, literally in a box somewhere..., and then I was invited to come to the Computer Music Conference in Paris, and... there was so much talk about all these circuits that they had, that needed to be controlled. And you know people were unhappy with the piano, so they were using all these little fader boards. And I suddenly realised, there was David Wessel doing a speech and he was like: "We need a real instrument..." and I realised suddenly that that crazy instrument that I had used in the Concertgebouw... I suddenly realised that I should really work on that, so this is where I immediately hooked up with David, and I got to know Joel (Ryan) through George (Lewis) and I started learning about MIDI, I started learning programming a little bit, and we invited George and Joel to STEIM...

As the development of The Hands continued in the STEIM workshops, the process became a model for how artist-driven technology could be supported. This approach placed the initiative and the design of an instrument with the musician, who has both the knowledge and the vested interest in developing and playing with the resulting musical object.

> ...that was maybe '84-'85, I don't remember exactly, and so then suddenly I found myself programming on a small computer and making these first little programs and together with Johan de Biggelaar developing The Hands further. And I think about a year later I did these first concerts at the IRCAM with The Hands working.

At this point it could be argued that The Hands left the experimental phase and entered the standardisation phase.

2.2 *The Standardisation Phase*

The standardisation phase is characterised by an effort to engage a greater number of performers in the adoption of the design solutions found during the preceding experimental phase.

One of the technical challenges faced during the development of The Hands was the implementation of an analogue to MIDI converter that enabled the conversion of the data gloves' sensor reading to MIDI in order to interface with the DX7 synthesiser. This preliminary work in 1989 lead to the development of the analogue-to-MIDI converter Sensorlab. The year after, a simplified version of The Hands known as "MIDI Conductor" was released and distributed amongst music practitioners and students.

The MIDI conductor hosted the same type of sensors that The Hands had (keys, pressure sensors, distance and tilt sensors) but just in smaller numbers. Six units were produced, and they were handed out to a duo (Frank Baldé and Michael Barker), a trio (BMBCon, i.e. Justin Bennet, Roelf Toxopeus and Wikke van't Hoof) and Edwin van der Heide. Waisvisz, too, wrote a piece for multiple MIDI

Conductor performers (i.e. the BMBCon trio) called Abracadabra. Baldé modified his MIDI Conductor to be able to use it with the second generation of the SensorLab, and as of today it is still operating. All this contributed to the making of a small, although unfortunately little documented, written literature. Yet, it provides historical evidence for the sharing of musical objects born by the need of a single performer.

2.3 The Customisation Phase

The customisation phase began in 1990 with the development of Version 2 of The Hands and ended in 2000 with the release of Version 3. The differences between these two versions are minimal compared to Version 1, thus present similar features and ergonomic solutions. Beyond an increased number of key switches and the addition of one potentiometer, the differences are found mainly in aesthetic features such as colour, case and circuit board layout.

After the finalisation of version 3 of The Hands, Waisvisz made the decision to stop developing and accept the physical layout as is. From this point onwards, he concentrated on refining the software settings and musical content of the performances. In this version, there were no longer big technical problems, thus energies were invested almost solely in learning to play The Hands. This lead to a long sustained practise of not only playing and developing music, but also developing physical practices and movements of playing. It was time invested in developing a personal relationship with his musical instrument.

> Some movements are also really to connect. I think there's a lot shoulder movements that I have that I do that are totally unnecessary for, you could say for the actuation of the sound, for the triggering of the sound. But they will steer my hands, move my hands in a way by moving the shoulders so that they are at the right moment. It's a kind of gel that connects.

This process of learning and re-learning the possibilities and limitations of the instrument is one that is well-known to anyone who has ever practiced to play an instrument. By freezing the design modifications and extensions when The Hands were physically stable and durable, it became possible to focus on the musical intent beyond the novelty of the devices and engage in the aesthetic and musical considerations, rather than the technical details that lie behind the interface. In the case of The Hands, this required a whole new set of skills to engage with the physical challenge of playing and mapping each moment to a continuous physical movement through each piece.

> It is a kind of internal connection of notes that is being helped by connecting them physically by changing your movement and posture. Then suddenly it is like almost like a real connection in time. And I think that if you would analyse all the movements of what I do, it's interesting because my instrument doesn't require that much specific movement, but I would easily guess that more than half of the movement is to connect and to help to shape time, rather than is being functional to the triggering like interacting with the sensor or so to say.

3 Conclusions and Notes

> But of course with The Hands, you can use your whole body. You can stand there as a boxer, or you can be there like trying to be fragile like a cloth in the wind or something.

When it comes to playing an instrument, whether it is traditional or a DMI, the body of the performer is in the centre of things, with the control and sound properties of an instrument intimately linked to its acoustical properties, technical execution, size, weight, construction and, more generally, to what the object affords (Gibson 1977). In this article we have argued that the way a demo-object becomes an instrument is through the practice of learning to play it, and the perseverance in seeing it through the three phases of instrument design: the experimental phase, the standardisation phase and the customisation phase.

It could be said that Waisvisz in a sense stumbled upon the initial notion of The Hands, but it was through his commitment to build a practice around them over a long period of time and develop a unique and personal set of expressions and virtuosic skills, which instituted them as instruments for musical expression.

This was made possible by deliberate intentions and goals. The architect Sabien de Klein, who designed the enclosure for the Crackle Synth, describes the way Waisvisz explained his intentions to her:

> He said: "I want it to be like a piano… I want to make it recognisable as an instrument of the old times, that it is, not after old instruments, but between old instruments."[2]

This is also in line with a statement by Sally Jane Norman, Michel Waisvisz and Joel Ryan in the catalogue for the STEIM Touch festival:

> Touch advocates an idea of performance in which the physicality of the encounter between artist and audience is essential. Touch vindicates the central position of the human element in the electronic arts, and the necessity to place technology at the service of the creative individual. (Norman et al. 1998)

By addressing new technologies for creative expression through the performing body of the musician, the process of allowing time and effort to move an interface towards becoming an instrument closely mirrors the process of learning to play. As the idiosyncrasies and quirks of an interface are either improved and eliminated or learned, the third phase of instrument building is also the phase where the musician has the opportunity to intimately engage with the emerging instrument. This may ultimately allow us to see the potential design and solution space of a musical interface through the framework of a much larger history of artistic objects and practices. If so, it seems that the time invested and the artists' perseverance may be a key to an instrumentality that is achieved by favouring long-lasting ageing processes over a habit of fast-prototyping and fast-dismissal.

Throughout the process of making and playing The Hands, Waisvisz considered each new modification in the light of a desire for making a new kind of music. Joel

---

[2]In personal communication with Kristina Andersen.

Ryan, who was present at STEIM for most of this period, expresses this stance well when he writes:

> The musical instrument is a vehicle for the desire to make music. It is both something that must be internalised, incorporated and made flesh and something other, without which we could not get to that sound from the world beyond (Ryan and Andersen 2014).

References

Cance, C., Genevois, H., & Dubois, D. (2009). What is instrumentality in new digital musical devices? *A Contribution from Cognitive Linguistics and Psychology*. https://hal.archives-ouvertes.fr/hal-00430397. Accessed April 15, 2016.

Cook, P. (2001). Principles for designing computer music controllers. In *Proceedings of the NIME Workshop at CHI 2001*, Seattle, Washington, April 1–2, 2001.

Cook, P. (2004). Remutualizing the musical instrument: Co-design of synthesis algorithms and controllers. *Journal for New Music Research, 33*(3), 315–320.

Dobrian, C., & Koppelman, D. (2006). The 'E' in NIME: Musical expression with new computer interfaces. In N. Schnell, F. Bevilacqua, M. J. Lyons, & A. Tanaka (Eds.), *NIME '06: Proceedings of the 2006 Conference on New Interfaces for Musical Expression*. Paris.

Gibson, J. J. (1977). The theory of affordances. In R. Shaw & J. Bransford (Eds.), *Perceiving, Acting, and Knowing: Toward an Ecological Psychology* (p. 127). Hoboken, NJ: Wiley.

Jordá, S. (2004). Digital instruments and players: Part I—Efficiency and apprenticeship. In M. J. Lyons (Ed.), *NIME '04: Proceedings of the 2004 Conference on New Interfaces for Musical Expression Hamamatsu, Shizuoka, Japan*.

Norman, S. J., Waisvisz, M., & Ryan, J. (1998). *Touchstone. Catalogue to the first STEIM Touch-Exhibition*. Amsterdam.

Ryan, J., & Andersen, K. (2014). 821 Words and 20 Images. In *No Patent Pending, Self-Made Performative Media* (3 ed., pp. 1–10). MER. Ghent: Paper Kunsthalle, Ghent.

Torre, G., Andersen, K., & Baldé, F. (2016). The Hands: The making of a digital musical instrument. *Computer Music Journal, 40*(2), 1–13.

Machine Learning as Meta-Instrument: Human-Machine Partnerships Shaping Expressive Instrumental Creation

Rebecca Fiebrink

Abstract In this chapter, I describe how supervised learning algorithms can be used to build new digital musical instruments. Rather than merely serving as methods for inferring mathematical relationships from data, I showhow these algorithms can be understood as valuable design tools that support embodied, real-time, creativepractices. Through this discussion, I argue that the relationship between instrument builders and instrumentcreation tools warrants closer consideration: the affordances of a creation tool shape the musical potential of theinstruments that are built, as well as the experiences and even the creative aims of the human builder. Understanding creation tools as "instruments" themselves invites us to examine them from perspectives informedby past work on performer-instrument interactions.

1 Introduction

The practice of building new musical instruments is predicated on the recognition that instruments matter: that the sort of music one can make with a xylophone is different than with a violin, which is different still from the music one can make with a computer. Instruments differ by more than just their sound qualities; acoustic instruments bring with them particular physical affordances, and these lead to idiomatic playing styles and repertoires.

The goal of many designers of digital musical instruments is to discover new idioms for expression, shrugging off old constraints of physical materials and acoustics. Each new configuration of sensors and sound synthesis algorithms patched together by software suggests a new way of being played.

Just as the instrument shapes the music that may be played, the tools for instrument creation shape the instruments that may be built. And just as each instrument demands its player develop a particular set of physical skills and musical knowledge

R. Fiebrink (✉)
Department of Computing, Goldsmiths, University of London, London, UK
e-mail: r.fiebrink@gold.ac.uk

T. Bovermann et al. (eds.), *Musical Instruments in the 21st Century*,
DOI 10.1007/978-981-10-2951-6_10

to become competent, each instrument creation tool demands the cultivation of certain technical skills and ways of thinking in its users.

In this chapter, I will discuss how machine learning algorithms can shape the design of new instruments. Machine learning algorithms can facilitate new types of design *outcomes*: they enable people to create new types of digital musical instruments. But, I will argue, they are also valuable in facilitating new types of design *processes*, allowing the instrument creation process to become a more exploratory, playful, embodied, expressive partnership between human and machine. And these qualities of the design process in turn influence the final form of the instrument that is created—as well as the instrument creator herself.

My aims in this chapter are: (1) to provide readers new to these ideas an introductory understanding of how supervised learning algorithms can be used to build new digital musical instruments; (2) to demonstrate that supervised learning algorithms are valuable as design tools, bolstering embodied, real-time, creative practices; and (3) to argue that, because the nature of any new musical instrument is intimately tied to the process through which it was designed, a closer attention to the relationships between instrument builders and instrument creation tools can deepen our understanding of new instruments as well as point to opportunities to improve design of both new instruments and creative experiences.

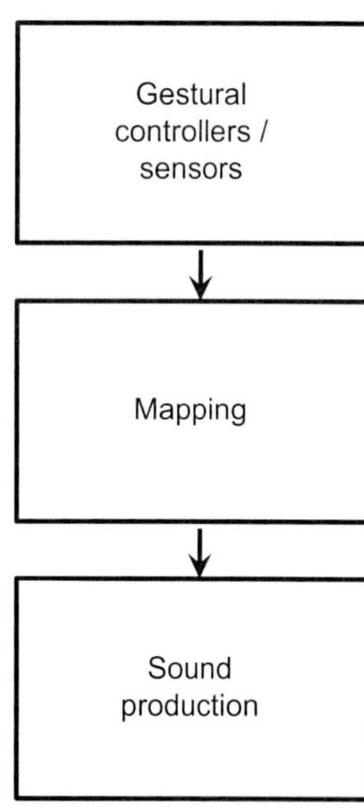

Fig. 1 Components of a digital musical instrument

2 New Instruments

2.1 Mappings and Mapping Creation Tools

Wanderley and Depalle (2004) use the following basic modular structure (illustrated in Fig. 1) to frame discussion of the design of digital musical instruments: First, a gestural controller (or other sensing component) senses the actions of the performer(s); this may include custom sensing hardware, a microphone, a camera, biosensors, and so on. These sensors pass a real-time stream of data to a "mapping" component, which is typically a software program. This component determines how to control the parameters of a sound production component, based on the values of the sensors. The sound-making component might be controlled with low-level musical parameters (e.g., amplitudes and frequencies of sinusoidal components, filter coefficients, or physical modelling parameters) or higher-level ones (e.g., determining the tempo or style of an autonomous agent).

In acoustic instruments, the relationship between a performer's actions and the sound of an instrument is dictated by physics, but there are few constraints on how digital musical instrument mappings might link these together. The design of the mapping determines, in the words of Hunt et al. (2002), "the very essence of an instrument": it defines the ways a performer may move or act, the dimensions of musical engagement that are possible, the means for an audience to perceive the relationship between a performer's intention and the music, and so on. Designing a mapping can thus be understood as designing a space of musical possibilities, and a number of instrument builders see this process as one of musical composition, where the outcome is a system that "carr[ies] as much the notion of an instrument as that of a score" (Schnell and Battier 2002).

Currently, computer programming is the de facto tool for creating an instrument mapping. Programming allows the creation of any imaginable mapping, in theory—just as a Theremin allows one to play nearly any imaginable melody, in theory. However, the practice of programming strongly encourages the creation of certain types of instrument mappings and discourages others. It is easiest to program mappings in which each sensor input controls a single sound synthesis parameter, and in which each synthesis parameter value is likewise impacted only by this single sensor; Hunt and Wanderley (2002) term such configurations "one-to-one mappings." Furthermore, it is easiest to program mapping functions that are simple (e.g., linear) and deterministic. The easiest instrument to build is therefore often analogous to a mixing desk: a set of independent sliders, each with an easy-to-reason-about control mechanism wherein the usable range of the sensor is mapped onto the useful range of a single sound control parameter.

This type of mapping naturally supports particular types of interactions between performer and instrument at the expense of others. Problematically, Hunt and Kirk (2000) found evidence that such simple, "one-to-one" mappings may present barriers to effective musical use when compared with more complex mappings. They found that mappings in which multiple dimensions of input affected multiple sound parameters simultaneously—"many-to-many" mappings—were more engaging to the user, offered more effective control over complex tasks, facilitated more effective learning of the interface over time, allowed people to think about sound gesturally, and were sometimes even considered to be more fun.

Researchers have developed various approaches to facilitate mapping creation through means other than programming, and a number of these approaches make it easier to create complex and many-to-many mapping functions. This work includes a variety of mathematical approaches to function generation, including matrix operations (Bevilacqua et al. 2005), interpolation (e.g., Garnett and Goudeseune 1999; Bencina 2005), and machine learning, which I discuss in the next section.

2.2 *Human-Computer Interactions with Digital Instruments: Control Versus Partnership*

The idea that "mappings" are a useful concept for framing the design or analysis of digital musical instruments is not without its detractors. For instance, Chadabe (2002) is critical of the paradigm, as it assumes a one-directional, simplistic relationship between human and instrument where the aim is control by the human over the sound. To employ a fixed, deterministic mapping function can be seen as ignoring the true potential for digital instruments to facilitate truly new forms of music making. Instead of taking advantage of computers' capacity for complex, non-deterministic processes, employing a static mapping function underutilizes the computer as simply a means of mimicking acoustic instruments, "to make the performer powerful and keep the performer in complete control" (Chadabe 2002).

In this chapter, I argue that the act of composing the instrument, like Chadabe's vision for the act of performing with an instrument, presents opportunities for new forms of relationships between humans and computers. The machine learning approaches I will discuss next create deterministic mapping functions that might be lacking interest on their own, at least in Chadabe's assessment; however, they support a rich dialogue and journey of co-discovery between human and machine throughout the process of *creating* a mapping. This process may unfold for months or years before a performance, or it may happen live on stage, making the mapping-building process a performative instrument in its own right. In either case, the quality of relationship between human and machine in the instrument composition process has significant aesthetic and practical consequences, as I will discuss.

3 Machine Learning and the Wekinator

Supervised learning algorithms are capable of learning functions from examples. An instrument mapping can be understood as such a function, whose inputs are sensor readings and whose outputs are sound synthesis parameter values. An

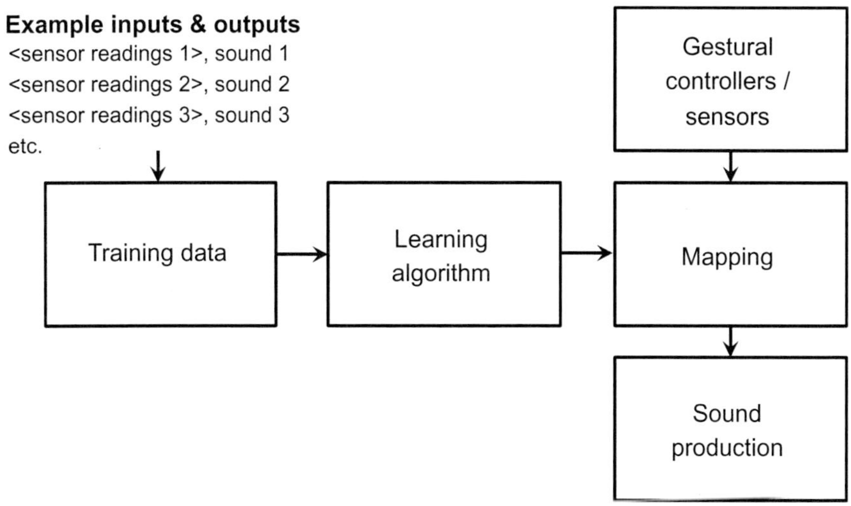

Fig. 2 A supervised learning algorithm can create a mapping from a set of training examples

algorithm can learn this mapping from a set of training examples, where each training example contains one set of sensor readings, paired with the set of sound synthesis parameter values that the designer would like to produce when those sensor readings are seen during performance (Fig. 2).

Different learning algorithms employ different strategies for learning a function from the training examples. However, the learning process can be roughly characterized as finding a mapping function which, upon seeing input values similar to those in a given training example, tends to produce output values similar to those in that training example.

Supervised learning has been used to create mappings for new musical instruments since the early 1990s. Neural networks—a type of supervised learning algorithm—were used by Lee et al. (1991) to control the timbre of synthesised sound using a MIDI keyboard, and by Fels and Hinton (1995) to control speech synthesis using a sensor glove.

In 2008, I began to build a general-purpose machine learning tool that could be used by composers[1] to create a variety of new digital instruments. By that time, seventeen years after Lee and Wessel's experiments, many composers had laptops

[1]In this chapter, I use the word "composer" to refer to people who build new musical instruments and create customized controller mappings, rather than referring to them as instrument builders or musicians. This word choice reflects an understanding of instrument building as an act of musical composition (cf. Schnell and Battier 2002, discussed above). It also accommodates the fact that there may not be a clear or consistent distinction between the notions of instrument, "preset" or mapping, and composition. For instance, at least two of the composers discussed here (Dan Trueman and Laetitia Sonami) have used the same controllers or sensors to play different musical pieces, but designed a different gesture-to-sound mapping for each piece.

which could easily train neural networks in a few seconds (or even faster). They had a wealth of sensors and game controllers, as well as fast audio and video feature extractors from which to obtain information about performers' actions. They had easy-to-use communication protocols such as Open Sound Control (Wright and Freed 1997) to patch these sensors to powerful, real-time sound synthesis software such as Max/MSP. However, composers did not have access to easy-to-use machine learning software tools. Outside of music, toolkits such as Weka (Hall et al. 2009) were beginning to make it easier for people without extensive machine learning expertise to experiment with off-the-shelf machine learning algorithms, using graphical user interfaces (GUIs) that did not require computer programming. However, general-purpose GUI toolkits such as Weka did not typically support real-time applications such as music performance.

I named my real-time machine learning toolkit Wekinator, in honor of Weka's achievements in making machine learning accessible to wider groups of users, and also because Wekinator used Weka's implementations of several learning algorithms. Fundamentally, Wekinator is a tool for building mappings like those in Fig. 1. In real-time performance, Wekinator receives input values from sensors or other input sources via Open Sound Control (OSC) messages, and it sends output control values to any sound synthesis program (or even animation program, game engine, etc.) via OSC. Wekinator provides a GUI for recording new training examples, training supervised learning algorithms (including neural networks and linear and polynomial regression for creating continuous mapping functions, as well as other methods), running trained models, and configuring various aspects of the machine learning process (e.g., specifying which sensor values will be used in computing each one of the synthesis parameters).

3.1 *Interactive Machine Learning as Design Tool*

In most conventional machine learning applications, the goal of using machine learning is to build an accurate model from the set of training examples. For example, the goal might be to build a model that predicts whether a medical treatment is likely to be effective for a new patient, using a training dataset with information about previous patients (the model function inputs) and the efficacy of the treatment on them (the model function's output). The set of training examples is often assumed to be fixed, and much of the human work of applying machine learning focuses on finding the algorithm that most accurately models the patterns in the given training set. Typically, the human practitioner relies on established quantitative metrics in order to compare alternative models and choose the best.

A composer using supervised learning to build a new instrument is faced with a very different type of application. She most likely does not begin the design process with an appropriate training set in hand—she must build a training set from scratch, creating examples that encode her understanding of how performer gestures or actions will be mapped to musical control parameters. While quantitative metrics may be helpful in assessing whether a model has accurately captured the patterns in the training data, these metrics cannot always reflect all of a composer's priorities for a trained mapping (Fiebrink et al. 2011). For instance, she might want the mapping to provide access to a range of sounds that fits the desired aesthetic of the piece, and to make these all accessible using a set of performer gestures that are comfortable to perform; or perhaps she wants to create a mapping that is easy (or difficult) for a performer to learn to play without making undesired sounds. The composer therefore cannot rely only on quantitative metrics of how well a model fits the training data to know whether a mapping is any good, or whether one alternative is better than another; she must use other means to evaluate a mapping, such as experimenting with it herself and listening to how it responds to her actions.

If a composer is dissatisfied with the mapping built by a supervised learning algorithm, changing the training examples is often an effective way to improve the mapping. For instance, if she wants a particular sound to be more easily playable using her mapping, she can provide additional training examples, pairing that sound with easy-to-demonstrate performer gestures, then retrain to build a new model. If she is unhappy with the outcome, she can delete those training examples and replace them with different ones. When supervised learning is used to build new musical instrument mappings, the training examples act as the conduit through which a composer communicates her intention to the computer. In more conventional machine learning applications, however, changing the training data is not a reasonable action to take to improve a model, because the training dataset is assumed to be a (more or less) accurate representation of some phenomenon in the world. This is the case with the medical treatment prediction example above, where the dataset recording treatment outcomes for previous patients is a valuable source of information about the problem domain.

For these reasons, Wekinator's user interface is designed to facilitate certain interactions between humans and supervised learning algorithms which are not part of more conventional machine learning processes: Users can create new training examples in real-time, by demonstrating performer actions along with the sound synthesis parameters they would like to be associated with those actions. Users can evaluate trained mappings by hands-on experimentation, observing how the mapping changes the sound as they change the input values. Users can iteratively add and remove training examples, and seamlessly move between these phases of editing data, re-training, and evaluating the effects of changes they make to the mappings (Fig. 3). This type of approach to machine learning in which a human user steers model behaviours through iterative and strategic changes to the training data is often called "interactive machine learning" (Fails and Olsen 2003).

Fig. 3 Interactive workflow with Wekinator

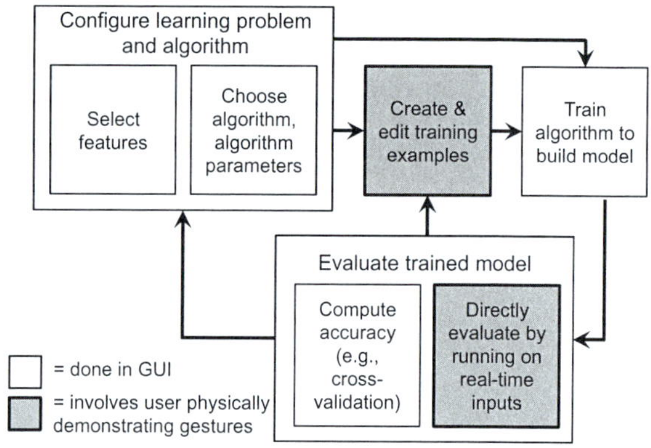

4 Machine Learning as Design Tool

In the eight years since developing Wekinator, I have observed it being used to create new instruments by dozens of professional composers, computer music and computer science students, "hackers" and "makers," and people with disabilities, and I have also used it in my own compositions and performances (Figs. 4, 5 and 6). Previous publications describe how I have used participatory design processes and surveys (Fiebrink et al. 2010), workshops (Katan et al. 2015), interviews (Fiebrink 2011), analysis of software logs (Fiebrink et al. 2011), and reflection on my own work (Fiebrink et al. 2009) to understand how people use Wekinator and why. This work all suggests that the most important benefits of Wekinator pertain to the way that it changes the design process, facilitating the creation of new kinds of instruments while also making design accessible to new people.

Fig. 4 Laetitia Sonami plays the Spring Spyre, an instrument she created with Wekinator (2015)

Fig. 5 The Sideband ensemble performs Anne Hege's composition *From the Waters*, in which Wekinator was used to create several GameTrak-controlled instruments

Fig. 6 *Nets0* was one of the first pieces written for Wekinator, and it requires performers to train new mappings for their own controllers live on stage

4.1 Speeding Up Implementation of Complex Mappings

One of the most immediately apparent benefits of using Wekinator to build mappings is the speed and ease with which composers can build a new instrument and modify it. Once the sensors or input devices are sending data to Wekinator via OSC, and a sound synthesis program is ready to receive control messages from Wekinator, the process of training a machine learning algorithm to create a mapping from input values to sound can take as little as a few seconds. This is true even for complicated, many-to-many mappings (the default type of mapping created by Wekinator) in which each sound control parameter is influenced by many input dimensions in possibly non-linear ways. Thus, using supervised learning encourages the creation of mapping types that have been shown to be more engaging, learnable, and controllable than those that are easiest to create using coding (see Hunt and Kirk 2000).

4.2 Supporting Prototyping and Exploration

Reducing the time it takes to create a viable instrument does not necessarily mean that composers using Wekinator spend less time building instruments. Instead, composers I have observed typically use their time to make many different variants of an instrument. They iterate many times, making slight or dramatic changes to the training data, as well as to the input devices and the sound synthesis software. Sometimes, these iterations are attempts to fix a problem with the mapping or otherwise improve the instrument according to a clear set of criteria. In these cases, changing a supervised learning model via changes to the training data can be a much faster way to fix a mapping or adapt it to a change in input or sound synthesis, compared to changing manually-written programming code.

However, these iterations are often the result of the composer intentionally exploring alternative designs in an effort to better understand what sort of instrument he really wants to make and how to make it. Prototyping and iterative refinement are recognized as activities that are critical to design in any domain (Resnick et al. 2005; Buxton 2010). Prototypes are physical manifestations of design ideas, and experimentation with a prototype helps a designer better understand the merits of the idea as well as potential ways to improve it. By reducing the time and effort needed to instantiate a prototype for a new idea, Wekinator encourages prototyping and allows composers to explore more ideas, and more refined ideas, over the process of building an instrument. In contrast, several composers I surveyed described how creating instruments by writing code often led to them using instruments they were unhappy with: changing a design using code incurred enough time and effort that they were discouraged from exploring new ideas, and they chose instead to accept instruments that limited them in problematic ways.

4.3 Supporting Surprise and Discovery

Creating an instrument can be understood as an example of what design theorist Horst Rittel (1972) described as a "wicked" design problem: the definition of the problem (What sort of instrument should I make? How will it be played, and what sort of sounds will it produce?) is not known in advance. It is only by designing the instrument that the problem becomes clear: the final instrument design embodies both the composer's final understanding of what the goals of the design process are, as well as the method of achieving them.

Composers using Wekinator to build instruments have often intentionally used machine learning in ways that will help them refine this "problem definition," to evolve their understanding about what kinds of instruments are possible to build, and what kind of instruments they ultimately want. A common strategy for a composer creating a new mapping with Wekinator is to "sketch out" the rough boundaries of the gestural and sonic space using the initial training dataset, then discover what sounds and gesture-sound relationships the supervised learning algorithm builds into the mapping trained from this dataset. A composer can construct this first training set by choosing a set of sounds she thinks she might want to play using the instrument, and a set of different input actions that span a comfortable range of control, then pair these together in a small number of training examples. A mapping created from these examples immediately allows the composer to discover new sounds that might exist in between and beyond the input values (e.g., gestures) she placed in the training set. When using this strategy, experimenting with the resulting mapping is really a process of discovering unexpected sounds and behaviors, rather than "testing" whether the mapping has learned the "right" behaviors from the given training examples. One Wekinator user described his rationale for this process thus: "There is simply no way I would be able to manually create the mappings that the Wekinator comes up with; being able to playfully explore a space that I've roughly mapped out, but that the Wekinator has provided the detail for, is inspiring."

Wekinator's support of interactive supervised learning allows composers to edit training examples to modify the mapping in response to the discoveries they then make. When a composer discovers a new sound she likes, she can reinforce this sound in her instrument by adding new training examples with this sound into the training set. When she discovers a sound she doesn't like, she can change the training examples in that region of the input gesture space to correspond to a more favorable sound.

Having access to surprise and discovery can fundamentally change the way a composer understands their relationship to the computer as well as the qualities of the instrument that they build. In particular, professional composers who have used Wekinator in their work have described how it allows them to move away from a paradigm of control over a computer into one where the computer is a collaborator. Laetitia Sonami, who has been using Wekinator for five years in the development of the Spring Spyre (Fig. 4), says in a lecture about her use of machine learning:

> …in a way, you don't want the instrument to perform like a well-trained animal circus, you kind of want it to be a little wild, and you want to adapt to it somehow, like riding a bull… I think the machine learning allowed more of this…fun of exploring, instead of going 'I have to have a result right away, this thing is going to do that,' and then leaving it at that. This… allows for a kind of flexibility that I think is essential for artists and musicians to… open up some kind of unknown and really create… things that excite you. I'm not sure about exciting the audience, but actually hopefully exciting the person who's making it, at least! And then you hope that it gets conveyed (Sonami 2016).

4.4 Supporting Embodied Design Practice

Another critical difference between designing instruments using machine learning and designing instruments by writing code is that composers are able to use their bodies directly in the design process. Instead of reasoning about what sort of movement-sound relationships he might want in an instrument, then deriving a mathematical function that he thinks will facilitate those relationships in a mapping, a composer can simply demonstrate examples of movements and movement-sound pairs that feel and sound right to him.

The ability to draw on embodied understanding of movement and sound in the process of designing an instrument is vitally important to many composers who work with Wekinator; the use of the body changes both the experience of composition and the type of instrument that can be created. Composer Michelle Nagai used the Wekinator to create an instrument, the MARtLET, from a piece of tree bark with embedded light sensors. She describes her experience:

> I have never before been able to work with a musical interface (i.e. the MARtLET) that allowed me to really 'feel' the music as I was playing it and developing it. The Wekinator allowed me to approach composing with electronics and the computer more in the way I might if I was writing a piece for cello, where I would actually sit down with a cello and try things out (Excerpt from interview, published in Fiebrink 2011).

Composer Dan Trueman, who used Wekinator to create game controller instruments for his piece *CMMV* writes:

> With [the Wekinator], it's possible to create physical sound spaces where the connections between body and sound are the driving force behind the instrument design, and they *feel* right. It's very difficult to do this with explicit mapping for any situation greater than 2–3 features/parameters [i.e., inputs and outputs], and most of the time we want more than 2–3 features/parameters, otherwise it feels too obvious and predictable. So, it's very difficult to create instruments that feel embodied with explicit mapping strategies, while the whole approach of [the Wekinator], especially with playalong, is precisely to create instruments that feel embodied. I like to think of digital instrument building as a kind of choreography. Choreographers are hands-on—they like to push, pull, hold their dancers, demonstrate how things should go, in order to get what they want, and the resistance and flow of their dancers in turn feeds back into their choreography. This is quite similar to the approach that [the Wekinator] engenders, and radically different than what explicit mapping strategies [i.e., mappings created with programming] enable (Excerpt from personal correspondence, published in Fiebrink 2011).

4.5 Supporting Accessibility

Wekinator allows people to build new instruments without programming. In addition to making the instrument-building process faster for programmers, this means that non-programmers have the ability to create new instruments for themselves and others to perform. As an educator, this has been helpful in teaching students about computer music performance and interaction design. Students can easily explore different designs, start to reason about design trade-offs, and experience the satisfaction of building and performing with a new instrument even if they are not confident programmers (Morris and Fiebrink 2013).

5 Discussion: Wekinator as Meta-Instrument

I describe Wekinator as a meta-instrument: an instrument for creating instruments (Fiebrink et al. 2009). Like anyone learning a new instrument, users of Wekinator must begin by mastering the fundamental techniques of training, testing, and modifying models, but they soon reach a point where their attention is no longer on the algorithms but on using them to achieve a creative vision. Building an instrument with Wekinator then becomes, fundamentally, a real-time process of self-expression, sculpting a unique space of musical possibilities that will afford creative engagement by oneself and/or others. In designing this space, just like in performing an instrument, a creator draws on a foundation of established musical practices while also seeking to imbue his work with an individual style, all the while being influenced by affordances of the tool which subtly encourage certain idiomatic ways of working and not others.

Understanding composition tools as instruments—whose affordances are vitally tied to the musical potential of the instruments created with them—invites us to bring aesthetic and philosophical considerations pertaining to the role of computers in musical performance to bear on the analysis and creation of composition tools as well. Composers have written of the value of creating "potential for change in the behaviors of computer and performer in their response to each other" (Moon 1997), of interfaces in which "interaction transcends control" (David Rokeby as described by Rowe et al. 1993), becoming more "like conversing with a clever friend" (Chadabe 1997, p. 287) or "sailing a boat on a windy day and through stormy seas" (Drummond 2009).

My work with composers suggests that a meta-instrument that supports these interactive qualities, as Wekinator does, can make the process of composition more engaging and musically satisfying. A meta-instrument that encourages playful exploration and discovery can help a composer navigate the wicked design problem of instrument building, sculpting the instrument to better meet her goals while simultaneously evolving those goals in response to the instrument. When the process of exploration and engagement is physical, rather than abstracted into mathematical functions and programming code, composers are able to engage in tight, *enactive* (Wessel 2006) action-feedback loops which further inform their embodied understanding of the instrument and their own musical aims.

Supervised learning algorithms are not the only computational tools which might give rise to these interactive qualities during instrument building or other compositional activities, and Wekinator's user interfaces are far from the only way to link human creators to supervised learning processes. Alternative approaches might facilitate faster exploration of more diverse instrument designs, or take advantage of additional information that composers could communicate through the body (such as examples of comfortable movement sequences or evocative sounds) without requiring a composer to format these as supervised learning training examples. Particular interaction qualities might be intentionally designed into tools, for instance making the "seas" of interaction even stormier with algorithms that make it difficult for composers to build instruments similar to those they have built before, or that introduce indeterminacy into more aspects of the tool. Those of us who are composers of meta-instruments have many new ideas to explore, ourselves, as we design new spaces of musical interactions for the composers who use our tools.

References

Bencina, R. (2005). The metasurface: Applying natural neighbour interpolation to two-to-many mapping. In *Proceedings of the International Conference on New Interfaces for Musical Expression* (pp. 101–104).

Bevilacqua, F., Müller, R., & Schnell, N. (2005). MnM: A Max/MSP mapping toolbox. In *Proceedings of the International Conference on New Interfaces for Musical Expression* (pp. 85–88).

Buxton, B. (2010). *Sketching user experiences: Getting the design right and the right design.* Morgan Kaufmann.

Chadabe, J. (1997). *Electric sound: The past and promise of electronic music*. Upper Saddle River, New Jersey: Prentice Hall.

Chadabe, J. (2002). The limitations of mapping as a structural descriptive in electronic instruments. In *Proceedings of the International Conference on New Interfaces for Musical Expression* .

Drummond, J. (2009). Understanding interactive systems. *Organised Sound, 14*(2), 124–133.

Fails, J. A., & Olsen Jr, D. R. (2003). Interactive machine learning. In *Proceedings of the International Conference on Intelligent User Interfaces* (pp. 39–45).

Fels, S. S., & Hinton, G. E. (1995). Glove-TalkII: An adaptive gesture-to-formant interface. In *Proceedings of the SIGCHI Conference on Human Factors in Computing Systems* (pp. 456–463).

Fiebrink, R. A., & Cook, P. R. (2011). *Real-time human interaction with supervised learning algorithms for music composition and performance* (doctoral dissertation). Princeton University, Princeton, NJ.

Fiebrink, R., Cook, P. R., & Trueman, D. (2011). Human model evaluation in interactive supervised learning. In *Proceedings of the SIGCHI Conference on Human Factors in Computing Systems* (pp. 147–156).

Fiebrink, R., Trueman, D., Britt, C., Nagai, M., Kaczmarek, K., et al. (2010). Toward understanding human-computer interaction in composing the instrument. In *Proceedings of the International Computer Music Conference*.

Fiebrink, R., Trueman, D., & Cook, P. R. (2009). A meta-instrument for interactive, on-the-fly machine learning. In *Proceedings of the International Conference on New Interfaces for Musical Expression*.

Garnett, G., & Goudeseune, C. (1999). Performance factors in control of high-dimensional spaces. In *Proceedings of the International Computer Music Conference* (pp. 268–271).

Hall, M., Frank, E., Holmes, G., Pfahringer, B., Reutemann, P., & Witten, I. H. (2009). The Weka data mining software: An update. *ACM SIGKDD Explorations Newsletter, 11*(1), 10–18.

Hunt, A. & Kirk, R. (2000). Mapping strategies for musical performance. In M. M. Wanderley & M. Battier (Eds.). *Trends in Gestural Control of Music*. IRCAM—Centre Pompidou.

Hunt, A., & Wanderley, M. M. (2002). Mapping performer parameters to synthesis engines. *Organised Sound, 7*(2), 97–108.

Hunt, A., Wanderley, M. M., & Paradis, M. (2002). The importance of parameter mapping in electronic instrument design. In *Proceedings of the International Conference on New Interfaces for Musical Expression*.

Katan, S., Grierson, M., & Fiebrink, R. (2015). Using interactive machine learning to support interface development through workshops with disabled people. In *Proceedings of the SIGCHI Conference on Human Factors in Computing Systems* (pp. 251–254).

Lee, M., Freed, A., & Wessel, D. (1991). Real-time neural network processing of gestural and acoustic signals. In *Proceedings of the International Computer Music Conference* (pp. 277–277).

Moon, B. (1997). Score following and real-time signal processing strategies in open-form compositions. *Information Processing Society of Japan, SIG Notes, 97*(122), 12–19.

Morris, D., & Fiebrink, R. (2013). Using machine learning to support pedagogy in the arts. *Personal and Ubiquitous Computing, 17*(8), 1631–1635.

Resnick, M., Myers, B., Nakakoji, K., Shneiderman, B., Pausch, R., Selker, T., et al. (2005). Design principles for tools to support creative thinking. In *Report of Workshop on Creativity Support Tools*. Washington, DC, USA.

Rittel, H. W. (1972). On the planning crisis: Systems analysis of the 'first and second generations'. *Bedriftsøkonomen, 8*, 390–396.

Rowe, R., Garton, B., Desain, P., Honing, H., Dannenberg, R., Jacobs, D., et al. (1993). Editor's notes: Putting Max in perspective. *Computer Music Journal, 17*(2), 3–11.

Schnell, N., & Battier, M. (2002). Introducing composed instruments, technical and musicological implications. In *Proceedings of the International Conference on New Interfaces for Musical Expression*.

Sonami, L. (2016). Lecture on machine learning, within online class "Machine Learning for Musicians and Artists" by R. Fiebrink, produced by Kadenze, Inc.

Wanderley, M. M., & Depalle, P. (2004). Gestural control of sound synthesis. *Proceedings of the IEEE, 92*(4), 632–644.

Wessel, D. (2006). An enactive approach to computer music performance. In Y. Orlarey (Ed.), *Le Feedback dans la Creation Musical* (pp. 93–98). Lyon, France: Studio Gramme.

Wright, M., & Freed, A. (1997). Open Sound Control: A new protocol for communicating with sound synthesizers. In *Proceedings of the International Computer Music Conference*.

Interfacing Sound: Visual Representation of Sound in Musical Software Instruments

Thor Magnusson

Abstract This chapter explores the role of visual representation of sound in music software. Software design often remediates older technologies, such as common music notation, the analogue tape, outboard studio equipment, as well as applying metaphors from acoustic and electric instruments. In that context, the aim here will be study particular modes in which abstract shapes, symbols and innovative notations can be applied in systems for composition and live performance. Considering the practically infinite possibilities of representation of sound in digital systems—both in terms of visual display and mapping of gestural controllers to sound—the concepts of graphic design, notation and performance will be discussed in relation to four systems created by the author: ixi software, ixiQuarks, ixi lang, and the Threnoscope live coding environment. These will be presented as examples of limited systems that frame the musician's compositional thoughts providing a constrained palette of musical possibilities. What this software has in common is the integral use of visual elements in musical composition, equally as prescriptive and representative notation for musical processes. The chapter will present the development of musical software as a form of composition: it is an experimental activity that goes hand in hand with sound and music research, where the musician-programmer has to gain a formal understanding of diverse domains that before might have been tacit knowledge. The digital system's requirements for abstractions of the source domain, specifications of material, and completeness of definitions are all features that inevitably require a very strong understanding of the source domain.

T. Magnusson (✉)
Department of Music, School of Media, Film and Music,
University of Sussex, Brighton, UK
e-mail: t.magnusson@sussex.ac.uk

T. Bovermann et al. (eds.), *Musical Instruments in the 21st Century*,
DOI 10.1007/978-981-10-2951-6_11

1 Introduction

Musical instruments are amongst the earliest human technologies. Possibly preceding fire and weaponry, we could speculate how early humans used rocks or sticks to hit other materials in order to define territories, communicate, or synchronise movements. For a social animal like the human, music is clearly a sophisticated and multipurpose cohesion technique. Some of the oldest known musical instruments are flutes found in Germany (the *Hohle Fels* flute) and Slovenia (the *Divje Babe* flute), estimated to be between 35,000 and 42,000 years old. Preceding these flutes would be generations of forgotten instruments in the form of rocks and sticks that might not even "look like" musical instruments at all.

Today we talk about "music technology," a two-word coinage that conjures up the image of plastic- or metal-surfaced equipment offering interaction through rotating knobs, sliders, or buttons, which are mapped to functionality represented on a screen of some sort. However, a quick etymology of the word "technology" clearly demonstrates that we are not discussing plastic gadgets here, but rather an embodied knowledge, skill or craft. The root of the Greek word *technē* is "wood," but at the time of the early philosophers, it had begun to denote the craft of producing something out of something else. For Aristotle, *technē* (τέχνη) is an activity where the "origin is in the maker and not in the thing made" (Ackrill 1987, p. 419). In *Rhetorics*, Aristotle uses the word "technology" to signify the "craft of the word" (techne and logos) as used in grammar or rhetoric, which is an inverse meaning to the later use signifying the knowledge (logy) of craft (techne). The word is not used much until the 17th century, which is when it enters the English language (Mitcham 1994, p. 130). At no point did the word signify objects, but rather the skill of doing things, as evidenced in Marx's *Das Kapital*: "Technology discloses man's mode of dealing with Nature, the process of production by which he sustains his life, and thereby also lays bare the mode of formation of his social relations, and of the mental conceptions that flow from them" (Marx 2007, p. 402). Earlier in the same paragraph we read: "Darwin has interested us in the history of Nature's Technology," and it is clear that he means the ways nature goes about its business. Bernard Stiegler defines technology as "the discourse describing and explaining the evolution of specialised procedures and techniques, arts and trades" (Stiegler 1998, p. 94) and encourages us to use the word in the manner we apply the words "psychology" or "sociology".

We *do* music technology: we don't buy it, own it, or use it. Thinking, designing, discussing, performing and composing are all acts of music technological nature. Musical instruments are the tools of music technology and represent the musico-theoretical framework of the specific culture. However, let's not forget the Greek origins, where the technology was about shaping something out of something else: in contemporary music technological practice, we are applying hardware, code libraries, communication protocols, and standards that become the material substance of our design explorations. We are working with *designed materials*, not wood or skin, but entities that already are of an epistemic nature

(Magnusson 2009). The new materials are semiotic in that they are part of a complex organisation of protocols and standards, which are needed for the diverse code libraries and hardware to be applied in the complex ecosystem of wired and wireless inter-software and inter-hardware communication.

When we create digital instruments we operate like a Latourian ant: busily operating as a part of a larger whole, applying actor-networks consisting of other actor-networks, or, in short, inventions that have become *blackboxed* in other technological processes, to the degree that we lose the possibility of grasping any origins. Where would a technological object originate from anyway? For this reason the instrument often appears before we know its expressive scope or indeed rationale (how, why, where, etc.).[1] The history of the saxophone provides a good example of how undefined the role of a new instrument can be, slowly gaining diverse functions amongst different musical cultures. In this context it is interesting to behold Attali's statement that in "music, the instrument often predates the expression it authorizes, which explains why a new invention has the nature of noise; a 'realized theory' (Lyotard), it contributes, through the possibilities it offers, to the birth of a new music, a renewed syntax" (Attali 1985, p. 35).

2 Digital Music Technologies—Designing with Metaphors

Music is many things to many people. If we were to attempt at a general definition, one approach might divide music into two key categories: in the first, music is performed, where an instrumentalist, or a group of them, engage in an act of generating sound, either from a score, from memory, or by improvisation. The context of co-players, the audience, and the location plays an important role here, where the liveness yields a sense of risk, excitement and a general experience of the moment's uniqueness and unrepeatability. The second category is music as stored: in people's memory, as written notation, on disks, tapes, or digital formats. The music could even be stored as an undefined structure in the form of algorithmic code for computer language interpreters. Now, in the 21st century, things are a little more complicated. New developments in digital music technologies transcend the above categories, deriving their symbolic design equally from the world of acoustic instruments, performance, notation, and electronic technologies. These new technologies further complicate the relationships between instrument makers, composers, performers, and the audience. Who is what? And the work itself … is it an instrument? A compositional system? A piece?

[1]The 160 character text message is a good example: the SMS (Short Message Service), although invented as part of the GSM cooperation in 1984, was initially implemented in Nokia phones for their engineers to test mobile networks. The technology was quickly adopted by users who began enjoying this mode of communication. This became a protocol of sorts, and as of 2016, Twitter is still respecting this 140 char limit.

There is a real sense that the technologies of music making are undergoing a drastic change by the transduction into the digital domain. This can be explored by studying the divergent natures of acoustic vs. digital instruments. The sound of a traditional musical instrument is necessarily dependent on acoustics, or the physical properties of the materials it is built of. Instrument makers are masters of materiality, with sophisticated knowledge of material properties and how sonic waves traverse in and through diverse types of matter, such as wood, metal, strings, or skin membranes. The instrumental functions of an acoustic instrument are necessarily determined by millennia old practices of material design. This is clearly not the case with digital instruments, where any interface element can be mapped to any sound. The mappings are arbitrary, and can be intelligent, non-linear, non-determined, inverse, open, and more. The design of digital interfaces ranges from being directly skeumorphic[2] and functional to more abstract and representational. In either case, every interface element *signifies* a function resulting in a sound rather than directly *causing* a sound. With the mapping function inserted between a gesture and the sound, the interface becomes semiotic: with this arbitrary relation, the digital instrument begins to surpass the acoustic as an epistemic entity, and at times manifests as a vehicle of a music theory or even a score.

The idea of making music with computers has existed since they were invented, and we can boldly claim that computers are the ideal media for composing, performing, storing and disseminating musical work. A quick tracing of this symbiotic relationship takes us back to early computers, with Ada Lovelace speculating about the musical potential of Baggage's Analytical Engine in 1842 (Roads 1996, p. 822). In the early days of electronic computers, we find Lejaren Hiller and Leonard Isaacson applying Markov chains in 1957 for one of the first algorithmically composed pieces, the *Illiac suite*, and Max Matthews inventing notation languages for computer generated synthetic sound. However, if we look at the history of mass produced digital musical instruments and software, we see that the computers have been used primarily as bespoke microchips integrated in instruments, for example in a synthesizer or an electronic drum kit, where the hardware design has been primarily mimetic, aiming at imitating acoustic instruments.[3] In the case of music software, we are faced with multiple imitations of scores, piano rolls, magnetic tape, where the key focus has been on developing tools for the composition and production of linear music at the cost of live performance. From both business and engineering perspectives it is evident that hardware manufacturers benefited from a model where new synthesis algorithms were embedded in redesigned computer chips, and sold as new hardware.[4] Software developers in turn addressed another

[2]Skeumorphic design is where necessary features in an original objects are used as ornamentation in the derivative object. Examples in graphical user interface design could be screws in screen-based instruments, leather in calendar software, the use of shadows, and so on.

[3]The contrasting design ideologies between Moog and Buchla are a good example of the problems at play here. It is evident that Moog's relative commercial success over Buchla's was largely due to the referencing well known historical instruments (see Pinch and Trocco 2002).

[4]There are exceptions of that model of course, such as the discontinued Nord Modular Synth.

market, applying the "studio in your bedroom" sales mantra, which sparked the imagination of a generation in the late 80s, who used Cubase on Atari computers, starting a genealogical lineage that can be traced to the current Logic or Live digital audio workstations.

Specialists in innovation studies, marketing, science and technology studies, and musicology, could explain in much more detail how technologies gain reception in culture, the social and economical conditions that shape their evolution, and the musical trends that support the development of particular technologies. From the perspective of an inventor, it is less obvious why the history of musical technologies has developed this way, although inventions ultimately have to depend on market forces in order to enter public consciousness. Here, the history of failures is as, if not more, interesting as the history of successes. ("failure" is here defined in the terms of the market, economy and sales). One such "failed" project could be Andy Hunt's MidiGrid, a wonderful live improvising software for MIDI instruments written in the late 80s (Hunt 2003). An innovative system, ahead of its time, the focus was on performance, liveness and real-time manipulation of musical data. Written for the Atari, Hunt received some interest from Steinberg (a major software house), which, at the time, was working on the Cubase sequencing software. Only an alternative history of parallel worlds could speculate how music technologies had evolved if one of the main music software producers would be shipping two key software products: one for performance and the other for composition.[5] At the time of writing certain digital interfaces are being produced that are not necessarily imitating the acoustic, although inspired by them. It is yet to be seen whether instruments such as the *Eigenharp* and the *Karlax*[6] will gain the longevity required to establish a musical culture around the technology of composing and performing with them.

Since the early 2000s, developments in open source software and hardware have altered this picture. The user has become developer, and through software such as Pure Data, SuperCollider, CSound, Max, ChucK, JavaScript, and hardware such as Arduino and Raspberry Pi, a world has opened up for the creation of new music technologies. The ease of access and low cost of these technologies, together with strong online communities that are helpful and encouraging, make such DIY approaches fun, creative and rewarding. When music software has become sophisticated to the degree that it can almost compose the music without the input

[5]Hunt's software is of course no failure. It is a highly successful research project that has served its author and many others as musical tool, for example in education, and it has inspired various other research projects, mine included. But the context of this discussion is innovation and how a specific music technology instance might fare in the world of mass markets and sales.

[6]The manufacturers of both interfaces call them "instruments". Some might argue that they only become instruments when coupled with a sound engine, as familiar instrumental models indicate (e.g., Wanderley 2000 or Leman 2008), but I do believe it makes sense, in terms of innovation, longevity and spread of use, to call these instruments. Will there be a day when something like the Karlax will be taught in music conservatories? How would that even work? What would the training consist in?

of the user (who becomes a "curator" of samples or a "wiggler" of knobs and buttons), many find that real creative approaches happen when music technology itself is questioned and redefined. Gordon Mumma's ideas of "composing instruments" (see also Schnell and Battier 2002) are relevant here.

This chapter describes such questioning of music technology. Here the investigation regards interface and interaction design, i.e., how the visual elements in music software can affect musical ideas, composition and performance. Considering the practically infinite possibilities of representation of sound in digital systems—both in terms of visual display and mapping of gestural controllers to sound—the process of designing constraints will be discussed in relation to four systems developed by the author that engage with visual representation of sound in music software.

3 Interfacing Sound with Screen Interfaces

Interfacing sound in screen-based music software is no simple task: traditionally the software tends to either follow linear scoring metaphors (piano rolls, traditional notation, tape tracks) that are useful for composition, or imitate hardware (sliders, knobs, buttons, cables, screens), allowing for a real-time manipulation of sound which is eventually "bounced down" to a fixed file. There have been myriads of other, more experimental approaches, that investigate how we can perform with screen-based musical interfaces. However, designing two- or three-dimensional representation of sound, where the physical interfaces might consist of a mouse, keyboard or touch screens, comes with some complications. Some of the design patterns that we find in the material world cannot easily be abstracted and represented in the digital domain. Such translations often become a process of *transduction*, where sounds or actions are transformed in the digital. Even when we attempt mimesis and aim to be true to the original object, we lose some of the unique (non-universal) characters of the individual instrument, the entropic qualities that often manifest in its behaviour, as well as the history and use of the particular object itself. A copy of software does not have a history in the same way as an individual object.

On the other hand, elements not found in acoustic instruments present themselves as natural properties within the digital, for example the possibility of timelines, looping techniques, learning mechanisms, or diverse mechanics of mapping gesture to sound. Here, screen-based instrument designers apply techniques from computer games, interface design, HCI, apps, installations, and computer networks. The metaphors abound, but they can be found in diverse areas of development, where techniques and user skills are *reused* in the new design. Commensurate with how the maker of hardware musical instruments seeks to enact the skills developed and incorporated into the motor memory of instrumentalists over the years, the designer of screen-based interfaces will apply techniques from diverse fields, such as drag and drop, shift-click for multiple selection, swipe for new screens, right click for menus, etc. Blackboxed design patterns are applied in the design of new instruments, consciously or not, and the user intuitively performs the system, learning its conceptual nature through interaction design that is already familiar.

3.1 ixi Software

The dozen or so applications that Enrike Hurtado and myself developed around 2000, and uploaded onto our *ixi audio* website (ixi 2000), were all experiments in sonic interaction design (Magnusson 2006). We wanted to create non-representational graphics that would control sound, through both real-time interaction (using the mouse and keyboard) and automation. The user would create visual objects that had associated sound which could be manipulated by moving them around, changing the shape, connecting them with other objects, and so on. We explored diverse design patterns, each represented by a small "app" (as we called the software—this was before the days of mobile media). Examples of the mapping of visuals to sound include: size for amplitude, vertical location for pitch, horizontal location for panning, shape or colour for timbre, blinking or rotating for automatic triggering of sounds, movement as a type of panning (perhaps a moving microphone that randomly navigates a space of sound), and so on (Fig. 1).

This software is now "abandonware," as we have no time to translate it to new operating systems, indeed it is a good example of how transient digital systems for musical production can be. However, what *is* of lasting value are the ideas developed, the use of metaphors, the interaction design, the idea of automation, computational creativity, and real-time playfulness. These ideas become a design language, a set of interface and interaction patterns that are learned, embodied, and easily implemented in new software. Clearly not unique to ixi software, they are design discoveries, often of personal—as opposed to historical—nature,[7] that have been reapplied in later software by us, and, indeed, inspired other software.

3.2 ixiQuarks

The ixiQuarks (Magnusson 2007) continued the research of ixi software, but here with a more coherent research agenda. They were developed in SuperCollider between 2004 and 2006 as an investigation into alternative screen-based interfaces, where non-linearity, performativity and real-time control of sound were the key design considerations. Discarding common concepts like timelines or linear

[7]See Boden (1990) on creativity - although her P-creativity and H-creativity stand for psychological and historical creativity (where the former is always included in the latter), in this case we use the term personal creativity.

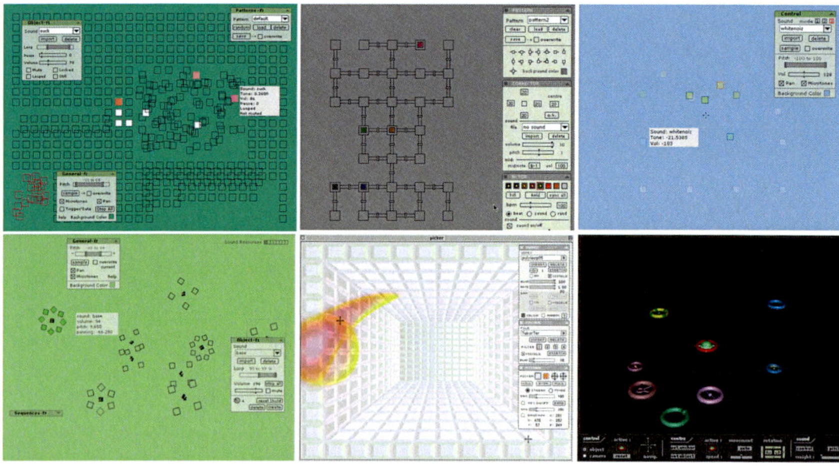

Fig. 1 A screenshot of six individual ixi software applications. Each of them served as an investigation into a different mode of interactive design

notation, most of the instruments developed as part of the ixiQuarks package were aimed at direct control of sound, where the mouse, they keyboard, pressure tablets and other cheap and common interface devices are used for control.

At the time of development, there were no multi-touch screens or trackpads, which resulted in more limited design decisions. However, it is not clear that multi-touch would be of a drastic benefit here since the lack of tangible interface elements makes the instrument less embodied and the user focus becomes more conscious on particular visual elements (think selecting a bespoke element on the screen with the mouse arrow). Furthermore, touching a visual element with your finger on a screen hides it (under the finger), prevents overlapping elements (as fingers can't be in the same space at the same time), and the anatomy of the hand also provides some expressive limitations. These would be interesting constraints to design around, but they simply didn't exist at the time (Fig. 2).

The ixiQuarks interfaces are non-representational, or, at least, they do not primarily derive their metaphors from physical instruments or music technological hardware. The interface and interaction metaphors were rather influenced by traditional HCI, computer games and web design. Creative audiovisual coding was a much more inspirational context than acoustic or digital music technologies. A central question to be explored was how visual interfaces and alternative interaction design would result in different music. We were equally interested in how the design itself inspired musicians, and also how the limitations of expressive actions would provide affordances and delineate constraints that would be navigated by users through a process of exploration (Magnusson 2010). For this reason, there were no manuals written, no demo videos created, no sound banks provided.

An element of ixiQuarks was that the user would be able to redefine the sound, create new sounds and change the function of the interfaces during performance.

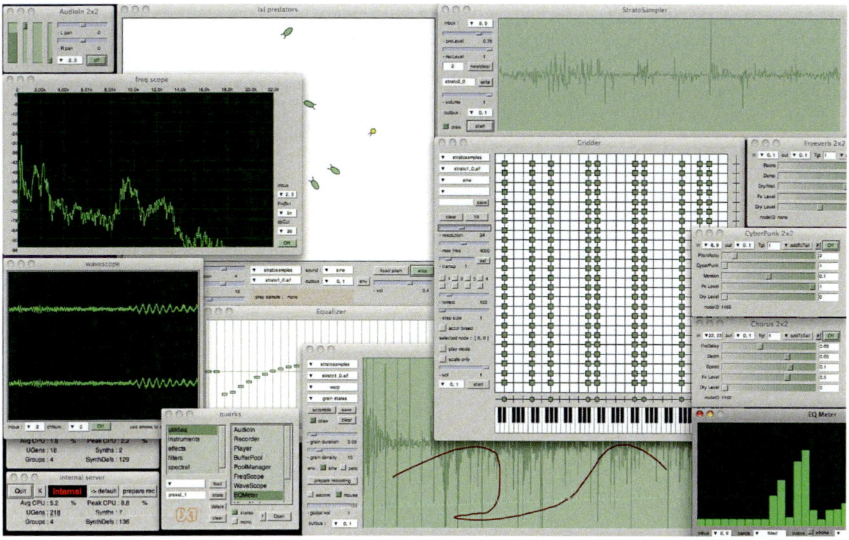

Fig. 2 A screenshot of ixiQuarks. Each of the instruments and widgets are independent from the other but work well together. The sounds from one can be used as input into another. Some of the instruments can be live coded and changed in real-time

This was an early version of live coding, where the interface could be altered in real-time. As a performer in improv ensembles I became more interested in the live coded aspects of musical performance, sometimes only using the graphical interfaces to trigger patterns, whilst changing the SuperCollider synth definitions. For this reason I decided, in 2009, to attempt at creating a live coding system that would continue some of the explorations of the early ixi work, but here through the use of language or code as opposed to graphical design.

3.3 *ixi Lang*

SuperCollider is an ideal platform for live coding. It is a real-time system where synths can be created and stopped without affecting other running synths, their parameters changed, and musical patterns can be written to control the synths. This is ideal for live exploration of sound synthesis and electronic music. Indeed, rumour has it that the term "live coding" was first used on the SuperCollider mailing list by Fabrice Mogini when describing his compositional process, sometime in the late 20th century. However, when performing live, speed and simplicity of syntax becomes important, as the performer ideally wants to be focusing on the music and not the code. With ixi lang some of the main design goals were: to create a simple, fast, and forgiving live coding system with a syntax that makes mistakes unlikely (no commas, brackets, or semicolons); a language that was easy to learn, and

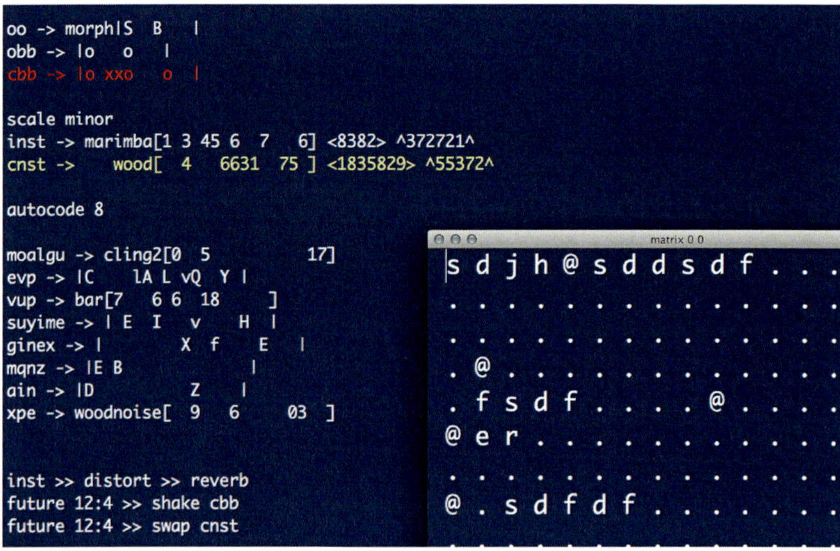

Fig. 3 A screenshot of ixi lang, with the Matrix grid-based event system

understandable by the audience, and where the use of visual elements was part of the language syntax (Fig. 3).

In many ways, ixi lang continues ideas from the graphical explorations but here using what I call a CUI (Code User Interface). Even though ixi lang is a textual interface, it perhaps more shamelessly applies design patterns that ixiQuarks tried to ignore: there is a temporal score where characters represent sounds and spaces silence. There is an underlying temporal grid, which syncs the musical events to a default tempo clock. There is a clear timeline-based design in ixi lang, but the flexibility of the language and specific features make both polyrhythmic and polymetric explorations easy. In terms of graphic design, we find diverse instances of visual elements in the language, for example where effects are applied with symbols such as "≫" and "≪", a visual reference to how guitar pedals are connected with jack cables (Fig. 4).

ixi lang has been described in detail elsewhere (Magnusson 2011), but in retrospect one could characterise it as a system of language elements that try to move away from typical programming language commands embracing playfulness and simplicity. For this reason the system has strong limitations and it does not extend very well. However, it is used for teaching children from the age of seven live coding and music, and anyone can learn it in about 30 min. The language has various hidden features that can be discovered by eager users, it is quirky, and it contains an autocoder, where the language writes its own code, often resulting in fine music.

```
ubu -> guitar[1   5 2 3   4  ]

ubu >> distort >> reverb

ubu << distort

shake ubu

ubu + 7
```

Fig. 4 This is a typical ixi lang score. First we create an agent called "ubu" and give him a guitar instrument. The guitar is the name of a SuperCollider synth definition, and the user can use any of their definitions. Ubu's score ([1 5 2 3 4]) are the notes in the selected scale, and the spaces between the notes are silences. This is effectively Rousseau's system of notation from the 18th century. Ubu is then given two effects (distortion and reverb), but in the next line the performer removes the distortion from the effect chain. Then ubu is shaken, which scrambles the numbers in the score, leaving it the same length, but with the notes in different places. Finally ubu's score is transposed up by a fifth. One of the system's innovations is to update the code in the text document when a function has been applied to it (such as "shake" or +7)

3.4 The Threnoscope

More recently I have been developing an instrument I call *Threnoscope* (Magnusson 2014). As the other software, it is a live coding system, split into three views: a notational view, a code view, and a console for system output. The notational view and the code view serve as a dual visual representation of the music: the former is a visual description of the sound, whilst the code is a form of prescriptive notation. Both aid the audience in understand the sonic events but they also serve as interfaces for further control by the performer.

The visual system contains circles that represent the harmonics of a fundamental frequency, often tuned to a 55 or 54 Hz A note. In the former case, the second harmonic is then 110 Hz, the third 165 Hz, fourth 220 Hz, and so on. Notes (or drones) can be created on these harmonic circles or anywhere in between through a structure of tonic (the harmonic, ratio, degree, or frequency). The interface has crossing lines that represent the loudspeakers. Any combination from two to eight speakers can be used. The speakers serve as static timelines where notes travel across the circular space. This creates an unusual looping structure, heavily influencing the music created with the system (Fig. 5).

The Threnoscope was initially planned as a musical piece, to be performed by the author and anyone who wanted to play the piece.[8] After two decades of interest

[8]However, further development and user experience shows that the system is more of a compositional tool, an instrument, and not a musical piece. Admittedly, the boundaries are not very clear here and the author has had interesting discussions with users who are of different opinions of what might constitute a musical piece.

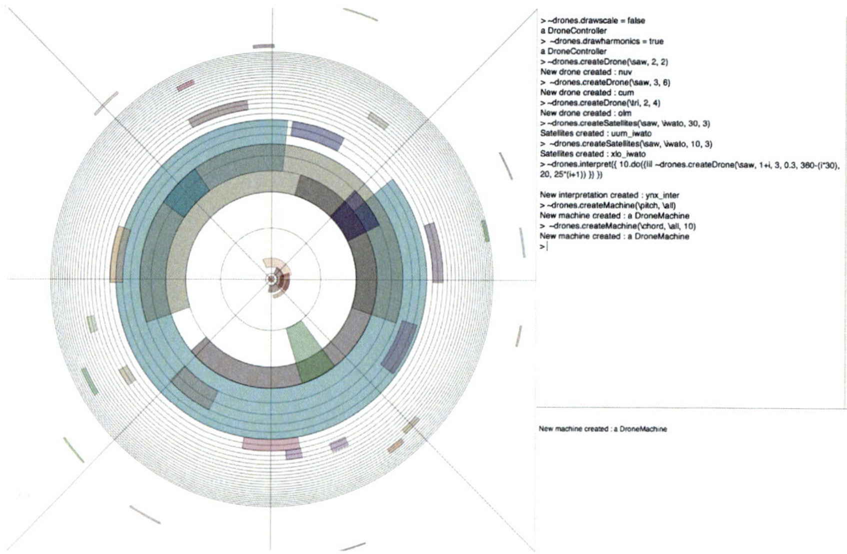

Fig. 5 A screenshot of the Threnoscope microtonal live coding system. The circles represent the harmonics of a fundamental tone. The crosshair lines represent the speakers, here an eight-channel surround system. Notes are the *coloured* wedges that can move around the space, or stay static. In the middle we see a machine that can affect the drones in different ways, supplementing the performer's actions

in microtonal music and alternative tunings and scales, I felt the need to study these areas more formally and the best way to do that is to develop a digital instrument implementing these concepts. My aim was also to move away from notes as events that happen in time, but rather conceive of them as spatial phenomena that move around without a set duration. Although the system can be used to play staccato notes that disappear immediately, the system is designed as an encouragement for long duration musical events, where Morton Feldman might be considered a restless character, but La Monte Young and Phill Niblock on a similar wavelength.

The Threnoscope is created for live improvisation, as a live coding instrument. The textual interface is considered the most expressive and free interface for this compositional system. A key problem with graphical user interfaces is that their elements take up screen estate, where the biggest elements call for the attention of the user. Designing a graphical user interface for musical software therefore resembles the writing of a musical score. With text, on the other hand, it is only the imagination and vocabulary of the performer that sets the limitations of what could be possible within the language framework. As an example, at the spur of the moment, some performer might want to create 100 sine waves randomly on the first six harmonics at different degrees in a chromatic scale. This could easily be written as following:

```
100.do({ ~drones.createDrone(\sine, rrand(2, 7), degree: rrand(1, 9))}).
```

Obviously, it would not make sense to create an interface for such musical acrobatics. Here the coding language is a much more appropriate and simple interface than buttons, drop down menus or sliders would ever be.

There is a score format for the Threnoscope, which enables composers to write linear or non-linear pieces. This format is a timed array with code instructions in it. The code score can be visually represented and manipulated during its playback. This can be useful when composing, but often the scores are short scores that are played during an improvisation, almost like a "lick" or an incorporated musical phrase in jazz or other improvised music.

4 Conclusion

The development of musical software has been described above as an experimental activity that goes hand in hand with sound and music research. A redefined boundary between the instrument maker and the musician is forged, a practice which requires a strong knowledge of both music and materiality. Musical composition at this level requires music technological research. We find that, in order to develop the music or instrument that one works towards, one has to understand key concepts of the source domain—such as human gestural patterns, or the resonant properties of physical materials. Just as a composer for acoustic instruments needs to understand the acoustics of the instruments, music theory, harmony, and rhythm, the composer of digital systems will need to comprehend the physics of sound, digital signal processing, software engineering, human-computer interaction, as well as music theory. For composers and software developers the question is how to represent these new features of sonic control. A new notational language is needed, and it is in that context that the work above is presented.

For some musicians, it might feel convenient to buy off-the-shelf products that perform many of the things we might want to do, but at some stage the software will limit compositional ideas and performance options. This inspires musicians to conduct their own compositional work through research and development of their own music technologies, designing affordances and setting constraints relevant to the particular musical work (Magnusson 2010). The best practitioners in this field are the ones who can manage their time on instrument building, software development and other engineering tasks, whilst still keeping a focus on their compositional intention. People who work this way report that time spent on learning, researching and experimenting will result in novel musical output that is unique, personal and of a strong musical and technical identity.

References

Ackrill, J. L. (1987). Nicomachean ethics, book 6, chapter 4. In A. New (Ed.), *Aristotle reader*. Oxford: Clarendon Press.
Attali, J. (1985). *Noise: The political economy of music*. Minneapolis: University of Minnesota Press.
Boden, M. A. (1990). *The creative mind: Myths and mechanisms*. London: Weidenfeld and Nicolson.
Hunt, A. (2003). MidiGrid: Past, present and future. In *Proceedings of new interfaces for musical expression 2003*. Montreal, Canada.
ixi. (2000). http://www.ixi-audio.net. Accessed September 8, 2016.
Leman, M. (2008). *Embodied music cognition and mediation technology*. Cambridge, Massachusetts: MIT Press.
Magnusson, T. (2006). Screen-based musical instruments as semiotic machines. In *Proceedings of new interfaces for musical expression*. Paris: IRCAM.
Magnusson, T. (2007). The ixiQuarks: merging code and gui in one creative space. In *Immersed music: The ICMC 2007 conference proceedings*. Copenhagen: Re:New.
Magnusson, T. (2009). Of epistemic tools: Musical instruments as cognitive extensions. *Organised Sound, 14*(2), 168–176.
Magnusson, T. (2010). Designing constraints: Composing and performing with digital musical systems. *Computer Music Journal, 34*(4), 62–73.
Magnusson, T. (2011). ixi lang: A SuperCollider parasite for live coding. In *International computer music conference*. Huddersfield: University of Huddersfield.
Magnusson, T. (2014). Improvising with the threnoscope: Integrating code, hardware, GUI, network, and graphic scores. In *Proceedings of new interfaces for musical expression 2014*. London: Goldsmiths College.
Marx, K. (2007). *Capital: A critique of political economy—The process of capitalist production*. New York: Cosimo.
Mitcham, C. (1994). *Thinking through technology: The path between engineering and philosophy*. Chicago: University of Chicago Press.
Pinch, T. J., & Trocco, F. (2002). *Analog days: The invention and impact of the moog synthesizer*. Cambridge, MA: Harvard University Press.
Roads, C. (1996). *The computer music tutorial*. Cambridge, MA: The MIT Press.
Schnell, N., & Battier, M. (2002). Introducing composed instruments, technical and musicological implications. In *Proceedings of new interfaces for musical expression 2002*. Limerick: University of Limerick.
Stiegler, B. (1998). *Technics and time, 1: The fault of epimetheus*. Stanford: Meridian. Stanford University Press.
Wanderley, M. M. (2000). Gestural control of music. Paris: Ircam. http://recherche.ircam.fr/equipes/analyse-synthese/wanderle/Gestes/Externe/kassel.pdf. Accessed February 2016.

Digital Media and Electronic Music in the Classroom—The Loop Ensemble

Marten Seedorf and Christof Martin Schultz

Abstract The production, performance, storing, processing, dissemination and reproduction of music are and always have been technologically determined. The continuing digitalization benefits and accelerates this development. At present, children and adolescents grow up in this cultural environment, their perception and handling of music is pervaded by digital technology. But despite their cultural relevance and several impulses from the academical discourse, these aspects of music culture are still marginalized in the educational practice in German classrooms. Younger research focuses the adequacy of music software for educational purposes, formulating the need of educationally suited software. Research shows, that main obstacles with integrating digital music media into education are the high cost and the deterrent complexity of music software. In the context of the interdisciplinary research project 3DMIN, we developed the loop ensemble. It consists of three virtual instruments created in the open source software Pure Data and is designed for the practical dissemination of electronic music culture and its technical basics in a pedagogical context. As an Open Educational Resource it is designed as didactic material for an action-oriented music education in combination with autonomous learning. We evaluated the instruments' usability in three ways. The results show a distinct practical suitability of the ensemble, yet further empirical research is needed for a profound evaluation.

M. Seedorf · C.M. Schultz (✉)
Audio Communication Group, Technische Universität Berlin, Sekr. EN 8, Einsteinufer 17c, 10587 Berlin, Germany
e-mail: christofmschultz@gmail.com

M. Seedorf
e-mail: marten.seedorf@campus.tu-berlin.de

T. Bovermann et al. (eds.), *Musical Instruments in the 21st Century*,
DOI 10.1007/978-981-10-2951-6_12

1 Introduction

This article deals with electronic music in classrooms in Germany. The *loop* ensemble, which consists of three virtual instruments built in Pure Data, is presented as an approach to integrate electronic music into education. Even though, as didactic material, the ensemble is focused on specific aspects of digital music culture, in particular the technical function and aesthetic of electronic sounds, it will be discussed in the wider context of digital music media in the classroom. In German music education, we do not see a specific consideration of electronic music and its instruments. In the following, we refer to a wide concept of electronic music, covering a variety of styles such as electronic dance music or the electronic sounds of the post-World War II avant-garde, including the aesthetic influence electronic music and its technology has taken on other styles of music. We understand electronic music as every sound, that has been produced or edited electronically. The article attempts to integrate perspectives from musicology, music education, cultural science and artistic practice to finally present a practical approach towards an exemplary integration of digital media into German music lessons with a focus on electronic music.

Music and technology have always formed a strong symbiosis. The production, performance, storing, processing and the dissemination of music as well as listening to it is shaped by technology (cf. Enders 2013). Therefore, technological shifts constantly lead to innovations of sound, change its cultural context and most of all, affect listeners (cf. Smudits 2013). The proceeding digitalization further intensifies the symbiosis and thus defines the current music culture (cf. Tschmuck 2013), which children and adolescents are confronted with. Their comprehension and handling of music is strongly influenced by an omnipresent digital technology (cf. MPFS 2014, p. 58f) and, more importantly, by using it they actively contribute in many ways to the music culture and thus shape it (cf. Gall 2012, p. 11f). But in doing so they are endangered to be overwhelmed and frustrated by the vast mass of medial information and the complexity of a medially globalized world. This confrontation illustrates the importance of media literacy in a digitalized music culture and the need of educational institutions to consider these aspects (cf. Muench 2013).

The *loop* ensemble arose from the 3DMIN project. 3DMIN's basic concept formulated an urgent need for new digital instruments that provide an access to the world of electronic- and computer music in the context of music education, since in comparison to classical music and its instruments modern electronic music still is neglected here and thus this innovation is urgently needed (Bovermann et al. 2014). This objective constituted the starting point of our work. We understand electronic instruments as the contemporary continuation of the historical development of musical instruments and therefore they do not conflict, but relate. Learning about technical procedures of digital sound-synthesis means acquiring instrumental, musical skills. Music education should consider the potential of teaching these skills at a young age analogous to the training of a classical instrument. Not only is this an opportunity for music education to catch up on recent cultural development, but it could assist young people in their confrontation with the contemporary music culture. It could contribute to their media literacy and offer them skills to actively shape their digitalized music culture.

2 Digital Media in German Music Education—Tradition, Stagnation and Progress

German music education has a tendency to favor the preservation of cultural traditions over the integration of cultural changes, particularly if they concern medial or technological aspects of music (cf. Gies 2001, p. 6f). In the early phases of digitalization, the emerging fundamental cultural consequences, regarding for example the empowering character of the new media and the speed of its development, had a reinforcing effect on the discourse in music education (cf. Stubenvoll 2008, p. 109). As a result, proponents and critics of electronic or digital media in musical education lead an emotional and partly ideological debate in the 80s and 90s. Knolle summarized the arguments of the critics in 1995:

1. Thinking and acting become algorithmized by the computer […]
2. Musical material and its production becomes standardized […]
3. Sensual experience is diminished—escape into a synthetic, virtual world […]
4. The new technologies are attributed with the ability to generate an ideology […] (cf. Knolle 1995, p. 42f)

The following theses of Rheinländer supply an exemplary summary of the proponents' point of view:

1. The Computer is to be regarded as a musical instrument […]
2. The Computer is a medium of teaching and thus a didactical instrument of musical education. […]
3. The use of the computer in education is established by a precise assignment of teaching methods. […]
4. The Computer is a new medium of teaching and thus, it generates new teaching methods. […]
5. Besides writing, reading and math, the use of the computer has become the fourth basic cultural technique in the western world. […] (cf. Rheinländer 2002, p. 10f)

At the end of the millennium a pragmatic shift could be observed. The socio-cultural impacts of digital culture became increasingly obvious, for example through the globalizing effects of the internet. Subsequently, the topic gained more relevance and acceptance among experts and practical consequences became necessary. Research projects were sending impulses towards educational policy and the academical training of the teachers, with examples given by Auerswald (2000), Eichert (2004), Muench (2005) and Roth (2006). A great number of didactical concepts dealing with theoretical and practical aspects of digital musical technology were developed to integrate the topic into music education (cf. Strasbaugh 2006, p. 5f). There were, however, only few actual consequences in the music classes, and in the middle of the first decade of the new millennium the discourse lost its dynamic, leading into an acceptance of the status quo (cf. Ahlers 2012, p. 127). Several studies show, that digital media is rarely been used in the classroom in spite of all guidelines and sufficient equipment of the schools with the necessary devices (Maas 1995; BMBF 2005; Sammer et al. 2009). In these studies, teachers referred to several aspects as reasons for this avoidance, which will briefly be discussed in the following.

Generally, it seems that digital media is perceived as incompatible with certain established teaching methods. Teaching approaches based on direct instruction or lecturing look back on a long tradition (cf. Meyer 2010, p. 185)[1] and they still are the prevailing style in German music classrooms (cf. Riedl 2010, p. 173). Digitalized music culture, on the other hand, calls for active participation. In conjunction with the contemporary significance of self-socialization (cf. Ahlers 2011, p. 225f), it would benefit from modern teaching methods, where the teacher no longer is the center of attention in a sense of direct instruction or lecturing, but the learning subject moves into focus in the course of action-oriented and autonomous learning concepts (cf. Schläbitz 2002, p. 40ff). Throughout the discourse, the available software was perceived as too expensive and didactically unsuitable with respect to complexity and usability. Teachers complained about technical difficulties and an inappropriate amount of time spent on preparation. Although "there are enthusiastic, curious or adversary positions amongst music educators and researchers concerning technology's benefits or problems… [, some] of the problems that were already raised about 20 years ago still cannot be eliminated" (cf. Ahlers 2012, p. 131). In view of this situation, in the second half of the first decade of the new millennium German researchers shifted their focus on educational software, its production and evaluation (Stubenvoll 2008; Ahlers 2009) and (Weidler 2014). Still, there is a need for software designed for educational purposes to include digital media and its cultural context on a reflective and practical level (cf. Ahlers 2012, p. 130f). This is where our project, the *loop* ensemble, wants to propose an exemplary solution.

The ambivalence of the present situation can be illustrated by an analysis of current curricula. Since in Germany these are a matter of the federal states, there are large differences in the extent of integration of digital media in general and electronic music in particular. In the curricula of secondary level I and II in Berlin/Brandenburg, for example, the integration of digital media in combination with media literacy is only vague and superficial. Even though the central aims of digital media literacy are formulated, the didactic content does not include the topic, and electronic music culture seems marginalized (LISUM 2006a, b). A positive example are the curricula on the secondary level of Nordrhein-Westfalen, where several aspects of electronic music culture are represented, even as a topic of the

[1]These teaching styles are often referred to as "chalk-and-talk".

school leaving exam. Under the title *New ways of sound and expression: The effects of new technologies on musical design* Karlheinz Stockhausen, Kraftwerk, techno and house music, sampling culture, etc. are included (MFSW 2013).

Recently, the topic is gradually gaining more relevance in training programs of teachers, in the curriculum as well as in an obligatory part of the teaching material. The recent generation of music teachers has grown up with an increasingly digitalized culture themselves and thus are able to bring a more natural handling into the classrooms. Still, in comparison to its cultural relevance, especially in the living worlds of the students, the presence of digital media and electronic music culture in music classrooms seems weak. Its potential for a self-determined, contemporary education is rarely put into practice (cf. Schläbitz in preparation, p. 5ff).

3 The *Loop* Ensemble

With the development of *loop* we tried to create a music software for use in the classroom that tries to fulfill the various demands mentioned above. As a collection of computer-based instruments *loop* attempts to offer the possibility to integrate digital media with a focus on electronic music culture and its technology. *Loop* consists of three independent but connectable electronic instruments made in Pure Data: ADD, DRUMBO and JERRY (see Figs. 1 and 2). They use different

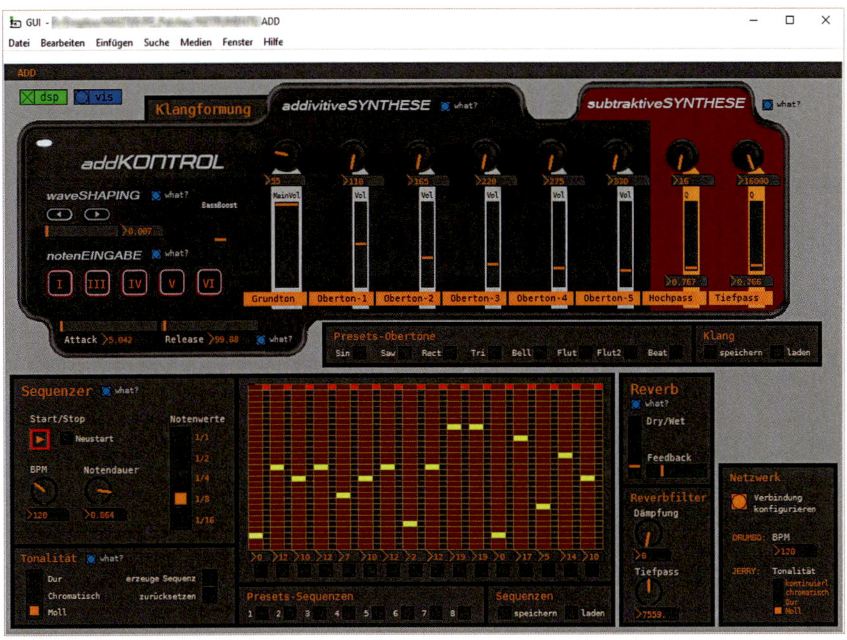

Fig. 1 Main interface of ADD

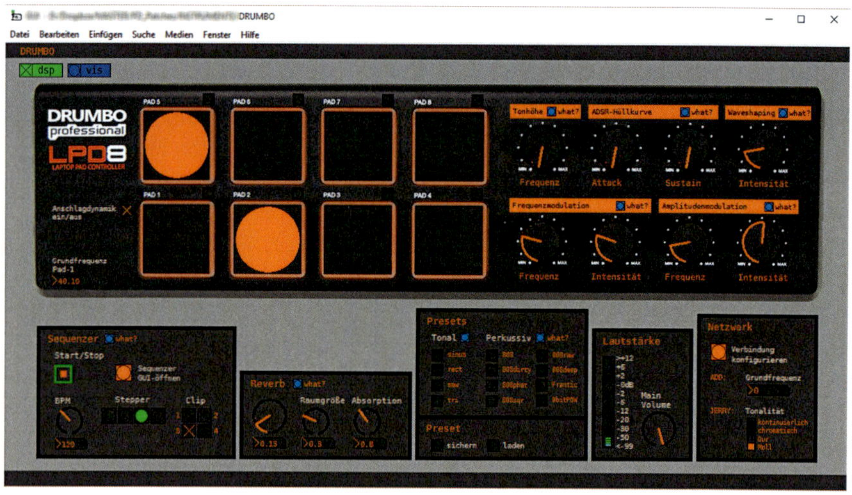

Fig. 2 Main interface of DRUMBO

controllers, are based on different methods of sound synthesis and are different in their musical roles.

Firstly, the issue of cost had to be taken into account. The *loop* instruments completely rely on open source software (Pure Data) and are released under the General Public License (GNU GPLv3), which guarantees end users the freedom to run, study, share and modify the software. The absence of licensing fees results in a zero purchase price for the developers and end users. Additionally, the optional hardware controllers are available for a relatively low price of approximately 50 Euros each. Open Source also means flexibility and freedom in using, customizing and sharing the software. Thereby the *loop* ensemble meets the requirements of the current call of the German Federal Ministry of Education and Research for Open Educational Resources (cf. BMBF 2016).

Another demand is the reduction of software complexity (cf. Sammer et al. 2009, p. 168). It is the basic approach of the three instruments that they are able to self-describe their technical principles through interaction. *Loop's* so-called illustration patches (see Fig. 3 on p. 7) are subroutines, small interfaces within the interface, that use the instruments main engine but focus on a particular functionality such as pitch, sequencing, reverberation, frequency modulation, amplitude modulation, ADSR envelopes, etc. With interactive minimal examples and brief explanation texts they try to explain the function of the specific modules, audio-technical basics and special phenomena. The patches can be opened directly from the main interface and allow the users to playfully and interactively experience individual features.

This experience is meant to be exploited creatively when using *loop* as a music instrument. Thus, the learning process is closely tied to a creative musical practice, making the *loop* ensemble a suitable tool for action-oriented learning.

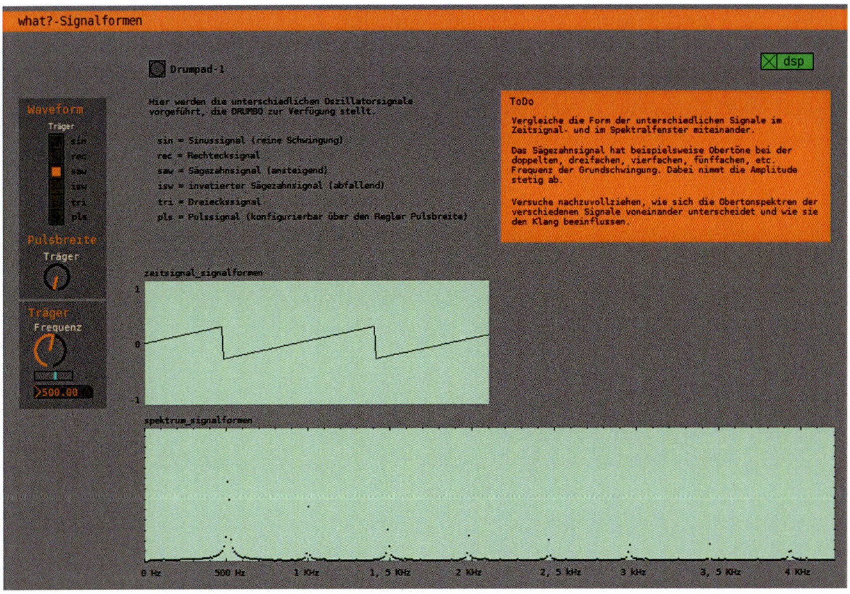

Fig. 3 Illustration patch visualizing and interactively explaining waveforms within the instrument DRUMBO

To further facilitate the use of *loop*, the system requirements are kept to a minimum. Even ten-year-old computer hardware with any major operating system should be able to run the instruments smoothly. For an optimal experience active loudspeakers with sufficient frequency bandwidth or quality headphones are recommended.

During the development we had to find a balance between restricting the technical complexity in terms of the educational objectives and yet implementing interesting functionalities to expand the musical possibilities. Due to the frequently mentioned problems expressed in the pedagogical discourse we often decided to choose simplicity over functionality. The complexity of the software should be reduced as far as possible and focus on the target group. As a design principles for software tools in education, we followed the *low threshold, high ceiling* guideline, supporting an easy access for novices and powerful facilities for more experienced users Resnick et al. (cf. 2005, p. 25ff). For beginners the first steps should be easy and motivating. Getting started with *loop* is particularly low-threshold. Due to the automation provided by the sequencers, the instruments can be played without a long period of training. On the other hand, advanced users have to be kept in mind and motivated with interesting and more demanding possibilities. With experience or after exploring the patches the instruments have the potential to pad/lead to a more complex and elaborate style of play. Connecting the three instruments with each other using the network functionality creates the ensemble, providing a rhythmic and harmonic synchronization that enables small groups to play together.

This way *loop* allows intuitive cooperation and supports collective musical improvisation, making the ensemble suitable for action-oriented group lessons. ADD, DRUMBO and JERRY can loosely be assigned to musical roles: Bass, drums and lead. However, these boundaries are soft due to their ambivalent sound production. The instruments are designed to support diverse traditions in electronic music. The interconnection of the three instruments enables the students to play different styles of electronic dance music, depending on the adjustable aesthetics of the sounds and the settings for the rhythmic parameters. Also, *loop* is meant to motivate the students to experiment freely, for example by designing unusual sounds and arrangements beyond conventional rhythmic or harmonic structures.

Optionally, the instruments can be controlled with low-cost hardware controllers, such as a KORG nanoKONTROL2 and an Akai LPD8. The third instrument JERRY is controlled entirely with keyboard and mouse. The graphical interfaces are adapted to the appearance of the associated controllers to help understanding their layout. All instruments can be fully controlled without the controllers, just using mouse and keyboard.

Due to the educational context, we tried to use appropriate language, e.g. with everyday analogies, that still includes the technical terminologies. At the moment the ensemble only exists in German. An English version is planned.

In summary, the didactic concept of the *loop* ensemble provides interactivity and a self-explanatory approach, putting an emphasis on self-determined and action-oriented learning. Due to its capacity to be used as an ensemble via network connection, it is also suitable for group lessons in the classroom. To also support teacher-centered teaching, *loop* is released in an additional version, which is optimized for lecture and presentation situations. In this version the interface got adjusted and small text boxes were removed. In general, the ensemble follows the "low threshold, high ceiling" concept (Resnick et al. 2005, p. 2). On the one hand *loop* offers an easy access towards electronic music and allows beginners or even non-musicians to express themselves musically. On the other hand, it is also capable of complex musical actions and offers a deeper insight into the technical principles behind electronic sound-synthesis, for example through the exploration of the code of the instruments.

The used framework Pure Data is a visual programming language that uses data flow of objects connected by patches. This is related to analog synthesizer patches and quickly enables users to get started and provides an intuitive way of programming. Pure Data is development and application environment at the same time. Building the instruments logic and actively using the instrument both happens in the same window. Every change is compiled and executed in real-time, which provides direct feedback. Main advantage of Pure Data is its visual character that makes it suitable for educational use. Additionally, it comes with an easy to learn and easy to use interface, plenty of libraries and a strong community. A noticeable negative effect on the usability arises from the rudimentary and limited possibilities that Pure Data offers developers for designing the user interfaces. Also the proximity of development and application environment is risky, since users can unintentionally damage primary functionalities.

For *loop* we chose the *Pd-extended* distribution, which includes a thorough assembly of additional libraries, extensions and documentation. Unfortunately, it is no longer maintained and the last update was released in 2013. This disadvantage still gets compensated by the easy-to-install setup and the support of all major operating systems. During our development we discovered the *L2Ork* distribution of Pure Data. With its growing popularity, an active community and major interface improvements *L2Ork* is evolving into a viable alternative. The developers have indicated that they plan on supporting other operating systems besides Linux in the near future. The migration of *loop* from *Pd-extended* to *L2Ork* has already been tested and seems possible.

4 Evaluation

We put the instruments through three phases of evaluation. First we applied a rating system for music software utilized in educational contexts that uses basic ISO norms on the subject usability (Ahlers 2009). The results show that *loop* positively stands out in exploration, self-descriptiveness and suitability for learning. However, it shows deficiencies in fault tolerance and controllability in comparison with commercial products.

The second evaluation was exploratory and began while the instruments were still under development. We used them in workshops to evaluate their usability and their suitability in educational contexts. For an easy access to the target group we got in contact with university support programs for girls provided by the Technische Universität Berlin and the Freie Universität Berlin. The *Techno-Club* and the *MINToring* program both attempt to encourage girls to study Science, Technology, Engineering and Mathematics (STEM) disciplines by offering workshops and trial lessons in these specific fields. The chosen target group for the instruments, students in the upper secondary, was approached in three independent workshops (N = 10). In those two-hour sessions the students could freely experiment with the instruments (see Fig. 4). We cautiously assisted their exploration of the ensemble by answering questions and providing suggestions.

Every workshop ended with a 20-min-long group interview. The applied guideline covered the subjects: Innovation, fun, usability and integration potential. The results were quite promising and provided us with valuable feedback to optimize the usability. The majority of the students considered the instruments as desirable for them and their music classes. They especially valued the visual presentation and the activity-oriented possibility to experiment. The automated sequencers integrated in ADD and DRUMBO seemed to help restrained students to begin playing. In contrast to the other two instruments the early version of JERRY used in the workshops was lacking a sequencer. It seemed that the students treated it with more reservation and caution. After the workshops we integrated this crucial feature to improve accessibility.

Fig. 4 Explorative workshop evaluation with students of the *Techno-Club*

The third accompanying evaluation examined typical problems of technology development in the field of computer science from the perspective of the Gender and Diversity Studies and brought social aspects into account. For this we used the Gender Extended Research and Development (GERD) model, which tries to encourage developers to reflect their design choices at all critical sections of the research process and the development (Draude et al. 2011). With its support we became aware of excluded user groups (like people with visual and auditory disabilities or without access to the required hardware), reflected about the main beneficiaries (educational institution, teachers, students) and realized how our personal background affected the development (like our own experiences as high school students and own preferences in electronic music).

Ultimately, the various evaluations helped us to adopt varied perspectives and thereby improved the development of the instruments. With the evaluation model of Ahlers that uses ISO usability norms we could identify innovative strengths and shortcomings that came with the rudimentary open source environment Pure Data. The possibilities of configuring the GUI are very limited and the overall performance and load distribution are far from being efficient. To eliminate these disadvantages, we would have had to refrain from Pure Data, but this would most likely have resulted in instruments that are limited in their openness, flexibility and accessibility.

The exploratory workshop evaluation confirmed that the instruments appear to be usable in a classroom context. A focused evaluation of the *loop* ensemble in school-based practice which captures its actual suitability remains still pending.

Finally, we would like to encourage teachers and students to freely use, distribute and modify the *loop* ensemble. *Loop* and its manual can be downloaded free of charge at the PD Community Portal.[2]

References

Ahlers, M. (2009). *Schnittstellen-probleme im Musikunterricht: Fachhistorische und empirische Studien zum Einsatz und zur Ergonomie von Sequenzer-Programmen* (vol. 89). Forum Musikpädagogik: Augsburger Schriften (Wißner, Augsburg).

Ahlers, M. (2011). Neue Medien im interdisziplinären Musikunterricht—Geschichte, Chancen und Beispiele. In D. Meister (Ed.), *Schule in der digitalen Welt. Medienpädagogische Ansätze und Schulforschungsperspektiven* (pp. 221 237). Wiesbaden: VS.

Ahlers, M. (2012). Information communication technology as creativity support tools?: On German music educations history in ICT: Selected research and recent developments. In M. Gall, G. Sammer, & A. de Vugt (Eds.), *European perspectives on music education: New media in the classroom* (pp. 125–134). Innsbruck: Helbling.

Auerswald, S. (2000). *Der Computer im handlungsorientierten Musikunterricht. Didaktischer Stellenwert und methodische Konzeptionen* (vol. 40). Augsburg: Forum Musikpädagogik.

Berliner Landesinstitut für Schule und Medien LISUM. (2006a). Rahmenlehrplan für die Sekundarstufe I—Musik. Senatsverwaltung für Bildung, Jugend und Sport Berlin. http://www.berlin.de/imperia/md/content/sen-bildung/schulorganisation/lehrplaene/sek1_musik.pdf. Cited August 01, 2016.

Berliner Landesinstitut für Schule und Medien LISUM. (2006b). Rahmenlehrplan für die gymnasiale Oberstufe—Musik. Senatsverwaltung für Bildung, Jugend und Sport Berlin. http://www.berlin.de/imperia/md/content/sen-bildung/unterricht/lehrplaene/sek2_musik.pdf. Cited August 01, 2016.

Bovermann, T., Egermann, H., Foerstel, A., Hinrichsen, A., Hildebrand Marques Lopes, D., Pysiewicz, A., et al. (2014). 3DMIN—Challenges and interventions for the design, development and dissemination of new musical instruments. In *Proceedings of ICMC/SMC*, Athens: National and Kapodistrian University of Athens.

Bundesministerium für Bildung und Forschung. (2005). IT-Ausstattung der allgemein bildenden und berufsbildenden Schulen in Deutschland—Bestandsaufnahme 2005 und Entwicklung 2001 bis 2006 https://www.bmbf.de/pub/it-ausstattung_der_schulen_2005.pdf. Cited August 01, 2016.

Bundesministerium für Bildung und Forschung. (2016). Richtlinie zur Förderung von Offenen Bildungsmaterialien (Open Educational Resources—OERinfo) Bundesanzeiger 15 Jan 2016 https://www.bmbf.de/foerderungen/bekanntmachung.php?B=1132. Cited August 01, 2016.

Draude, C., Maaß, S., & Wajda, K. (2011). Gender-/diversity-aspekte in der Informatikforschung: Das GERD-modell. In N. Marsden & U. Kempf (Ed.), *Gender-UseIT. HCI, web-usability und UX unter gendergesichtspunkten* (pp. 67–78). Berlin: de Gruyter. doi:10.2791/83798

Eichert, R., & Stroh, W. M. (2004). Medienkompetenz in der musikpädagogischen Praxis. In H. Gembris, R.-D. Kreamer, & G. Maas (Eds.), *Vom Kinderzimmer bis zum Internet. Musikpädagogische Forschung und Medien* (vol. 59, pp. 36–65). Augsburg: Wißner.

[2]https://puredata.info/Members/loop2016/ accessed: August 01, 2016.

Enders, B. (2013). Vom Idiophon zum Touchpad. Die musiktechnologische Entwicklung zum virtuellen Musikinstrument. In B. Flath (Ed.), *Musik/Medien/Kunst—Wissenschaftliche und künstlerische Perspektiven* (pp. 55–74). Bielefeld: Transcript.

Gall, M., Sammer, G., & de Vugt, A. (Eds.) (2012). Introduction to new media in the classroom. In *European perspectives on music education: New media in the classroom* (pp. 11–30). Innsbruck: Helbling.

Gies, S., Jank, W., & Nimczik, O. (2001). Musik lernen. Zur Neukonzeption des Musikunterrichts in den allgemein bildenden Schulen. *Diskussion Musikpädagogik—Konzeptionen des Musikunterrichts, 9*, 4–22.

Knolle, N. (1995). Zur Ideologiekritik der Neuen Technologien in Schule und Gesellschaft und ihre Konsequenzen für die Musikpädagogik. In *Musiklernen und Neue (Unterrichts-) Technologien* (vol. 16, pp. 41–59). Essen: Die Blaue Eule.

Maas, G. (1995). Neue Technologien im Musikunterricht. Eine Erhebung zum Stand der Verbreitung und zur Innovationsbereitschaft von MusiklehrerInnen. In *Musiklernen und Neue (Unterrichts-)Technologien* (vol. 16, pp. 96–123). Essen: Die Blaue Eule.

Medienpädagogischer Forschungsverbund Südwest. (2016). JIM 2014: Jugend, Information, (Multi-)Media: Basisuntersuchung zum Medienumgang 12–19 jähriger in Deutschland. http://www.mpfs.de/fileadmin/JIM-pdf14/JIM-Studie_2014.pdf. Cited August 01, 2016.

Meyer, H. (2010). *Unterrichts-Methoden II: Praxisband* (13th ed.). Berlin: Cornelsen Scriptor.

Ministerium für Schule und Weiterbildung des Landes Nordrhein-Westfalen. (2013). Vorgaben zu den unterrichtlichen Voraussetzungen für die schriftlichen Prüfungen im Abitur in der gymnasialen Oberstufe im Jahr 2016. Vorgaben für das Fach Musik. https://www.standardsicherung.schulministerium.nrw.de/cms/zentralabitur-gost/faecher/fach.php?fach=19. Cited August 01, 2016.

Münch, T. (2013). Medien im Musikunterricht. In W. Jank (Ed.), *Musik-Didaktik: Praxishandbuch für die Sekundarstufe I und II* (pp. 220–228). Berlin: Cornelsen.

Münch, T., & Knolle, N. (2005). *Abschlussbericht des BLK-Modellvorhabens Me[i]Mus im Rahmen des BLK-programms, Kulturelle Bildung im Medienzeitalter (KUBIM)*. Bonn: ACRult Mediastra.

Resnick, M., Myers, B., Nakakoji, K., Shneiderman, B., Pausch, R., Selker, T., et al. (2005). Design principles for tools to support creative thinking. In B. Shneiderman, G. Fischer, M. Czerwinski, B. Myers, & M. Resnick (Eds.), *Creativity support tools—A workshop sponsored by the National Science Foundation* (pp. 25–35). National Science Foundation. http://www.cs.umd.edu/hcil/CST/Papers/designprinciples.pdf

Rheinländer, M. (2002). *Der computer—Instrument im Musikunterricht—Instrument des Musikunterrichts*. Oldershausen: Lugert.

Riedl, A. (2010). *Grundlagen der Didaktik* (2nd revised ed.). Stuttgart: Franz Steiner.

Roth, J. (2006). Click to learn. Umgangsweisen mit computerbasierten Lernumgebungen zum Erwerb musikalischer Kompetenzen. In N. Knolle (Ed.), *Lehr- und Lernforschung in der Musikpädagogik* (vol. 27, pp. 253–267). Essen: Die Blaue Eule.

Sammer, G., Gall, M., & Breeze, N. (2009). *Using music software at school: The European framework NET MUSIC project, 1*, 155–177.

Schläbitz, N. (2002). The 'winAmp' takes it all. Zeit nehmen für einen zeitgemäßen Musikunterricht. In R.-D. Kraemer (Ed.), *Multimedia als Gegenstand musikpädagogischer Forschung* (vol. 23, pp. 27–71). Essen: Die Blaue Eule.

Schläbitz, N. (in preparation). Musik-Medien und die Musikpädagogik. Eine Romanze mit manchmal tragischen Momenten und ungewissem Ausgang. In R. Großmann & E. Ungeheuer (Eds.), *Musik und Medien*. Laaber-Reihe: Kompendium der Musikwissenschaft, Basiswissen Musik.

Smudits, A. (2013). Musik in der digitalen Mediamorphose. In B. Flath (Ed.), *Musik/Medien/Kunst—Wissenschaftliche und künstlerische Perspektiven* (pp. 75–96). Bielefeld: Transcript.

Strausbaugh, L. G. (2006). *Digitale medien im Musikunterricht: Ansätze zur Didaktik und Methodik des computergestützten Musikunterrichts*. Saarbrücken: VDM.

Stubenvoll, M. (2008). Musiklernen am computer: Zur Qualität von Musik-Lernsoftware und ihrer empirischen Überprüfung. In R.-D. Kraemer (Ed.), Multimedia als Gegenstand musikpädagogischer Forschung. (vol. 81) Essen: Die Blaue Eule.

Tschmuck, P. (2013). Elektronische Musik—von der Avantgarde-Nische zum paradigmatischen Musikstil. In B. Flath (Ed.), *Musik/Medien/Kunst—Wissenschaftliche und künstlerische Perspektiven* (pp. 97–109). Bielefeld: Transcript.

Weidler, B. (2014). *Computersoftware im Musikunterricht*: *Am Beispiel von "Band-in-a-Box"*. Hamburg: Diplomica. Publisher = Diplomica.

The Birl: Adventures in the Development of an Electronic Wind Instrument

Jeff Snyder

Abstract This article reflects on the markedly distinct development stages of an electronic wind instrument called the Birl. Stemming from an early idea for an electro-mechanical oscillator inspired by the sounds of pen plotters, the Birl was formed through the connection of that oscillator prototype to a rough wind instrument body. Originally intended to fulfill the role of the wind section in an ensemble of instruments built for the author's doctoral dissertation composition, the instrument took on a new life after the completion of the piece. The development of a "cello-like" resonator body and refinements to the electro-mechanical aspects had brought the instrument to a performable state, but several limitations suggested further development. A desire to make the instrument more conducive to exploratory improvisation pushed the Birl in new directions, toward open-holed fingering systems and embouchure sensors with neural net mapping structures and physical models of dynamically configurable toneholes, resulting in an instrument that bore little resemblance to the original electro-mechanical concept. The author discusses the design challenges that arose as the instrument evolved, the solutions that were found along the way, and the ways in which user feedback informed the design as the needs of the instrument changed.

1 Introduction

This is the story of how the instrument I call the Birl morphed from a large, strange electromechanical contraption into a miniature wind controller. The current version of the instrument is arguably completely unrelated to the original design. Only the name has carried over, and the explanation of the name no longer makes sense with what the instrument has become. The convoluted story of the instrument's development gives some insight behind the scenes at the various design problems, creative inspirations, and unplanned discoveries that guide the creation of new instruments.

J. Snyder (✉)
Princeton University, 310 Woolworth Center, Princeton, NJ 08542, USA
e-mail: josnyder@princeton.edu

T. Bovermann et al. (eds.), *Musical Instruments in the 21st Century*,
DOI 10.1007/978-981-10-2951-6_13

I approach instrument design with a few things in mind I want to achieve, but many of those ideas do not end up in the final product. I don't consider this a failure of the design goals, but a gift of the process. One of my favorite parts of instrument design is when ideas emerge from accidents and surprises along the way. The Birl is an example of how sometimes the resulting object evolves from the process, as much as—or even more than—vice versa.

2 Origins of the Birl (2008)

In 2008, I formed a band with fellow composer and technological adventurer Victor Adan called the Draftmasters. We had both gotten excited about the musical and visual possibilities hidden in 1980s pen plotters, those large mechanical drafting machines that print images by moving a real pen around on a page. Victor and I were collecting plotters via Ebay bids, and, encouraged and guided by fellow plotter enthusiast Douglas Repetto, we were experimenting with controlling the plotters live, treating them as musical instruments as well as drawing tools. We found an X/Y plotter (meaning the paper stays stationary while the pen moves in both X and Y dimensions) that seemed perfect for the job, called the Roland DXY-1100. It was big but still portable, and quick to respond to serial commands sent from Python or Max/MSP over a USB-to-serial converter, so we could control it live without much trouble. We wanted our stage act to integrate the visual and audio elements of the plotter. As we drew an image, the sound produced would be an amplification of the motor noises generated by the instrument as it followed our drawing instructions. We experimented with placing electromagnetic pickups against the stepper motors inside the plotter to get a stronger audio signal by capturing sounds directly from the electromagnetic field the motors gave off as they turned. It worked beautifully, and produced a gritty, intense sound that combined the bass frequencies of the rotary motion with the digital hissing and white noise of the drive signal being sent from the motor controller ICs. A contact microphone on the pen-up/pen-down solenoid completed the instrument, and we drilled holes in the plotter bodies to install 1/4″ jacks so we could simply show up with our plotters as though they were electric guitars. Video of our performance is available Iglesia (2009).

While I had one of our plotters open to repair a pickup, I accidentally pushed the plotter arm while the pickup on the motor was connected to an amplifier, and was surprised by the beautiful, clear glissando that erupted from the speakers. The tone color of the plotters in our live performance was naturally harsh, evoking a sort of robot apocalypse, but this sound was sweet and subdued. The difference was that there was no power applied to the plotter, so I was hearing only the electromagnetic waveform generated by the motor's motion (indirectly through the body of the motor), without the interference from the noisy PWM drive signal. I immediately began to ponder how I could harness that sound in a new instrument and be able to control it musically (Fig. 1).

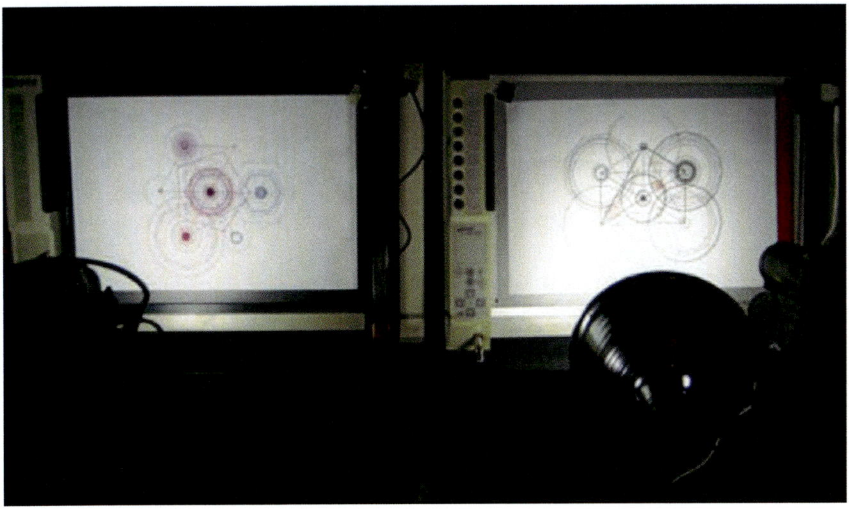

Fig. 1 The Draftmasters pen plotter band

The primary challenge was how to turn the motor at a precise speed without driving it electronically. I built a small test rig with two stepper motors mounted on an aluminum plate, and experimented with ways to mechanically couple the rotors. If I drove one stepper motor electronically, I could use a friction belt and pulleys to drive a second passive stepper motor at the same speed. I soon realized that in this configuration I could dispense with the electromagnetic pickup, since the passive motor was acting as a generator and I could simply connect the unused leads from one of the electromagnets inside the motor itself. This resulted in an even more pure signal, approaching a sine wave in timbre. Soon, I had a working prototype that allowed me to accurately produce desired pitches within a range of a two octaves. Going above the usable pitch register resulted in a loss of torque, stalling the motors. Going below the usable pitch register produced a waveform that got more rugged as the steps in the motor became audible and the rotor inertia could no longer smooth out the tone. Swapping out the motors for different stepper models moved this pitch range around, but didn't manage to expand it much. I did find I could raise the register easily by varying the pulley sizes, for instance a pulley size ratio of 4:1 produced the expected pitch shift of two octaves up.

I later noticed a description of a similar idea in Handmade Electronic Music by Nic Collins (2006), although he uses a DC motor instead of a stepper. In terms of historical precedent, the Hammond organ is also based on a related principle Aldridge (1996), with a spinning metal tonewheel being sensed by an electromagnetic pickup, however in the organ the pitch changes are produced by switching between several pickups pointed at tonewheels with different numbers of teeth, rather than by changing the speed of rotation.

In my early experiments, I was controlling the driver motor with MIDI signals sent to an AVR microcontroller brain. I found myself wondering how this new instrument should most naturally be controlled. Was there an instrumental interface better suited to this sound production method than any other? The sound properties of the stepper motor synthesizer were the following:

- It was monophonic. There was no possibility for polyphonic sound without creating multiple identical mechanisms, and this seemed unnecessary to me at the time.
- The tone color was very dark, with a strong focus on the fundamental and first harmonic.
- The amplitude could be controlled electrically, with a VCA, and had no "natural" envelope (such as a plucked string or percussion envelope).
- The amplitude was also somewhat coupled to the pitch, as higher frequencies produced higher amplitude signals, perhaps through inertia of the rotor.
- There was a natural vibrato to the sound, caused by slight inaccuracies in the pulleys.
- There was a natural portamento to the sound, due to the need for speed ramping in the motor control to avoid stalling. When moving between nearby pitches it was inaudible, but when going from a very low note to a high note a ramp of more than 10 ms was usually necessary.
- There was brief but noticeable overshoot to the pitch contour when changing speeds, due to the stretching of the rubber friction belt.

Several of these properties suggested that a control paradigm based on a wind instrument model would make sense. The most obvious was its monophonic nature, which is commonly a property of wind instruments not shared by most string instruments, keyboard instruments, or percussion instruments. Also, the dark tone color immediately reminded me of a recorder or flute, and many people commented on its "birdlike" character, which brought to mind whistles, ocarinas, and other wind-powered instruments. I quickly started working on a wind-style interface for the new electromechanical oscillator.

3 The First Birl (2009–2010)

While I worked on the interface, the instrument took the name "the Birl". The word "birl" seemed appropriate in multiple ways: It is an old English or Scottish word for "rotate with a whirring sound," a type of bagpipe ornament, and slang for "to carouse" "Birling" is also the name of the sport where lumberjacks run on logs in a river, a connotation that delighted me.

The first Birl was a large instrument, held between the legs and connecting to the floor with a cello pin (see Fig. 2). The top of the instrument was a wind controller, with mechanical momentary push buttons arranged for the fingers and thumbs of

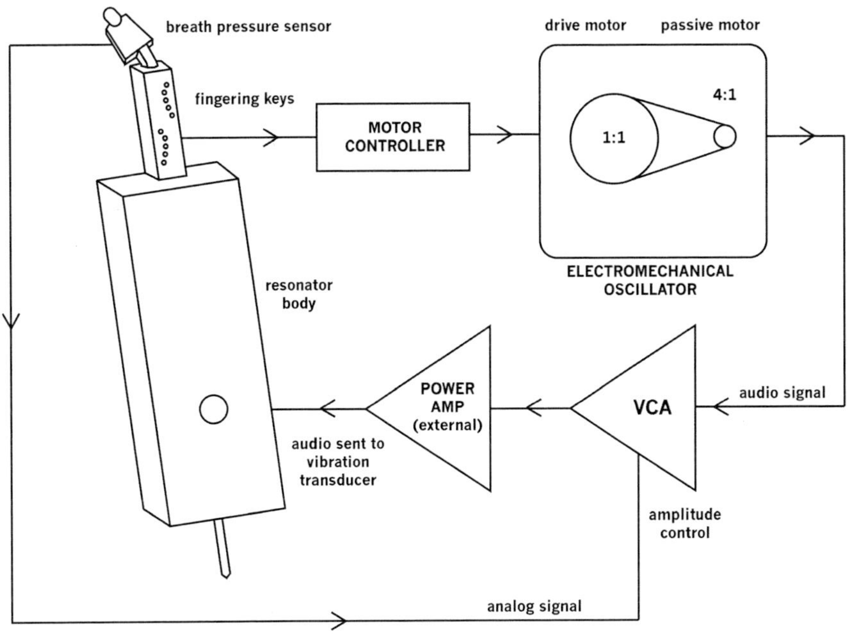

Fig. 2 The first Birl, diagram

the left and right hands. The base of the instrument was a large wooden resonator, in keeping with the concept of "acoustic electronic music" I had developed in my dissertation work Snyder (2011). The body of the instrument was made from 1/4″ birch plywood cut on a laser cutter, and the front was made from a thin spruce board with internal spruce X-bracing, like the top of a guitar or harp. The body housed the motor and pulley system, which was screwed into the inner right side of the instrument, and a hole in the center of the body was decorated with a laser-etched lute rose. Screwed into the spruce front from the inside was a Rolen Star vibration transducer, which resonated the top-plate to create the instrument's acoustic sound. By resonating the electronic sound through an acoustic body rather than from a speaker cone, I could achieve both a more natural radiation of the sound in space, and I could get an individualized color for the instrument. Each wooden resonator imparts a unique sonic filter onto the electronic sound passed through it, emphasizing certain frequencies and attenuating others. This idea extends back to instruments like the Ondes Martenot, an early electronic instrument with several acoustic resonators, and in my case was influenced by David Tudor's installation piece, ***Rainforest IV***. As for breath input, by 2010, a mouthpiece with a breath pressure sensor was fitted to the top of the instrument, but for the first performance in 2009, a Yamaha BC1 breath controller was used, since a more tailored custom solution hadn't yet been completed.

The instrument was self-contained, except for the power amplifier needed to drive the vibration transducer. Due to the inefficiency of the wood top when compared to the paper cone in a standard speaker, more watts were needed to get stronger sound levels than a small amplifier that could fit inside the body would have been able to provide. Therefore, the signal path was:

- The pattern of pressed pushbuttons for the keying system is sensed by a microcontroller, resulting in a frequency being sent to the motor controller.
- The motor controller controls the "drive" motor, which then spins the passive motor via a friction belt.
- The electrical signal generated by the passive motor is sent to a voltage controlled amplifier (VCA). The amplitude of the VCA is directly controlled by the voltage from the breath pressure sensor.
- The audio signal from the VCA is sent to a power amplifier, which sends an amplified signal to the vibration transducer.
- The vibration transducer mechanically vibrates the spruce top-plate on the front of the instrument, producing the acoustic sound the performer and audience hear.

This was the system used for the first public Birl performance, a Wet Ink Ensemble Wetink concert where my dissertation piece, ***Concerning the Nature of Things*** Snyder (2011), was premiered. I had written two parts for the Birl in the composition, having determined the most useable pitch range and knowing the basic timbre the instrument would have, but not having actually finished the instruments' construction. About a month before the performance, worried about getting both Birls functional in time, I decided to focus on finishing one instrument, and cut the second Birl part from the score. Erin Lesser, the flautist for Wet Ink, learned to play the new instrument and provided feedback during the design and development phase. I had built the fingering system with only four buttons per hand, one for each finger (not counting the thumb buttons), so we had to work together to design non-standard fingerings for the pitches that weren't well served by the simple recorder-based keying system (such as a low C, C#, and F#). On the back of the keying system, I had added two buttons for the left thumb, for octave up and octave down, and three buttons for the right thumb, allowing for maneuvering within my Adaptable Just Intonation Snyder (2011) tuning system. The fingering-to-pitch mapping was implemented as a lookup table, with specific patterns of open and closed buttons resolving to a particular note. Lesser tackled the unfamiliar instrument with enthusiasm and managed a very expressive performance even with the limited rehearsal time resulting from the instrument being completed barely a month before the premiere. However, after the performance, I was left with a considerable list of design problems Lesser had discovered with the instrument. It should be noted that I don't play any wind instruments, so I was heavily reliant on information from Lesser and other musicians who tested the prototypes (Fig. 3).

First, there were tuning problems in the upper octave. I had switched shortly before the concert from metal pulleys to plastic pulleys, since the reduced weight

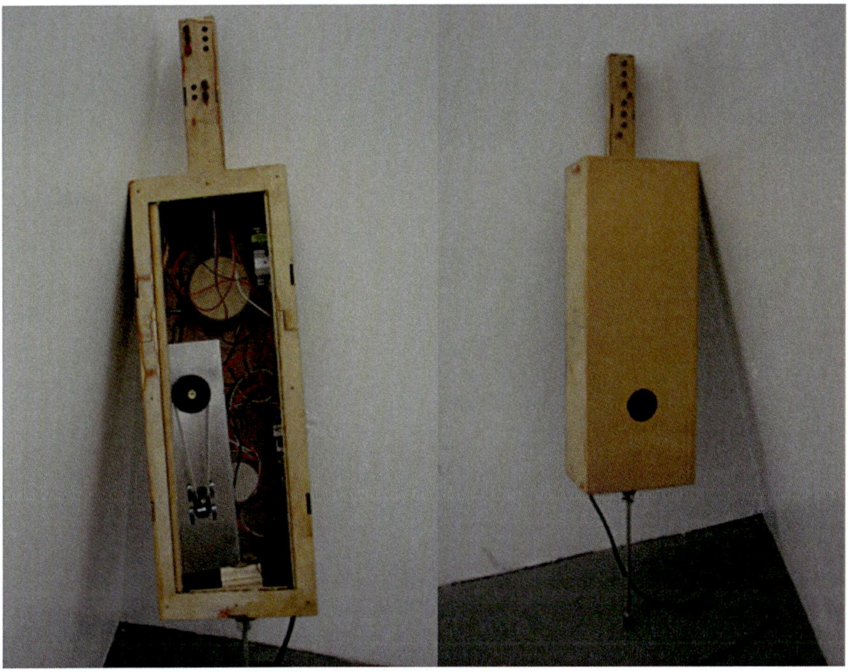

Fig. 3 The first Birl, prototype

allowed me to lower the ramp times for the motors. The plastic pulleys were not as precisely sized, though, and the difference caused seriously flat pitches in the high register that I didn't recognize until the day of the performance. In my music, this is especially problematic, since a great deal of attention has gone into precise Just Intonation Doty (2002). This was easily fixed by switching to precision steel pulleys from SDP/SI. But the added weight meant I needed to find motors that could handle more torque.

Another serious problem was the acoustic sound caused by the keys. I had used tactile pushbuttons because they had a satisfying click response when actuated, but when the pushbuttons were mounted in the resonator body, they were acoustically amplified to an unacceptable level. I liked the key click sounds in principle, but they made truly quiet playing untenable. Lesser also noted that the actuation force required for the pushbuttons was far above what was normal for a flute or other wind instrument, and was tiring for her fingers. I decided to redesign the button system (Fig. 4).

Most problematically, the electromechanical tone generator system produced unintended acoustic vibration noise in the resonator body in addition to the intended electrically amplified signal. This sound was not unpleasant in itself, as it was in tune with the electrical signal and changed pitch with the notes being played, but since it was mechanical in nature it could not be electrically attenuated by the VCA.

Fig. 4 Stina Hasse plays the first Birl

Therefore, whenever the motor was spinning, the instrument was humming, even if the VCA had attenuated the volume to "off." Since the motor in the Birl changed speed with every new pitch, the humming sound seemed unnatural, as though the instrument didn't really stop sounding the note when the performer ceased to blow into the instrument. The motor couldn't be stopped between notes because ramping up from a standstill would create a dramatic glissando on every attack. This definitely had to be solved, and a solution was not immediately obvious.

There were also more minor issues I hoped to address in the next iteration. The higher pitches from the motor were naturally louder for reasons I didn't entirely understand. This is also the way many woodwind and brass instruments operate—it is difficult to play quietly in the high registers as it takes more breath to overblow the notes—and Lesser was able to compensate by reducing her breath pressure for higher pitches, but it was very difficult in faster passages with leaps. It seemed worthwhile to build a more automatic compensation system into the instrument. It was difficult to minimize pitch glitches when changing many keys at once, such as going over the "break" in the instrument, where the fingering changes most drastically from one note to the next. This is also a problem for acoustic wind instruments, but it seemed much less forgiving in this instantaneously calculated digital version. Lesser told me that the majority of her practice time was spent working to minimize these glitches.

4 The Second Birl (2011)

The first issue I dealt with in the second iteration was the unintended acoustic vibration noise. I removed the stepper motor system from inside the resonator and made a prototype board that combined a custom power amplifier with the stepper motors and motor controllers, the analog VCA, and the power supplies for all the circuitry. This worked well and sounded much better—I recorded a studio version of ***Concerning the Nature of Things*** Snyder (2011) using this prototype, with the stepper motor board in an isolation booth to keep the mechanical noise away from the resonator and microphones. However, it was very messy, fragile, and not really useable in live performance. Having a board with the motors on it backstage seemed impractical, so I decided to try to build an enclosed box to acoustically isolate the motors.

I designed a box about 12 inches square, made from birch plywood. The front was an aluminum panel for controls and jacks, and the back had an acrylic window that made the motor system visible. When the first Birl had hidden the pulleys inside the resonator I had been disappointed that they weren't part of the visual signature of the instrument, and, thinking back to the Draftmasters, I wanted to give the audience more of a view into the unusual workings of tone generator. Inside the box, the motors were mounted on a thick aluminum plate, and the plate was suspended off the base of the box with rubber vibration isolation mounts. The box itself was isolated from the table or floor with large rubber feet. I lined the inside of the box with vibration damping rubber-lined foam, intended to muffle sounds from boat engine rooms.

I designed a printed circuit board (PCB) with the stepper motor driver circuitry, and another PCB for the audio processing of the electrical signal from the passive generator motor, designed to stack with a PCB for the control panel components. The goal was to get everything that had been on the messy prototype board into a nice, neat box that could be on stage next to the controller/resonator body and connected with a short MIDI cable. I left out the power amplifier once again, due to space, heat, and weight considerations, but I decided to expand upon the shaping of the audio signal.

In the original instrument, the audio path was simple. The waveform generated by the passive motor went directly through a VCA for amplitude control and was converted into acoustic sound through the driver transducer. In the time since I designed the first Birl, though, I had started to see the tone generator as an oscillator for a system that could be a more complete synthesis voice. Therefore, I chose to build into the new Birl some extended functionality that allowed further shaping of the sound (Fig. 5).

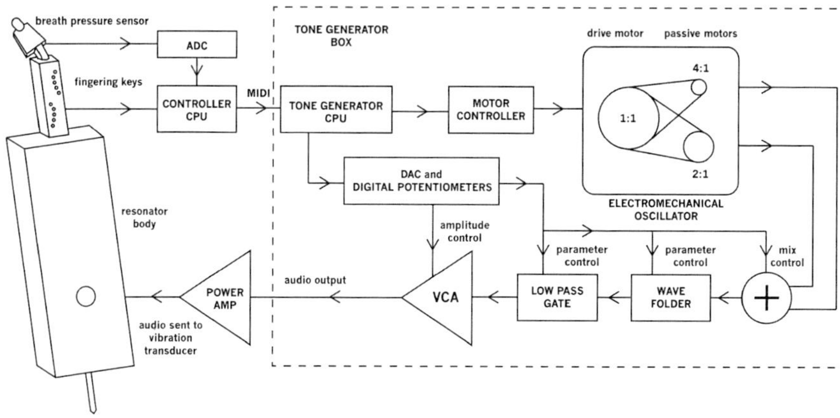

Fig. 5 The second Birl diagram

First, I added to the motor mount the ability to drive two simultaneous passive generator motors. The drive shaft of a single driver motor was affixed with two pulleys, and these pulleys drove the two passive motors with different pulley size ratios, one at a 2:1 ratio and the other at a 4:1 ratio. This meant I could mix two resultant oscillator signals, one an octave higher than the other. On the audio PCB, I made an "oscillator" section that allowed for crossfading between these—like an 8′ and 4′ stop on an organ. After the oscillator crossfader, the mixed signal went through a waveshaper that could add high harmonics to the signal, essentially a distortion circuit. This allowed for more timbral possibilities than the original "natural" waveform. After the waveshaper, the signal passed through a Low Pass Gate, based on Don Buchla's design from the 1970s Parker (2013)—a vactrol-controlled lowpass filter acting as a VCA. The signal then went through a final VCA and finally to an output jack on the box. Aiming for maximum flexibility, I designed the whole "voice" as a semimodular system, with patch points for each input and output, and voltage control inputs for all parameters. I also added digital-to-analog converters (DACs) and digital potentiometers to allow computer or MIDI control of the analog functionality.

Part of the reason for adding comprehensive digital control was the need to compensate for the higher volume in the upper octave of the instrument. With the final VCA controlled digitally, I could easily program curves to apply to the amplitude based on the frequency of the oscillator, allowing for a more even response.

Once I had assembled the audio and motor control PCBs, I installed them in the box that could now be controlled from the Birl wind controller or using MIDI from a computer (Fig. 6).

While the new instrument had a unique sound, there were design problems preventing it from being as useable as I had intended. The mechanical noise had been solved—the box was very quiet and no longer caused any acoustic issues, but

Fig. 6 The second Birl PCBs and motors, taken out of the box enclosure

electrical noise problems arose. The noise from the motor controllers was audible in the audio circuits, despite carefully isolated ground planes and independent power supplies. The physical proximity of the circuits was just too close to expect the high currents of the motors to not interfere with the audio. This problem was similar to the acoustic sound problem, in that the injected motor noise continued even when the VCA was off, and the sound was related to the frequency of the spinning motors, so it couldn't be easily ignored.

Even more seriously, the acoustic insulation had come at the price of heat insulation. The motors generated heat, and despite heatsinks and the 1/4″ thick aluminum mounting plate, once I had sealed off the air transfer from the inside of the box to the environment, the heat had nowhere to go. After about a half hour, the box was hot to the touch and needed to be turned off to cool down. This also limited the instrument's ability to be used in a full-length concert. I've since discovered the existence of heat pipes, a technology designed to handle this very problem, but I have yet to try that solution (Fig. 7).

Testing this system with composer and flautist Natacha Diels, we decided that despite the problems, the tone generator sounded quite good (Fig. 8). The controller, however, was still awkward, and the discrete nature of its pitch control was not ideal. The pushbuttons needed to be replaced with something more comfortable

Fig. 7 The second Birl tone generator box control panel

Fig. 8 Natacha diels tests the second Birl

and mechanically quiet, ideally something that sensed a continuous change on each key rather than simply an on/off event. Also, travelling to events in far away locations convinced me the controller needed to be separated from the resonator for portability. This led me to focus the next phase of research on improving the controller portion of the instrument.

5 The Third Birl (2012–2015)

Now that I had decided to break the instrument into three separate pieces—controller, resonator, and tone generator—the first order of business was improving the finger sensing on the controller. I wanted the instrument to be more like open-holed wind instruments such as the bansuri and the tin whistle. I built prototypes to test both infrared (IR) reflectance sensing and capacitive sensing, and found the response for the capacitive sensing fit my needs better, so I moved forward with that.

With the resonator no longer being a necessary part of the controller, I had to reconsider the visual design of the instrument. The resonator on a cello end pin had given the instrument a striking and unusual visual presence, and with the breath controller and key system removed from the resonator base, I worried about the instrument taking on a "typical" soprano saxophone visual style, like the Akai EWI Akai and the Yamaha WX Yamaha instruments. I decided to try to keep the vertical orientation of the instrument, and enforce that by creating a mouthpiece to angle the instrument more like a bass clarinet or tenor saxophone. I imagined the instrument would be played seated, and tried to design it so that it would be comfortable to play with the base of the instrument resting against the player's leg. I found the laser-cut plywood design of the first Birl somewhat ugly, so I aimed to create a design that could be milled out of solid wood, to have a more beautiful presentation.

I had only limited time access to a 3-axis CNC mill, so I designed the wooden body to be millable without flipping the part, to avoid wasting time with realignment during the milling process. I made it with a clamshell type of design, so that I could mill both parts from only one side, and then screw the two parts together around the circuit board to house the electronics. Before I milled the wooden version, I tested the design using a 3D printed white plastic model. The shape of the new enclosure was highly influenced by Scandinavian design; I had recently visited the Danish Museum of Art & Design Museum DK and I found the sharp corners and clean lines of Jacob Jensen's Jensen designs for Bang and Olufsen to be particularly inspiring. This led me to a relatively boxy design, somewhat reminiscent of a 1970s Volvo automobile, but unusual for wind instruments, which are typically cylindrical or conical in shape. In a way, the melodica is a closer visual reference than a recorder or flute.

The controller's circuit board had undergone several revisions since the first Birl. Now that continuous data from the fingers was going to be possible, I wanted to improve the data throughput from the device, so I switched from standard serial MIDI to OSC over Ethernet. I designed the brain of the controller around an AVR32 AVR microcontroller because it could easily send Ethernet information. The capacitive sensing for the keys was handled by a Cypress PSoC Cypress microcontroller using the CapSense CSD library. The keys themselves were simply aluminum standoffs, since any metal object can be a sensor using that technology. The initial revision of the board added multiple capacitive sensors for embouchure sensing; I was hoping I could retrieve some reasonable data on lip position by doing machine learning on the data from sensors touching the top, bottom, and sides of the mouth (Fig. 9).

While I still at this point considered the electromechanical oscillator to be an important part of the instrument, it was slowly beginning to seem more optional, rather than essential. As I made plans to lend Birl prototypes to musicians for "in-the-field" user feedback, the impracticality of the motor system for anyone's use but my own became more and more apparent. For instance, in transporting the Birl for to show at an event at the Mass MOCA museum, the motor connections were damaged, and I had to do some emergency surgery on the instrument that a musician couldn't be expected to do. I decided that the instrument needed to have the option of a simpler digital voice for musicians who weren't as interested in the strange electromechanical oscillator, but found the wind controller useful. I designed an internal digital synthesis circuit to live inside the Birl controller itself, which could be used instead of an external computer, although in practice those who used the instrument in this next iteration always used it with an external computer, to allow themselves more flexibility in synthesis options.

One major hurdle was the question of how I would map data from multiple continuous key sensors into a single pitch output. I wanted expressive pitch bends to be intuitive for players to execute with their fingers, but it wasn't immediately clear how that mapping should work. The first idea I had was to use machine learning. My colleague Rebecca Fiebrink had written a program called Wekinator Fiebrink (2011), which allowed for easy experimentation in applying machine learning techniques to digital music making. It could take OSC data in and send OSC data out, which was perfect for my new Ethernet-ready instrument. Fiebrink, who happens to be a flautist as well, joined me to try some tests where we trained a neural net on a simple flute scale. The results were astounding. The trained model made choices that were surprisingly intuitive, and the resulting system allowed Fiebrink to bend each pitch up to the next. With this encouraging solution to the pitch-mapping problem showing the way, I started testing the Birl with other professional performers (Fig. 10).

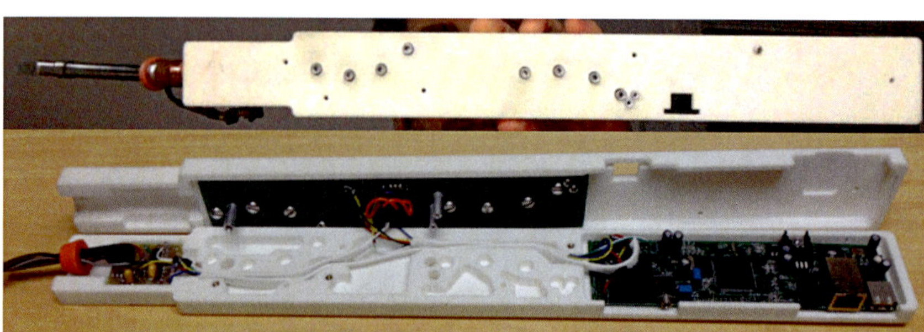

Fig. 9 Third Birl early prototype, 3D printed with vacuum cleaner mouthpiece

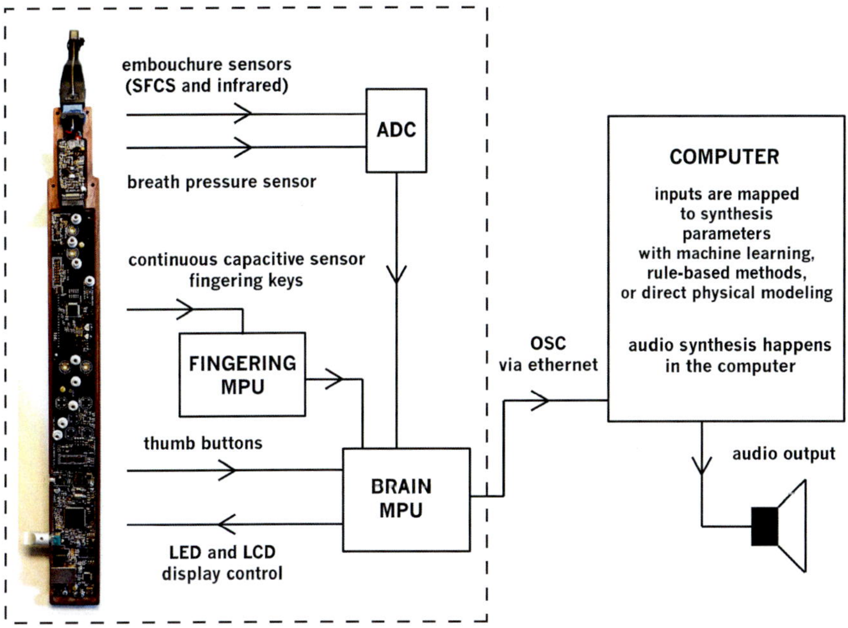

Fig. 10 Third Birl diagram

The primary testers of the third Birl in 2013 and early 2014 were jazz saxophonist David Schnug, and avant-rock saxophonist Sam Hillmer. Both were most interested in using the instrument as a controller for digital synthesis. At the time they first tried it, it had a vacuum cleaner attachment for a mouthpiece, and the embouchure sensors were pieces of copper wire covered in heat-shrink tubing and wire-tied to the mouthpiece. I tried using neural nets to map data from the embouchure sensors, but the values drifted significantly during testing, possibly due to the loose wire-ties allowing the sensors to move. I still got a surprisingly useful result, as can be seen in an online video Schnug. I mapped the embouchure sensors to timbre parameters on both a simple FM synthesis patch and a wind instrument physical model (the "blotar" by Dan Trueman and Perry Cook).

Both Schnug and Hillmer found the pitch bends reasonably easy to control. Interestingly, Hillmer was entirely focused on the pitch bending possibilities—in the software I had left an option to turn it off to allow easier discrete fingering, and he commented that he would never use this feature. He enjoyed the strange, yet controllable bends that were possible, and even invented an extended technique in his first performance with it—wearing rubber gloves while playing to reduce the sensitivity of the sensors and make the bends more intense, eliminating the ability to achieve exact pitches but exaggerating the slippery weirdness of the pitch mapping. He used this technique in a live installation performance in NYC called Apparition Hillmer.

Unlike Hillmer, Schnug was interested in both the continuous and the discrete fingering options. What Schnug found most exciting was when I mapped the embouchure sensor data to physical modeling synthesis parameters. He was extremely interested in the possibilities opened up by this mapping, which allowed for very unusual sound transformations controlled entirely with the mouth. As a musician who specializes in experimental and free jazz, he wanted a way to have a wild range of sounds under predictable embouchure control, to provide a world of timbres that could rival the saxophone's variability (Fig. 11).

The feedback from these performers started me thinking in another direction. When the original idea for the Birl was forming, the choice of a wind control paradigm was simply because I wanted a "winds" section in the ensemble of invented instruments I was building, and the properties of the electromechanical oscillator I was experimenting with suggested that wind control would be a good match. It had a wooden resonator because it was intended to be used for my own chamber music, specifically a suite of pieces I am writing in just intonation for electronic instruments with acoustic resonators. While the electromechanical oscillator and wooden resonator of the earlier Birl designs made sense in my original context, they weren't ideal for the needs of these professional musicians. Though these musicians were very experimental and open-minded, touring internationally would not be easy with those contraptions in tow. However, both of these performers found the possibility of a more expressive wind controller aimed at the

Fig. 11 Leila Adu and Dave Schnug test the first prototype of the third Birl

needs of an experimental musician to be very interesting. While existing commercial wind controllers were certainly useable in those contexts, there seemed to be room for something that sought to fill those needs more directly—a wind controller designed with a player like Evan Parker in mind.

Rather than considering the Birl controller as just part of a whole that I would later reconnect to the oscillator and resonator, I began to imagine what the controller could be without those more mechanical components. For the past few years, the controller has become separated from electromechanical ideas that initiated the design process, and both the oscillator and the resonator have been placed on the back burner while I solve the problems posed by these new design goals. They may rejoin the Birl someday, but it will be as accessories rather than as the heart of the instrument.

After a few adjustments to the dimensions of the body and the shape of the mouthpiece, I milled several bodies out of walnut and maple during a residency at the Haystack Mountain School of Crafts. I also made a new CAD design for a mouthpiece that I had 3D printed (Fig. 12).

I wrote a paper for the New Interfaces for Musical Expression (NIME) conference Snyder (2014) about the application of neural net machine learning to the pitch-mapping problem (Fig. 13). While I was showing the Birl at NIME, a colleague questioned why machine learning was really an improvement over a rule-based approach, and my answer didn't really satisfy me. I was interested in the fact that machine learning would easily allow the user to modify or completely alter the mappings by entering new training examples, but it was true that a rule-based approach to the mapping would be significantly more efficient, and, although it would be more difficult for the user to change the behavior to their preferences, the inner workings of the algorithm could be understandable to an expert user, unlike the more black-box internals of a neural net. The following year I had a student develop a rule-based pitch mapping algorithm that worked well for situations where user-defined fingerings were not necessary.

In addition to the neural net and rule-based pitch mapping options, I was curious about using a physical modeling approach. Gary Scavone had written several papers describing physical models of woodwind toneholes Scavone (1997, 1999), Scavone and Cook (1998), Scavone and Smith (1997) and Scavone and Van Walstijn (2000) that could be continuously varied from open to closed. Since I was already exploring digital synthesis options as an alternative to the electromechanical oscillator, that suggested another idea for the mapping: rather than trying to get a "pitch" parameter from the array of floating-point values from the key sensor readings, one could instead generate the synthesis directly using a digital physical model of a tube with holes of the right size in the right places. A student and I managed to create a model of a full tube with continuous toneholes for every key on the Birl. Once that was completed, I needed to figure out how to put the virtual toneholes in the right places, and with the right radii. The student wrote a solver

Fig. 12 Dave Schnug and Pedro Eustache try later prototypes of the third Birl

allowing the user to enter a scale in cents. From this user-defined scale, the solver designs a virtual tube with the correct tonehole placement. It worked great, although the tuning was not completely accurate—more work needs to be done on that front. One interesting advantage of this method is that extended techniques like multiphonics arise naturally out of the system, and the results of half-holing will usually be intuitive. One downside is that, unlike the machine learning or rule-based systems, the available pitches are limited by the number of keys, making this approach better for creating instruments similar to those with only a few open holes, such as the shenai, rather than instruments with complex keywork, like the oboe.

Throughout the 2014/2015 school year, I was lucky to have four wind players in PLOrk, the Princeton Laptop Orchestra, which I direct. We were working up a

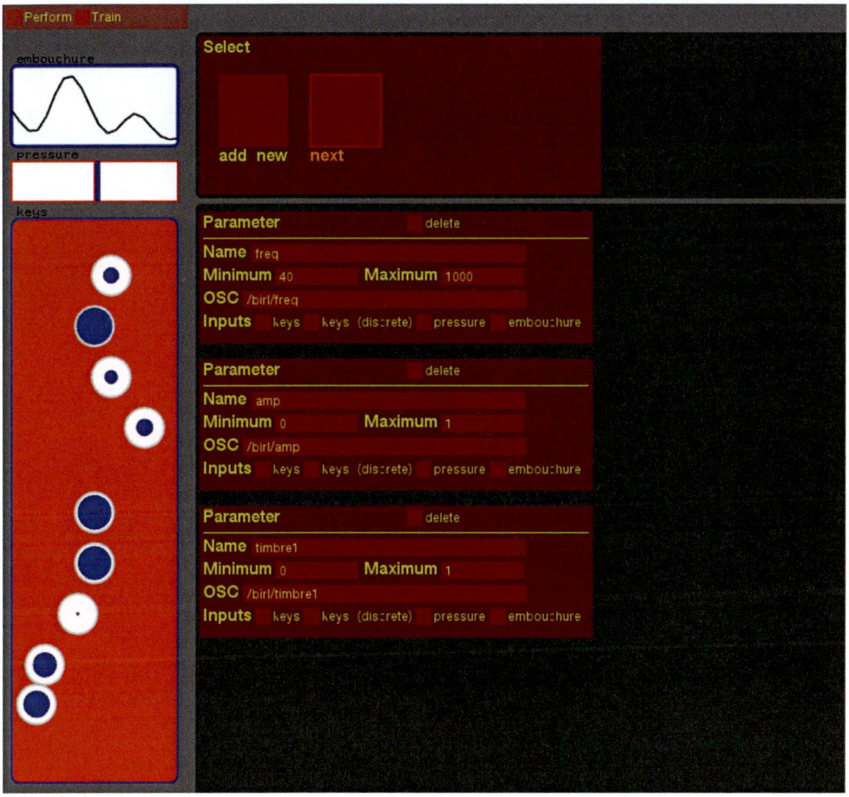

Fig. 13 Neural net training software for the third Birl by Gene Kogan

program of 15th and 16th century music arranged for electronic instruments, so the Birl was a perfect addition to our electronic orchestra. This was a fantastic opportunity to get real-world test data on the problems of the current prototype (Fig. 14).

The PLOrk musicians made several requests. First, they found the finger holes (at the time just holes drilled in the enclosure with standoffs inside the holes) to be too small and hard to locate without looking at their hands. I experimented with inserting thumbscrews into the standoffs to make physical touch plates for the keys, similar to the capacitive keys on an EWI. With that change, some of them noted that it was now hard to avoid accidentally activating keys when you just wanted to rest your finger near them. I added 3D printed plastic guards around the keys, intended to give fingers a place to rest when not pressing the keys.

We found that the sensitivity curve on the keys wasn't ideal, as there was a jump in the value when the sensors went from touched to not touched. I experimented with applying clear vinyl stickers to the keys to eliminate this threshold, then eventually found a durable solution in epoxy spray. Some of the players didn't like

Fig. 14 PLOrk performs with the Birl

the custom mouthpiece, and instead attached a clarinet mouthpiece, which worked fine since the embouchure sensors were not yet functional in these prototypes anyway, but doing so did bring the instrument's profile back into soprano saxophone territory. I had made some versions of the circuit board with capacitive sensors for the thumbs as well as the fingers, but the PLOrk members greatly preferred mechanical switches for the thumb buttons. The PLOrk members also complained of the difficulty in avoiding glitches in the pitch output of the instrument when changing octaves. This problem remained from the first Birl, and it remains unsolved as yet. We performed several pieces using the Birls at the final concert of the year, which was themed around electronic arrangements of Medieval and Renaissance works; video is available Plork.

I worked on further developing the embouchure sensors, trying a swept frequency capacitive sensing technique Sato et al. (2012) as a way to get more reliable data from the lip sensors. It worked well, and I designed a new embouchure sensor circuitboard to integrate the new technology. However, I noticed that although the new lip sensors did a decent job of detecting lip position, they did not register the inside of the mouth, which began to reveal itself as an important element of embouchure. I tested several methods of sensing the space inside the mouth, including acoustic sensing (via a microphone) and infrared, and found infrared to be remarkably reliable.

I was now considering the controller as its own independent instrument, not needing either the resonator or the electromechanical tone generator. The new prototype was also designed in such a way that it wasn't out of the question to affordably make it in multiples. I began envisioning the release of a future version of the instrument as a product.

I had released the Manta, a hexagonal grid touch controller, as a product in 2008, but I wasn't as confident of an existing market for the Birl, since electronic wind instruments have a bit of a cheesy reputation, and there are several existing wind controllers on the market. However, the continuous key sensing and general feel of the instrument seemed different enough for a successful product. A friend suggested that I develop a simplified version of the instrument, focused on making it affordable. Thus I started the design of the fourth Birl, the MiniBirl.

6 The Fourth Birl (2015-Present)

In designing the MiniBirl, I decided to eliminate the most expensive parts: embouchure sensing and embedded sound synthesis. I needed to reduce the cost of manufacture, simplify the user's experience of communicating from the instrument to a computer, and improve the portability.

I reduced the cost by stripping the design down to a single 2-layer circuitboard and removing all the complex parts. Instead of a separate brain microcontroller, the key sensor microcontroller would also be the brain. In addition to removing the embouchure sensing and the internal synthesis, I removed the LCD display. I replaced the Ethernet jack with a USB jack. Using USB as the only communication meant that I no longer needed a separate power cable, which made the instrument more elegant. What remained were the continuous keys sensing, the breath pressure sensor, and an indicator LED for the breath pressure. I added two new features that didn't add significant cost: an X/Y touchpad for the right thumb and an accelerometer/gyroscope IC to detect orientation of the instrument. I replaced the combination of standoffs and thumbscrews that I was using as keys with Chicago screws, which integrated several parts into one and lowered the cost of each key (Fig. 15).

Once I had removed the Ethernet jack and the LCD screen, I realized that I could reduce the thickness of the instrument. I simplified the design from the wooden clamshell to a single piece of wood, milled from the back, with a fiberglass panel covering the cavity. With the reduction in circuit complexity, I was also able to shorten the length of the instrument considerably.

After switching to USB I had to decide on a protocol, and I chose USB-MIDI so as to be compatible with most music software. Sending the data as 7-bit packets would be limiting, but pairing bytes to create 14-bit messages made for reasonable resolution. I'm still working on the problem of how to encode the continuous pitch information in a way that standard music software can understand. Continuously varying pitch over a seven-octave range does not easily translate to note-ons, note-offs, and pitch bend data.

The new MiniBirl prototype had its premiere in April 2015, played by Sean Mac Erlaine of the group This is How We Fly. The following year, it was tested by professional wind players Pedro Eustache, Noah Kaplan, and Steve Lehman, who also provided valuable feedback.

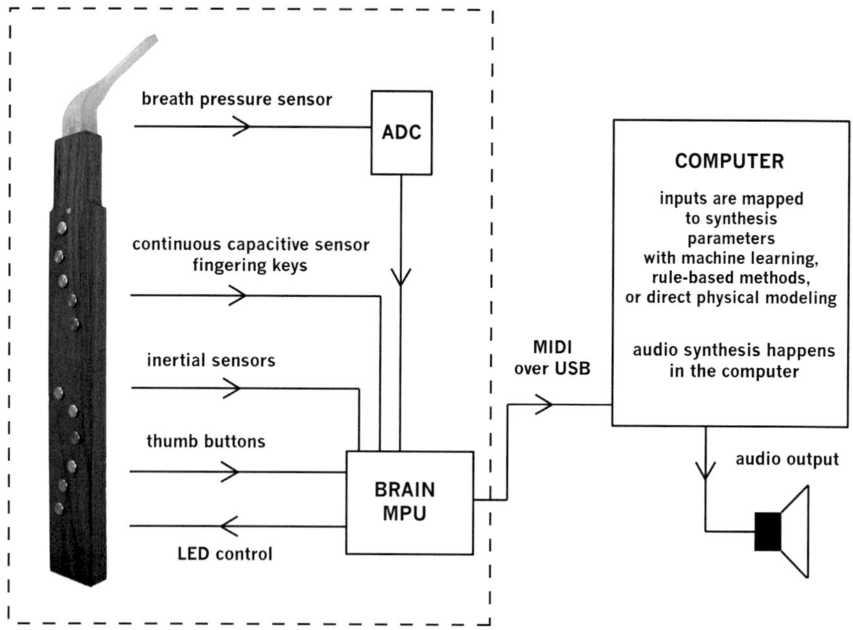

Fig. 15 Diagram of the fourth Birl

Lehman was interesting in that he was almost entirely uninterested in the continuous finger sensing possibilities. His style emphasizes clean, precise, accurate playing—as he told me "I've worked for years to sound like a computer on the saxophone!" To this end, he also found the lack of tactile feedback on the touch sensors to be a significant downside, not balanced by the continuous sensing possibilities. On the other hand, he really liked the X-Y touchpad for the right thumb and the unusual look of the instrument.

Kaplan is a saxophonist and composer who is deeply fascinated with alternate tunings, so he was excited by the possibilities in the software for using machine learning with the Birl that Gene Kogan had built. We spent a half hour training the system to recognize several unconventional fingerings as being either quarter-tone deflections or comma alterations. This was interesting to him, since the existing commercial wind controllers output only "cooked" pitch information, not allowing access to the actual key-press data to reprogram unusual fingering options (Fig. 16).

Eustache came to visit my New Jersey lab from California, where he works as a film soundtrack session musician, with credits on several films, including Pirates of the Caribbean. His feedback was incredibly detailed, since unlike any of my other test users, he has been regularly playing commercially available wind controllers for many years. He had very precise and helpful ideas about how the thumb buttons should be situated, and other physical layout details. He was most excited about the continuous fingering option, since he is a multi-instrumentalist (most of my test users had been either saxophonists or flautists) and he is regularly called upon to

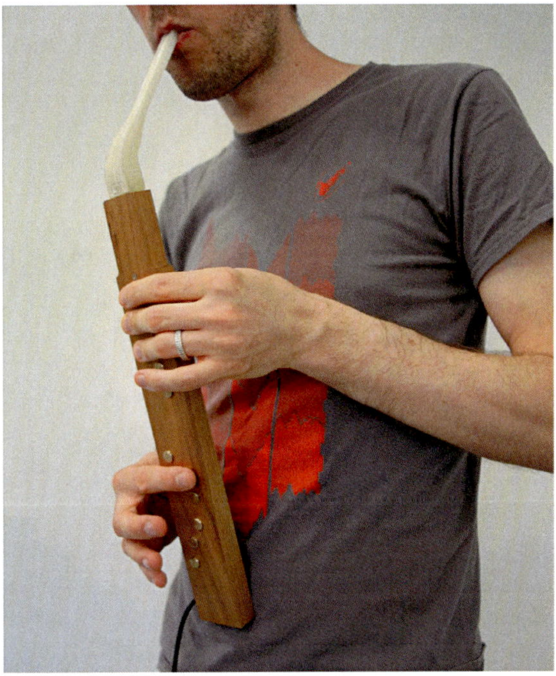

Fig. 16 The author testing out the fourth Birl

play a huge variety of wind instruments from various cultures, many of which use open-holed fingering techniques. He found the open-holed fingering options to function very well, although he wanted the keys closer to the sides of the instrument so that he could roll his fingers off of them as one does on a bansuri, and he thought the relatively sharp corners used in the design were counterproductive for this technique. He was also overjoyed about the small size of the instrument, especially the fact that it could fit in a backpack. He was theoretically very excited about the embouchure sensing, although it wasn't working when he visited, and he loved that the breath sensor was fast enough for fluttertongue effects. I was surprised by his enthusiasm for the visualization tool I had built with Gene Kogan, which allowed the user to view the data from the fingering sensors in real time. He wanted to be able to record and slow down the visualization so that he could analyze the synchronization mistakes made when leaping across octaves, imagining a tool something like Duncan Menzies's P-bROCK training program Menzies (2013).

While refining the design, I started to see the MiniBirl not just as a simpler, cheaper cousin to the Birl, but potentially as the core of the Birl itself. Instead of making two Birls, one larger and one smaller, I could make the MiniBirl modular, so that if one wanted to add internal audio synthesis or embouchure sensing, one could do so by snapping on another piece, perhaps even a wooden resonator body or an electromechanical oscillator, bringing the instrument back to its conceptual roots. The upcoming circuit board iteration takes this idea into account, designing in connectors that can communicate with additional modules.

7 The Birl of the Future! (Present-Beyond)

I've been working on the Birl continuously for eight years. In this wandering adventure, I explored various things the instrument could be, and followed performer needs and my own varied interests where they took me. The future of the project is still very open, and I hope to continue to use it as a platform to follow whatever intriguing paths present themselves.

8 Reflections on the Process

It might seem unusual to call all of these instruments "the Birl" when each has distinct design characteristics. However, in my mind they are one: I consider the Birl to be a gradually changing, continuous project. It has been exciting to undertake a project where the focus of the research is about the process, where continual reinvention flows from experiments I undertake and feedback from performers.

References

Akai. http://www.akaipro.com/product/ewiusb. Accessed May 30, 2016.

Aldridge, H. (1996). Music's most glorious voice: The Hammond organ. *The Journal of American Culture, 19*(3), 1.

AVR. http://www.atmel.com/products/microcontrollers/avr/32-bitavruc3.aspx. Accessed May 30, 2016.

Bang and Olufsen. http://www.bang-olufsen.com/en. Accessed May 30, 2016.

Collins, N. (2006). Handmade electronic music: The art of hardware hacking. Taylor & Francis.

Cypress. http://www.cypress.com/products/32-bit-arm-cortex-m-psoc. Accessed May 30, 2016.

Design Museum Denmark. http://designmuseum.dk/en. Accessed May 30, 2016.

Doty, D. (2002). *The just intonation primer*. San Francisco: Other Music.

Fiebrink, R. (2011) Real-time human interaction with supervised learning algorithms for music composition and performance (Doctoral dissertation from Princeton University).

Hillmer. https://vimeo.com/86461028. Accessed May 30, 2016.

Iglesia. 2009. https://vimeo.com/4611451. Accessed 30 May 2016.

Jensen. https://en.wikipedia.org/wiki/Jacob_Jensen. Accessed May 30, 2016.

Menzies, D., & McPherson, A. (2013). A digital bagpipe chanter system to assist in one-to-one piping tuition. In *Proceedings of the 2013 Stockholm music acoustics conference/sound and music computing conference* (*SMAC/SMC*).

Parker, J., & D'Angelo, S. (2013). A digital model of the Buchla lowpass-gate. In *Proceedings of the 16th international conference on digital audio effects (DAFx-13)* (pp. 278–285).

Plork. https://vimeo.com/album/3382506. Accessed May 30, 2016.

Sato, M., Poupyrev, I., & Harrison, C. (2012). Touché enhancing touch interaction on humans, screens, liquids, and everyday objects. In *Proceedings of the SIGCHI conference on human factors in computing systems* (pp. 483–492).

Scavone, G. (1997). An acoustic analysis of single-reed woodwind instruments with an emphasis on design and performance issues and digital waveguide modeling techniques (Doctoral dissertation from Stanford University).

Scavone, G. (1999). Modeling wind instrument sound radiation using digital waveguides. In *Proceedings of the ICMC 1999* (pp. 355–358).

Scavone, G., & Cook, P. (1998). Real-time computer modeling of woodwind instruments. In *Proceedings of the international symposium on musical acoustics (ISMA-98)* (pp. 197–202).

Scavone, G., & Smith J. O. (1997). Digital waveguide modeling of woodwind toneholes. In *Proceedings of the 1997 international computer music conference.*

Scavone, G., & Van Walstijn, M. (2000). The wave digital tonehole model. In *Proceedings of international computer music conference (ICMC-2000)* (pp. 465–468).

Schnug. https://vimeo.com/86167177. Accessed May 30, 2016.

Snyder, J. (2011). Exploration of an adaptable just intonation system (Doctoral dissertation from Columbia University). http://www.scattershot.org/full-dissertation.pdf

Snyder, J., & Ryan, D. (2014). The birl: An electronic wind instrument based on an artificial neural network parameter mapping structure. In *Proceedings of the new interfaces for musical expression (NIME2014)* (pp. 585–588). http://www.nime.org/proceedings/2014/nime2014_540.pdf

Wetink. http://www.wetink.org/. Accessed May 30, 2016.

Yamaha. http://usa.yamaha.com/products/music-production/midi-controllers/wx5/. Accessed May 30, 2016.

Case Study: The Endangered Guitar

Hans Tammen

Abstract The author describes his 15-year development of the hybrid interactive instrument "Endangered Guitar", and how it grew out of an already decade long practice of sonic performances on guitar, followed his aesthetic interests through hundreds of concerts, and influenced these interests in turn. The Endangered Guitar is an instrument is made to facilitate live sound processing. The software "listens" to the guitar input, to then determine the parameters of the electronic processing of the same sounds, responding in a flexible way. The instrument is interactive, in that it does not react in a fully predictable way to the input of the performer. The author makes a case that in order to truly improvise with electronics one has to program "uncertainties" into the machine. He uses weighted random functions, feedback strategies, and the fuzzy behavior of pitch tracking devices when presented with overtone-rich sounds, which the performer draws from the guitar with a variety of tools.

The "Endangered Guitar" is a hybrid interactive instrument meant to facilitate live sound processing. The software "listens" to the guitar input, to then determine the parameters of the electronic processing of the same sounds, responding in a flexible way. Since its inception 15 years ago it has been presented in hundreds of concerts; in 23 different countries on 4 continents; in solo to large ensemble settings; through stereo and multichannel sound systems including Wavefield Synthesis; in collaborative projects with dance, visuals, and theater; and across different musical styles. It is well developed, so it is time to describe the history of this instrument, and to look at the paths taken and the paths abandoned. The latter are equally significant, because some of them represent musical approaches that, however important, nevertheless ran their course at some point. One needs to remember that there was a musical concept behind every change in the instrument, and as I changed as a composer, my ideas changed as well. There is no difference between the Endangered Guitar or my works for chamber ensemble: some techniques stay at the core of my work, some others stay for a few years, others do not last for more than one piece (Fig. 1).

H. Tammen (✉)
New York, USA
e-mail: hta@mm.st

T. Bovermann et al. (eds.), *Musical Instruments in the 21st Century*,
DOI 10.1007/978-981-10-2951-6_14

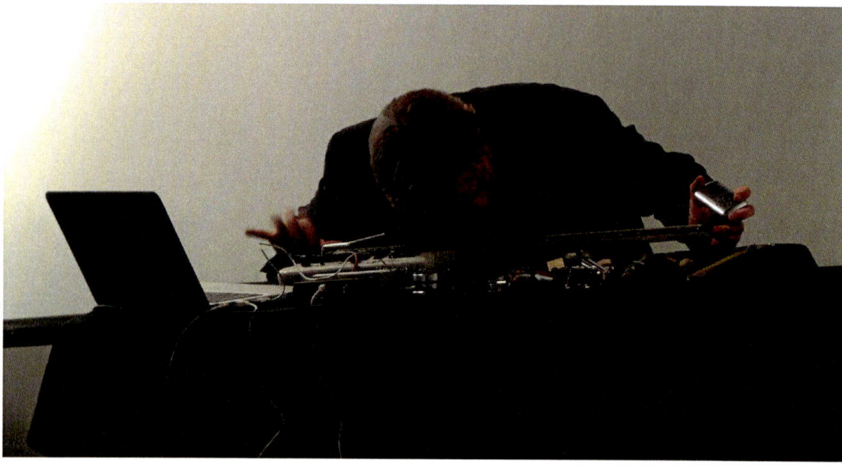

Fig. 1 Endangered Guitar performance photo by Matthew M. Garrison (from 2012)

Today I call my setup the Endangered Guitar when the following tangible components are present: (a) the actual guitar, of course; (b) the specific tools and materials that facilitate sonic progression; (c) a computer. However, there is more to the computer than mimicking a guitar effects board. Instead the software is part of an interactive instrument. 30 years ago, Joel Chadabe described the consequences as follows:

> An interactive composing system operates as an intelligent instrument—intelligent in the sense that it responds to a performer in a complex, not entirely predictable way adding information to what a performer specifies and providing cues to the performer for further actions. The performer, in other words, shares control of the music with information that is automatically generated by the computer, and that information contains unpredictable elements to which the performer reacts while performing. The computer responds to the performer and the performer reacts to the computer, and the music takes its form through that mutually influential, interactive relationship. The primary goal of interactive composing is to place a performer in an unusually challenging performing environment. (Chadabe 1984)

These are the concepts that constitute the Endangered Guitar:

- It is a **hybrid** instrument, in that it needs both parts, the guitar and the software, to function and sound the way it is intended.
- It is **interactive**, in that the software responds in a flexible way to the guitar input, and both performer and software contribute to the music.
- It is **unpredictable** in its responses, because it designed for an improviser.
- It is **process**, in that it is not conceived and built at a single point in history, but developed over 15 years, with no end in sight.
- It is **modular**, because programming in a modular way allows for rapid changes in the structure.

What follows is a history and description of its components as they relate to the concepts mentioned above.

1 The Instrument as Process

The Endangered Guitar is made for me as an improviser operating in a variety of musical situations.[1] I am interested in a variety of artistic expressions, collaborations and presentation modes. One cannot foresee what a specific musical situation requires the software to provide, so I cannot foresee what I had to program into the machine.

The Endangered Guitar is a continuing process, and even this account is just today's snapshot. It grew naturally out of a 25 year long musical development, and followed (and in turn also influenced) my musical paths for the following 15 years. It was not even planned to be "hybrid" or "interactive", in fact at the time I started working on it, I had no concept of these terms in relationship to my music. So how did it come to that?[2]

2 Background

I usually explain that I found my own voice in the 1990. I started playing guitar in 1972, performing rock music and studied classical guitar a little. In the 1980s I emulated a wide range of jazz styles. Then my ears opened up to sonic progression.

In the 1990s I moved towards improvising in the "British Improvisers" style,[3] which provided the right framework for my sonic interests. It requires quick reaction in group settings, transparency of sounds, and fast shifts in musical expression.[4] My guitar approach became very much similar to Fred Frith's: using sticks, stones, motors, woods, mallets, Ebows, measuring tapes, vacuum cleaners, violin bows, metals, screws, springs and other tools I could lay my hands on to coax sounds from the instrument. Not a single hardware store could escape my search for the perfect screw, the perfect spring. During this period the guitar was lying flat on my lap, and was treated as an elaborate string board, often equipped with multiple pickups. Adding a small mixer allowed for tricky routing of signals through the occasional guitar pedal (Fig. 2).

The sonic range I achieved can be best heard on my 1998 solo CD "Endangered Guitar", released in a metal box on the NurNichtNur label.[5] That's also when I

---

[1] I do not pretend that the instrument is designed for ALL cases, because there is music I am not interested in.

[2] I can't resist bringing up Joel Chadabe's observation: "I offer my nontechnical perception that good things often happen—in work, in romance, and in other aspects of life—as the result of a successful interaction during opportunities presented as if by chance" (Chadabe 1984).

[3] I listened to this music as soon as Music Improvisation Company's LP came out on ECM in 1970 (ECM Records 1970). But I started to understand its implications when I read Derek Bailey's account in the 1987 German translation of the first edition of his book "Improvisation: Its Nature And Practice In Music" (Bailey 1987).

[4] I can illustrate those musical interactions best with our quartet recording from 2000, Kärpf (Tammen 2000).

[5] CD Endangered Guitar (Tammen 1998).

Fig. 2 Pre-computer performance photo from 1999

coined the term—I just needed a title for the CD. I felt it was a break with the traditional notion of the guitar, but not so far as to call it “extinct”, the guitar being just “endangered”.[6]

3 Adding Max/MSP

Around the same time home computers became fast enough to allow digital audio, and laptops became affordable, so they could be carried around to gigs. I had worked already with notation and other music software on the Atari when I got one in 1986, so I was no stranger to using computers for music. When the prospect of bringing a laptop on stage instead of all the cables, pedals and mixer, the idea of moving everything onto the computer came to me quite naturally.

The software I used was Max/MSP. That decision was a natural one, too: my soon-to-be wife, Dafna Naphtali,[7] had already worked with Max for about a decade, controlling an Eventide H3000 to do live sound processing on her voice and

[6]In light of the guitarists who came forward since Rowe and Frith were using those techniques (especially the explosion in the 1990s), it is safe to say that playing the guitar with knitting needles or alligator clips is just another traditional guitar playing technique.

[7]Website Dafna Naphtali (2016).

acoustic instruments. She suggested to not look into ready-made applications, but to write my own. Since I had written business applications[8] some 10 years earlier, writing my own music application wasn't such a strange idea, and so she gave me a head start with node-based programming. It took about 2 years to get it comfortably working: I started programming Max in 1999, did the first performances with the software in 2000, and performed primarily with the guitar/software instrument by 2001.[9]

Originally the goal was just to transfer the physical setup into the digital realm. I have worked for 14 years at Harvestworks Digital Media Arts Center[10] in New York, overseeing projects of clients, students and Artists In Residence, and the wish of simplifying one's setup by moving it onto the computer was often the first goal.

While this was the plan, it was out the window as soon as I started experimenting. I did own LiSa already, Michel Waisvisz's Live Sampling Instrument[11] distributed by STEIM. I had some fun with it, but only when I connected my MIDI guitar to Max, which in turn controlled LiSa, it all of a sudden made sense. The basic idea of LiSa—cutting up audio, playing the material at different speeds and in different directions[12]—became the core of the software written in Max—and that had nothing to do with the original intent of using a computer. As for LiSa, after a few months I figured out how to move its capabilities into MSP.

I moved to New York in January 2000, and over the years my musical interests shifted towards longer musical developments, wider dynamic ranges, minimalistic concepts, and rhythmically a shift back from "pulse" to "groove"—all of it had its impact on the Endangered Guitar.

Next I will describe the major elements of the Endangered Guitar. Screenshots I took from various interfaces over the years suggest that the core approach to the software has been established by mid-2005. Meanwhile the guitar was usually lying flat on the table, Keith-Rowe-style. This allowed me to focus entirely on sonic progressions by completely obliterating traditional guitar playing techniques,[13] using both hands independently (including better access to the occasional sensor), or operating the laptop keyboard with one hand while simultaneously using the other on the guitar.

---

[8]Ask me about database applications for undertakers, orthopedic shoemakers and for those who sell colostomy bags!

[9]See my mini-CD release on NurNichtNur, recorded June 2001 during a tour in France: "Endangered Guitar Processing" (Tammen 2001).

[10]Harvestworks Website (2016).

[11]It is strange that I can't find a proper link to this groundbreaking software. STEIM has a successor to LiSA, and you may get some information about Waisfisz's approach from their RoSa page (STEIM 2016).

[12]It is not a coincidence that it sounded like Teo Macero's application of an echoplex to Sonny Sharrock's solo on Miles Davis' Jack Johnson album. I always wanted to sound that way, and Sonny Sharrock and the music of Miles Davis' Bitches Brew period belong to my earliest Jazz influences.

[13]The occasional tapping not withstanding.

I have to stress that the Endangered Guitar is still an instrument not designed to work with melody and harmony, instead to focus on sound and timbre, rhythm and dynamics. Pitch does concern me, since every sound has pitch (or numerous pitches). I do play with this, of course, but the main purpose is to organize sound in time.

4 Core: Sample Manipulation

For an overview of the signal routing and the components of the software see Fig. 3.

Audio is recorded into "revolving buffers" that are constantly filled, means recording starts again from the beginning when the end is reached. Of course, the buffer can be played in different directions and speeds, but pressing the TAB key syncs up both recording and playing "heads" in a way that one perceives it as the audio passing through. In fact, I had implemented a bypass routine for a while, but got rid of it after a few years when I realized I didn't need it.

Originally I used MIDI guitars to control the software, but they were abandoned by 2004. I figured out how to work with Miller Puckette's fiddle ~ object, allowing for direct pitch and velocity analysis of the incoming sound. This freed me from the need for specific guitars, pickups, and MIDI interfaces. Plus, advantages of MIDI guitars such as providing information about string bending did not yield significantly different results on a musical level.

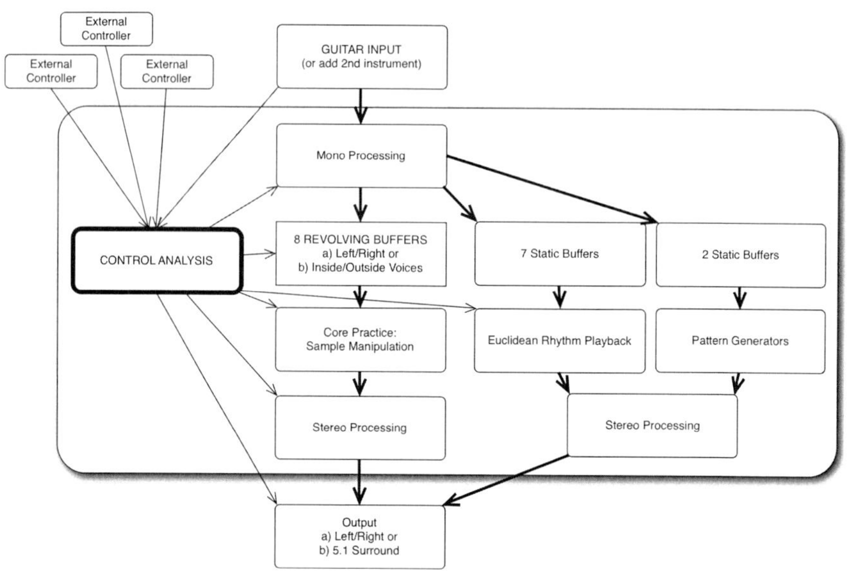

Fig. 3 Signal routing as of 2016

I use pitch and velocity information from fiddle~, e.g. to affect the speed of the "playhead". Higher notes than C (on the B-string) result in increasing speed in similar intervals, lower notes decrease speed. With arrow keys the "spread" can be increased from semitones (C# plays a semitone higher) up to a double octave (C# plays the buffer two octaves higher). To work with microtonality, the "spread" can be used to dividing the octave from 12 into 24 equal parts. Specific settings allow for working directly with granular synthesis approaches.

I use often a buffer length of 2 seconds, but it can be in a range from 1 ms to 15 s. When speed is changed, audience and other players often notice a considerable delay between playing and the audible result. What setting I choose depends on how fast I need to interact with others: if quicker interaction is required I go to shorter buffer lengths, but in solo situations I sometimes use the entire range.

In the very beginning I also experimented with other ways to use data from the analysis, such as moving average. However, if I didn't hear an immediate change as the result of my playing, I felt I lost control. I will later come back to this, because eventually "losing control" I consider today one of the main components of this hybrid instrument.

5 Audio Inputs

Originally the input was a single guitar sound coming in on the left channel. The right channel has subsequently been used for a second pickup mounted on the guitar's headstock, built-in piezos, audio from the second neck of a double-neck guitar, additional piezo strips on the table, or electronics such as Rob Hordijk's Blippoo Box. Eventually this lead to input from other live players (sound poets, hyperpiano, violin, percussion, string quartet, etc.[14]).

Each channel's analysis and sound can be turned off, to the extent that e.g. the guitar is just the controller for the violin processing, or the other way around.

The sound is then fed into a row of processes. As of this writing, I use bit and sample rate reduction, freeze reverb, ring modulation, plus an assortment of VST plugins.[15] As with many routines over the years, some processes ran their course and have been abandoned after a while, such as convolving different audio streams.

This section is "mirrored" after the sample manipulation unit, so I can apply processes to the input, the manipulated audio, or both.

---

[14]"Hyperpiano" is Denman Maroney's term for his approach to the piano (Maroney 2016). A few of my live sound processing projects can be explored on my website (Tammen 2014).

[15]I am fond of GRM Tools—I may have been able to replace some of those with my own programming, however, I am not sure it would ever sound that good! (Institut National Audiovisuel 2016).

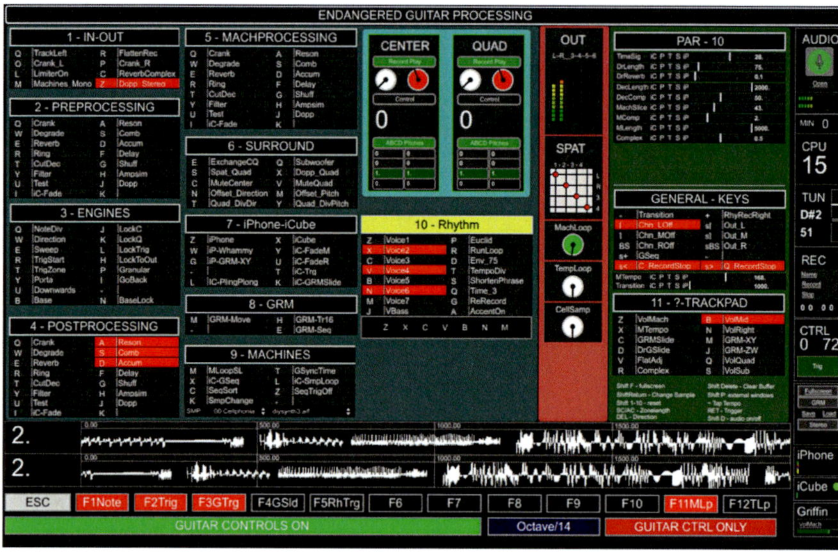

Fig. 4 Endangered Guitar software interface as of 2015

6 Modularity

The "mirroring" is easy to implement because Max/MSP offers many options to program in a modular way—each instance of the effects group can be seen as being recreated from a template.[16] Those files, once defined, can be used numerous times in the software, and this makes for a very modular system that can be changed rapidly. The Endangered Guitar software is constructed from over 200 files, nested within each other up to seven levels deep. What you see in the heavily color-coded interface is only the "container" patch.[17] This may sound complicated, but the hours spent on that kind of preparation have saved me weeks of hunting down multiple copies of the same routine (Fig. 4).

Where this system really shines is when I introduce new approaches. Once the main routine is written, it just needs a few arguments. It will be automatically included in the top interface, hooked up to the various levels of the reset system, and having its parameters connected to all the modes of control (values from audio analysis, external sensors and controllers, etc.). Hooking up an external sensor is easy as well, besides being included in the system as explained, its output can be routed to all parameters available by setting arguments.

As for controllable parameters a cursory check reveals that the system is currently set up to control approximately 100 parameters across the entire system. Any

---

[16]In Max/MSP parlance, I use bpatchers, poly~ objects and abstractions.

[17]The interface is entirely done with bpatchers—Max's presentation mode wasn't out yet.

incoming control value could be routed to any of these parameters, but of course nobody could actually control that amount of data. For the sake of modularity and standardization (and because I do not know what I may need tomorrow) these control channels are theoretically available. However, many of them I haven't used in years, some I certainly have forgotten about. In fact I barely use 10 % during performances, and what helps is that the interface only shows parameters that are actually in use.

This approach emerged from the notion of the "instrument as process"—in fact, there have been times in my life where I was making changes and adjustments every single day, even after performances late in the hotel room.

7 Multichannel Sound

The sample manipulation core is happening simultaneously on 8 buffers, to facilitate multichannel sound. After an initial residency at a multichannel sound gallery in 2002 I concentrated on 5.1 surround sound. My concept places the center-front speaker elsewhere in the room, but keeps the remaining quad system surrounding the audience. By using slightly different processing parameters on the quad, I create an immersive environment for the listener. That way I can set a mono (center) voice against the surrounding (quad) voice. Since both are processed differently, I have two distinct sonorities on the mono and the quad system that I play against each other (Fig. 5).

Fig. 5 Basic multichannel sound setup, established in 2006: outside versus center

I have used this approach in very different setups, with the single speaker moved into the geographical center, or to the back of the audience, into an adjacent room or hanging from the ceiling.[18]

8 Euclidean Rhythms

I was adding a "drum computer" setting from about 2005 on. There were various experiments over the years, even using drum samples, or convolving drum sounds with the actual guitar input. Eventually I settled on the Euclidean Rhythm concept as outlined by Godfried Toussaint, with sounds recorded on the fly during performance (Toussaint 2005). I like Euclidean Rhythms not because generating rhythms out of two integers should be appealing for computer programmers, but rather because my works tend to use ostinati, with phase techniques superimposing different odd meter rhythms. Toussaint's concept just yielded the best results. I create up to 7 different of these rhythms of different lengths, distributed in stereo or quad space. Sounds we hear are taken from various points of the signal chain, so the rhythms can be seen as another way of live sound processing. Materials are recorded into static buffers, so they only change when I need them to. Euclidean Rhythms are also used for two additional pattern generators, that I predominantly use in multichannel setups to create immersive environments.

These pattern generators are followed by another instance of the effects processing section, as described earlier, to process them the same way.

9 Sound Analysis as Control Source

The main tool for the analysis of incoming sound is still Miller Puckette's fiddle~ object. There may be better options for pitch tracking these days, but I have a decade-long experience "playing fiddle~". I have a good (albeit often sub/unconscious) understanding of the way it behaves, such as the chaotic data it provides when I use my steel pot scrubber. I know when it reports a new "event" depending on how hard I hit the strings, or which mallet I use. Replacing fiddle~ with something else feels I'd have to learn to play my instrument all over again (Fig. 6).

I have used FFT analysis at some point, to draw information from the different overtones, but this had been abandoned at some point since it didn't make much difference musically. Currently I only use pitch and velocity information, and "event" (which depends on how long the volume has to go down for the object to report a new action).

---

[18]A list of approaches can be found on my website (Tammen 2012).

Fig. 6 Tools, 2-channel guitar and computer setup (photo from 2007)

10 External Controller Input

Over the years I have experimented with numerous external controllers as well, some of them disappeared quickly, others were in use for years. The two most successful ones were an infrared proximity sensor and the iPhone's accelerometer. As of this writing I occasionally work with the Leap Motion controller.

The use of a proximity sensor had an interesting origin. Since I move body and hands in sync with the music, I often got asked what parameters I would control with my hand motions. But there weren't any sensors, it was just the music that moved me. However, it gave me the idea to use proximity sensors on the instrument. Alas, on the first test it became clear it wouldn't work - music moves my hands, not the other way around. What eventually made sense musically (and I have played with that setting for about 5 years) was to situate one sensor in front of me, and to move my head in and out of the infrared beam. I used it to turn specific routines on and off, which allowed me to create a third action when both my hands were busy working on the strings.

Using the iPhone accelerometer does come quite natural to a guitarist. I held it in my left hand, and a metal phone case allowed me to utilize it in similar ways to a pedal steel bar. As with other controllers, the data was routed to affect a variety of parameters. I have used this controller for about 3 years.[19]

11 Unpredictability

One of the main components of the software is to deliberately program unpredictable or "fuzzy" elements into the software. Before I was even working on the software I was aware that my improvisations became predictable if nothing surprising happened. I simply play better if I have to struggle with unforeseen situations, and when something happens that keeps me on the edge of my seat.

A "source of uncertainty"[20] can be just the way the room acoustics respond to specific frequencies, in that material played yesterday elsewhere doesn't sound good today. It could even be strings or sensors breaking, the machine crashing,[21] accidentally pressing the wrong key, etc. To facilitate some kind of change I was already limiting the tools I brought to the gig to whatever fit in my backpack, so sometimes it was a little arbitrary what I had available at the concert.

This is also the reason why there are no presets in my software. While it is easy in Max/MSP to implement a preset structure, I have thrown it out quickly. I noticed I used presets in situations where I wasn't sure what to do next musically—and jumping to some preset was an easy way out. I had already experienced the same situation 10 years earlier when I used loop pedals: if I didn't know how to get out of the current situation, I started a loop. The results, though, were never satisfactory, eventually the preset structure had to go (as well as the loop pedals a decade earlier).

That does not mean that what I play is "new" in the sense that you have never heard it before. Sometimes this is the case, but to characterize my performances I would like to offer Earle Brown's understanding of the term "open form composition", in that the parts of the composition are arranged differently with every new performance.[22] While the performances are not planned in a way that I determine how to start and end, and what to do in between, there is an enormous amount of practicing and planning behind each individual part when it is presented in

[19]You can see an example in the first few seconds of this concert video from 2010 (Tammen 2010).

[20]I came to like Don Buchla's use of this term to name a synthesizer module that produces unpredictable control values.

[21]It is baffling (but also attests to the stability of Max/MSP) that in 15 years the software crashed just four times live on stage.

[22]As a method in itself it does not have to be of value—a musician playing the same licks over and over does also, strictly speaking, "open form composition".

performance. The sounds I coax from the guitar draw on 25 years of experience, the use of the software rests on 15 years of practicing.[23] The improvisational element concerns the choice of the actual sonic material (especially with regards to the sound of the room and the sound system), order and timing of parts, how to transition from one part to the other, when to make sharp cuts, and dynamics and other elements of the form—and as well dealing with the occasional technical and musical failure.

Improvising is here merely the technique that helps us dealing with the unforeseen. Bruce Ellis Benson tries to depart from the usual binary scheme of composition/improvisation (as it is common in European discourse), by suggesting that improvisation is when one ventures into the unknown, something both composers and performers do[24] (Benson 2003). I do not want to put too much weight on this, because in 2016 it should be obvious that these terms can be used to describe cultural notions, but they are not useful to describe what we actually hear. However, I do feel the notion of improvising as a technique to deal with unpredictability does accurately describe one of the main ingredients of my performances.

But how does one achieve unpredictability when creating software? By deliberately programming it into the machine. First I use weighted random functions. The machine does not react the same way to input (pitch, sensors, etc.) every time. It responds within a range of values, depending on the desired musical outcome. It may go without saying that completely random (as in equal weight for each value) does not yield musically interesting results. Some parameters I have tweaked for years to produce the right amount of fuzziness without becoming arbitrary.

Recently I have made increasingly use of feedback systems. Using the analysis of the audio output instead of the input, plus feeding back the audio into the input provided good results. It is quite unpredictable which parts of the spectrum are emphasized and which not, and which provide for good material for the analysis. Here we can additionally benefit from the shortcomings of pitch trackers such as fiddle~, in that feeding material without a fundamental will results in erratic behavior.

From 2005 on I have increasingly made use of these strategies. I relinquish control to the machine instead of seeing it as an extension of my guitar. I play *with* the machine, not *the* machine. The amount of unpredictability varies with the project, it has less autonomy in ensemble situations, but in solo performances the Endangered Guitar tends to be this crazy thing I struggle with.

---

[23]"Practicing" is a good point: I am always amazed by people whipping up some patches up the night before the gig, and then getting lost on stage. Exceptions not withstanding, one does hear lack of preparation.

[24]… thus replacing one binary scheme with the next.

12 Outlook

What is the future of this hybrid interactive instrument? One development in recent years is that I branched out by using other sound sources, in that I do live sound processing of other instrumentalists (from single performers to whole ensembles[25]). Sometimes I process a tabla machine[26] or a room microphone. Technically it is not the "Endangered Guitar", but it is still "hybrid" (in that it is comprised of a software plus another instrument), and "interactive" (in that data from the audio input is interpreted in various and unpredictable ways to effect processing parameters).

I am currently extending the software by including an external data set, around 65,000 lines from my own DNA analysis. As before, I hope to struggle with the software's unpredictability, but this time I would use data from my DNA to influence the "fuzziness" of the machine. There is still some tweaking needed, the piece is aptly called "Conflict Of Interest". While working with my own DNA data makes certainly sense on a conceptual level, it allowed me to consider working in the future with other data streams as well, such as live input from the internet.

Lastly, the biggest challenge would be to develop the instrument into a true improvising machine. In its current stage the Endangered Guitar does not qualify as such, because to be an improvising machine it would need memory—making decisions based not only on the current input, but also based on numerous previous performances. If it is improvising, it needs to react to unforeseen situations, so it would need extensive pattern recognition and machine learning algorithms. This would be an undertaking that requires concentrating on nothing else for a few years.

13 Listening Example

An Endangered Guitar retrospective was released on the Danish label CLANG[27] in June 2016, containing excerpts from performances between 2004 and 2011. Called "Deus Ex Machina—Endangered Guitar Live",[28] the music is grouped into various sonic themes, with short pieces acting as interludes in between. All pieces are from the time when the Endangered Guitar became an interactive instrument. Another reason for choosing the works on this release is that these were new sonic universes —by then I had left behind the prepared guitar/noise formula that dominated my playing throughout the 90s, figured out which live sound processing approaches worked best, and incorporated Euclidean Rhythms.

---

[25]Tammen (2014).

[26]Tammen (2015).

[27]Clang (2016).

[28]Tammen (2016).

References

Bailey, D. (1987). *Musikalische improvisation: Kunst ohne Werk*. Hofheim am Taunus.

Benson, B. E. (2003). *The improvisation of musical dialogue: A phenomenology of music*. Cambridge.

Chadabe, J. (1984). Interactive composing. *Computer Music Journal, VIII*(1), 23.

Clang. (2016). http://clang.cl. Accessed August 8, 2016.

ECM Records. (1970). The Music Improvisation Company. http://v.gd/kJiVM7. Accessed August 8, 2016.

Harvestworks Digital Media Art Center. (2016). http://www.harvestworks.org. Accessed August 8, 2016.

Institut National Audiovisuel. (2016). GRM-Tools, http://www.inagrm.com/grmtools. Accessed August 8, 2016.

Maroney, D. (2016). Denman Maroney http://www.denmanmaroney.com. Accessed August 8, 2016.

Naphtali, D. (2016). Dafna Naphtali. http://www.dafna.info. Accessed August 8, 2016.

Tammen, H. (1998). CD Endangered Guitar. http://tammen.org/cd-hans-tammen-endangered-guitar/. Accessed August 8, 2016.

Tammen, H. (2000). CD Kärpf. http://tammen.org/cd-karpf/. Accessed August 8, 2016.

Tammen, H. (2001). CD Endangered Guitar Processing. http://tammen.org/endangered-guitar-processing/. Accessed August 8, 2016.

Tammen, H. (2010). Endangered Guitar at HERE Arts Center, YouTube video 2010. https://youtu.be/vFXIfgljMjw. Accessed August 8, 2016.

Tammen, H. (2012). Multichannel Endangered Guitar projects. http://tammen.org/multichannel-endangered-guitar-works/. Accessed August 8, 2016.

Tammen, H. (2014). Live sound processing projects. http://tammen.org/live-sound-processing/. Accessed August 8, 2016.

Tammen, H. (2015). Percussores for tabla machine and live sound processing. http://tammen.org/percussores-for-tabla-machine-and-live-sound-processing/. Accessed August 8, 2016.

Tammen, H. (2016). CD Endangered Guitar Live—Deus Ex Machina. http://tammen.org/cd-endangered-guitar-live-deus-ex-machina/. Accessed August 8, 2016.

Toussaint, G. T. (2005). The Euclidean algorithm generates traditional musical rhythms. In (Extended version of the paper that appeared in) *Proceedings of BRIDGES: Mathematical Connections in Art, Music and Science* (47–56). http://cgm.cs.mcgill.ca/~godfried/publications/banff-extended.pdf. Accessed March 8, 2016.

STEIM. (2016). RoSa Product Page, http://steim.org/product/rosa/. Accessed August 8, 2016.

Part III
Compose Play Perform

Interplay Between Composition, Instrument Design and Performance

M.A.J. Baalman

Abstract With electronics and code as an essential part of new musical instruments, the boundaries between composition, instrument design and performance are blurring. With code that can be changed and compiled on the fly, the design of an instrument becomes a fluid process, which can even be a performance in itself. Starting with an example from my own artistic practice, I explore the concepts of composition, instrument and performance and what role the design of electronics and software plays in these. What influences design decisions when developing instrument? How does the materiality of electronics and code inform these decisions? How do the knowledge and skills of the makers play their role in this?

1 Introduction

With electronic and digital media as key elements in musical instrument design, the boundaries between composition, instrument design and performance are blurring. Practioners do not need to make a clear distinction while developing their projects. However, when they present their work in the context of the established cultural scene, they need to deal with the common division of roles. This paper sheds light on the work process of such artists and how their practice plays in the field between those roles. The questions raised are motivated by my own practice as an artist.

In the following section I will describe my piece "Wezen—Gewording" to motivate the questions I raise in Sect. 3. In Sect. 4 I will look at the discourse about composition, instrument and performing in the context of electronic and computer music. In Sect. 5 I will look at live algorithms or interactive music systems, music software and tools and livecoding. I will conclude with a discussion reflecting back on the questions I pose in the beginning.

M.A.J. Baalman (✉)
Nescivi, Amsterdam, The Netherlands
e-mail: marije@nescivi.nl

T. Bovermann et al. (eds.), *Musical Instruments in the 21st Century*,
DOI 10.1007/978-981-10-2951-6_15

2 Becoming

Since August 2013, I have been developing the piece "Wezen—Gewording".[1] Within this piece, I am looking for a connection between the sonic gesture and the theatrical gesture. I explore this connection by exposing my method of creating the instrument during the performance, while at the same time playing it Fig. 1.

Over the course of two-and-a-half years, I performed the piece for 18 times and, both before and during each show, I developed it further. While performing, I go back and forth between body movements and livecoding the mapping between sensor data and its effect on the sound.

I wear gestural controllers at my wrists and hands with a number of buttons in each palm. The controls are designed such that they keep my fingers and hands free enough for me to type on my laptop's keyboard. These controllers started out as an assembly of elastic bands, perfboards, buttons and a Sense/Stage MiniBee (Baalman 2016). Over the course of 1.5 years of performing the piece, the design settled in that I figured out how many buttons I needed, where to place them, and how to mount them onto a wearable, an open glove, that now contains the buttons, the microcontroller and the battery (see Fig. 2).

The prospect of a performance acted for me as a kind of *showtime-deadline*, before which I did a series of rehearsals and spent time developing the code and hardware. The performance itself had a duration of 10–15 min.

Code development meant to create a framework, within which I can quickly create connections between gestures and sound during performance. Both rehearsals and performances informed me whether my design decisions worked, whether I needed to revert them, or in which direction I had to to further develop them. The mapping and sound synthesis algorithms that I wrote while playing were sorted out during the design and development time, based on whether I found they gave interesting results during the performance. Thus I assembled a collection of code snippets for use in future performances.

Furthermore, I wrote and rewrote mapping architecture, allowing for shorter ways to implement the same kind of behaviour during performance, abstracting concepts I worked out in previous performances. Another part of the work was reverting these abstractions, as I discovered during playing that they did in fact impede the improvisation, rather than support it.

Over time, I added new and substantial elements like interactive lights or additional sensors on various parts of the body to the performance. At the same time, I found it necessary to add less direct interaction layers, which resulted in the inclusion of a gesture recognition algorithm, triggering subtle changes in the background soundscape. Sometimes these additions would become a constant part of the system, at other times they disappeared after one or more performances.

My role shifted throughout the whole process, from being the actor and mover to being programmer, from electrical engineer to garment maker. This means that,

[1]"Wezen—Gewording" can roughly be translated to *Being—Becoming*.

Fig. 1 The performance "Wezen—Gewording" in May 2015 at the Kunst Achter Dijken festival in Pingjum, The Netherlands

since I embody movements and listen to many variations of their sonification, I am aware of the kind of data that particular movements will create. I can incorporate this knowledge into programming a sound and its mapping to the sensors. At the same time, I create new scenarios by code in which I can explore my movements and their effect on the sound. The sound informs my movement, which in turn informs me in my subsequent mapping decisions.

While rehearsing for a particular performance I develop a rough structure. I come up with different directions for the piece to develop, and if these were hard to reach before, I try to make them accessible by adapting the code framework. During the livecoding within the performance, I am aware of the framework I built and in which I make changes. I recall the limits of easily modified code and what possibilities it offers to the me as the mover. Thus, there is a tight connection between my embodied knowledge of moving and listening, and my engineering knowledge of how the technology enabling the performance works.

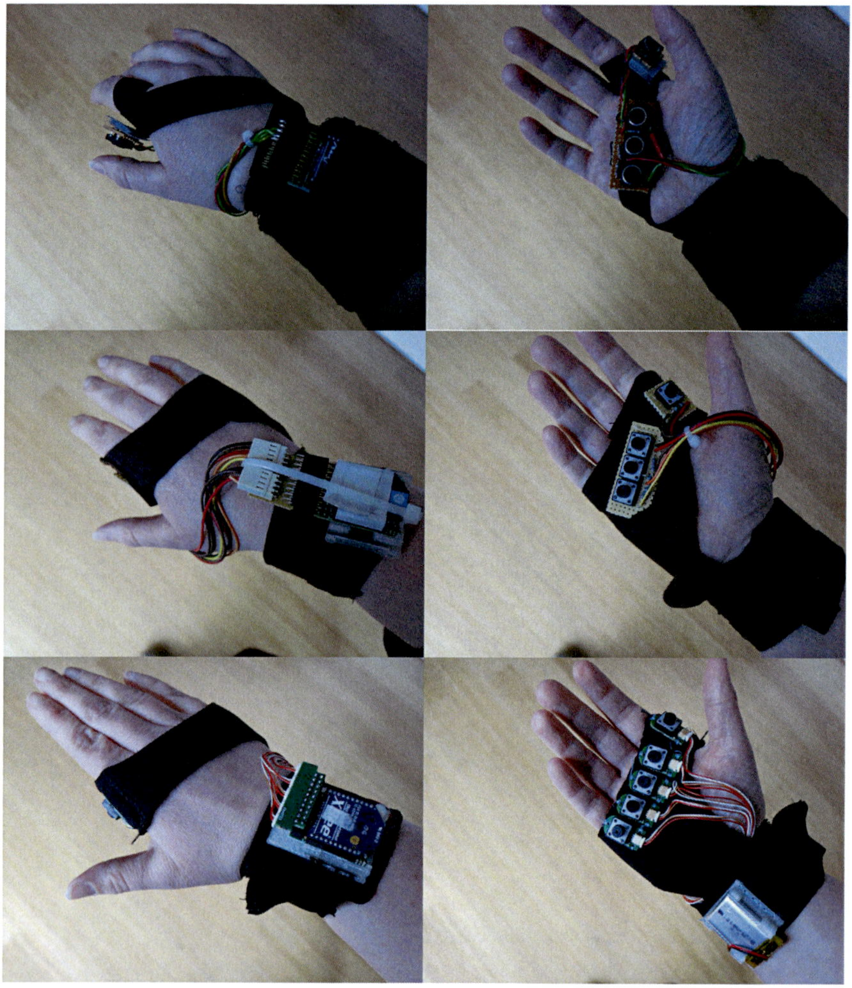

Fig. 2 The three stages in the development of the gestural controllers of "Wezen—Gewording". At the *top* the first version, *middle* the second, and *bottom* the final version

3 'Where Is the Piece?'

In the process described above, it is hard to determine whether the work I do is composing, building and designing an instrument, or performing. It is even hard to say what "Gewording" is. Did I compose it? Is it a performance? Is it an instrument? Can the different performances be considered the same piece?

John Richards poses similar questions in his article "Lead and Schemas" (Richards 2011) about his work on Dirty Electronics:

> 'Where is the piece? Is it in the process of building an instrument, the instrument itself, a notated score, the schematic, or the live performance? Another is, At what point does interpretation come into it?

For my piece "Wezen—Gewording", I have a vague understanding of which parts are the composition, the instrument or the performance:

- The *composition* is the concept of combining gesture, sound and (partially) livecoded software, and the (roughly) planned flow of procedure in which I perform with these elements.
- The *instrument* is the combination of the laptop, the software I wrote, the controllers I built, and the livecoding API[2] I implemented.
- The *performance* is the moment of playing the instrument in front of the audience. I call the preparation for a performance the *rehearsal*.

Yet, those elements are inseparable. My choices in the composition are informed by how I perform and my compositional needs inform how I (re)design the instrument. The instrument design again informs how I (re)do the composition and how I perform. While performing, I embody the possibilities both of the instrument as well as the composition.

The software or code seems to be what blurs the distinctions between the different elements. It is hard to distinguish if a particular segment of code is part of the instrument, of the composition, or even the performance, or perhaps all of these at the same time.

How do other artists relate to their practice in the field of electronic and computer music? What do they consider a composition, an instrument or a performance? What role does code play in these considerations?

4 Composition, Instrument and Performance

4.1 Composition

The basis of the word composition comes from the Latin verb *componere*, putting together (Oxford Dictionaries 2016), also used for ordering or arranging, i.e. putting an order to something. On Wikipedia (2016) musical composition is described as:

> Musical composition can refer to an original piece of music, the structure of a musical piece, or the process of creating a new piece of music. People who practice composition are called composers. "Composition" is the act or practice of creating a song or other piece of music.

---

[2]API stands for *application programming interface*; it defines which functions can be called from a library. In the context of livecoding it defines the set of commands that can be called during the performance, thus determining the vocabulary that is available.

The article gives a description of the elements that are present in different musical traditions, such as the use of music notation to communicate the composition to performers (who then interpret the composition), orchestration or arrangement for a set of instruments, and how improvisation is allowed or expected from the performers by the composers or by the musical tradition to which the pieces belong.

Maxfield (1963) writes about the contrast between instrumental music and "electronic music for want of a better term":

> Formal instrumental music is generally embodied in written score
> the notation of which is not to be taken as complete or exact;
> thus room is left its interpreter
> to decide nuance of detail
> anew for each performance.
>
> But although a recording of an instrumental work
> merely projects a *given performance*
> in this new art form
> it is the composer himself
> working directly with the recorded sounds
> who selects every nuance;
> there may indeed be no score
> and later interpretations not desired:
> the recording, instead, becomes the terminal object of creation.

Thus the composer is the interpreter or performer of his own music, and the music is not distributed by distributing its notation, but a copy of the recording. He further describes how the art object is then fixed, but that "the aesthetic experience it induces is never the same on two different occasions" and that the rigidity of its structure can only be heard upon second hearing. This second hearing can be avoided by never listening twice to the same recording of the work, or as he writes: "I frequently compose a new realization for each presentation of a given work."

While Maxfield (1963) referred to electro-acoustic tape music, for artists building their own electronic circuits to make music, the design of the circuit can be considered as the composition, the electronics as the interpreter, and its schematics as the notation. For example, Lewis (2007) writes:

> David Behrman and Gordon Mumma, implicitly advanced the radical idea of a musical composition that could exist purely and entirely in hardware. In this period, scores by the two composers, where they existed at all, often consisted only of a circuit diagram, accompanied by a set of sketchy instructions.

Holzer (2011) writes in his introduction to an issue of Vague Terrain:

> John Cage once quipped that Serge Tcherepnin's synthesizer system was 'the best musical composition that Serge had ever made', and it is precisely Cage's reformulation of the concert score from a list of deterministic note values to a set of indeterministic possibilities

> that allowed the blurring of lines between instrument-builder and music composer that followed.

Further, he writes about the artists' involvement in the issue:

> Their compositions take the form of systems which provide a map of what is possible, but lack a prescribed route on how to get there. The discovery—and the risk—is left to the moment of the performance.

For computer music, one can consider the actual writing of a music program as its composition, and the code as its representation (like a score). When the code is evaluated, it turns into an interpretation and is performed.

When looking at the process of composition, it is important to gauge the contributions of all people involved to the final (variant of the) composition. There may be cases where the composer has a concept, which a programmer translates exactly into code (or a circuit) and the code is executed and the composition is interpreted by the machine, unfolding over time, eventually perceived by a listener. However it is more likely that composer and programmer collaborate to fine-tune the code to the composer's intentions based on listening to the outcome. During this process it is likely that the very concept of the composition is adapted. Depending on factors like the skill-set of the programmer, the chosen programming language, or the complexity of the task, situations may occur in which decisions are made by the programmer that can be interpreted as compositional and which influence the composer in her conceptual choices. The composition becomes a joint artistic endeavour, hence single authorship can no longer be claimed, nor can the role of the programmer be seen as "just engineering". Hayles (2012) writes:

> Conceptualization is intimately tied in with implementation, design decisions often have theoretical consequences, algorithms embody reasoning, and navigation carries interpretive weight, so the humanities scholar, graphic designer, and programmer work best when they are in continuous and respectful communication with one another.

Although code may be considered an interpretation of the (idea of a) composition, it is rare that a composer asks several programmers to realize or interpret her composition, or that programmers reinterpret compositions from previous eras. As in the discussion above about tape music: the music is distributed by the code, which set the procedure how a machine will translate the composition to sound. In that sense, the programmer is also the performer.

In other cases, programmer and composer are the same person and roles shift between one and the other or, even more likely, boundaries between them blur into non-existence.

It seems that the act of composition within electronic and computer music is *conceptualising and building a system*. Both electronic circuit and code can be interpreted either as notation or instrument for a composition. Schnell and Battier (2002) proposed the term *composed instruments* to underline

> the fact that computer systems used in musical performance carry as much the notion of an instrument as that of a score, in the sense of determining various aspects of a musical work.

4.2 *"New" Instruments and Mapping*

In an acoustic instrument the link between the actor in the environment and the resulting output of the medium (sound) is direct or predefined (Miranda and Wanderley 2006). E.g., a performer playing a recorder blows into it and uses her fingers to close holes. These actions, combined with the physical shape of the instrument determine the sound that is heard. The performer adjusts her breath and finger positions trying to align what she hears with what she desires to hear.

Compared to such an acoustic instrument, its electronic equivalent introduces an "arbitrary" factor in its design: material and shape properties no longer determine the sound that can be emitted. Rather, there is a sense of freedom in determining how a gesture creates or modulates a sound. Waisvisz (1999) describes the search for a proper mapping as a very personal process, but that despite this

> one can analyse and create distinct relationships between the character changes of a gesture, and the change of musical content—and context— in a way that one's musical intentions are clearly grasped by listeners.

In the literature and research on the topic of mapping in the context of digital musical instruments (DMI) or new interfaces for musical expression (NIME), the process is mostly seen as and described as a technical area. Waisvisz (1999) on the other hand, believed that

> … the algorithm for the translation of sensor data into music control data is a major artistic area; the definition of these relationships is part of the composition of a piece. Here is where one defines the expression field for the performer, which is of great influence on how the piece will be perceived.

Rather than considering only the step of connecting sensor data to synthesis and music control as "mapping" [as Miranda and Wanderley (2006) propose], I consider the full process from physical gesture to (sonic) output "mapping" (Fig. 3), thereby following DelaHunta (2001)'s argument, that mapping constitutes the whole "invisible" part of the instrument. The steps in mapping (Fig. 4) encompass gestures and sensors to capture the gestures, electronic circuits to condition the sensor signals and digitize them, computational models to further condition the data, and eventually coupling of the conditioned data to parameters of the output system. Within the environment, the performer (as well as other spectators) perceives the effect and adjusts her gestures accordingly.

In the design process of an instrument at each of these steps in the mapping, one needs to make decisions. These are ultimately artistic decisions, but they are influenced by external factors, such as available technology, development time, and knowledge and practice of the artists and developers. This process consists of going back and forth between each of the different steps in the mapping process, until finally a satisfactory result is achieved.

Nilsson (2011) describes the design process of his DMI's in detail and divides the process into "design time" and "play time." He describes the iterative nature of the design process, alternating between design and play time, while the context of

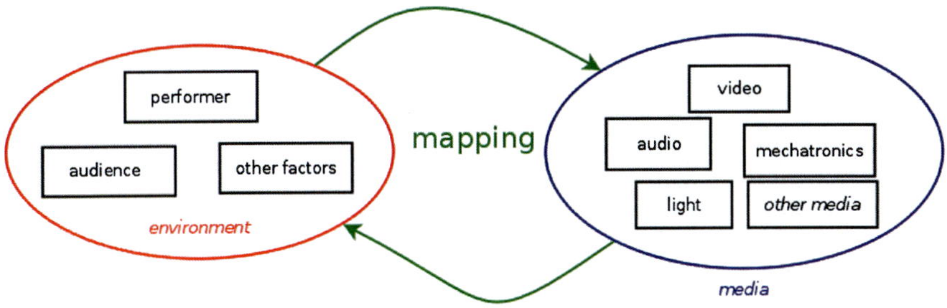

Fig. 3 Mapping between the environment and media—while we directly experience the environment and the media, the mapping is the "invisible" part between the two (DeLaHunta 2001)

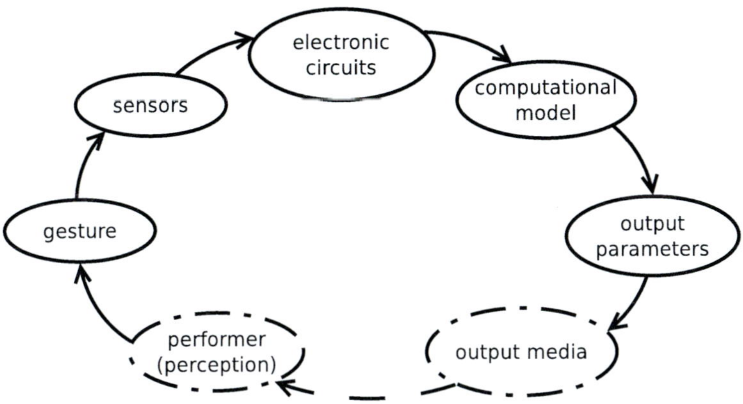

Fig. 4 The different steps involved in mapping from gesture to output media. The steps drawn in a continuous line make up the "invisible" part of the instrument. These are the steps that are adjusted over the course of the building of the instrument

the work grows larger—from private space to the context in which the instrument will be used, to public presentation in concert.

"Design time" is compared to composition, a process taking place "out of time", where design and implementation decisions are taken. "Play time" is the actual time playing with the instrument, where the instrument is evaluated in real-time in terms of how the playing of the instrument feels, what possibilities and sense of freedom it gives. Experience from playtime is then used again to change aspects of the instrument in design time, with multiple iterations until the performer is satisfied. Even after one or more concerts, one can take a step back to design time to adjust characteristics of the instrument.

Performing the instrument is thus an important part of the process of designing the instrument—performing is the exploration of the possibilities of the instrument and gives inspiration on how to improve the instrument for the performance that the creator has in mind.

Waisvisz (1999) advocated to stop development at some point and take a step back from the building process and start playing with the instrument as it is: to compose, perform and explore and exploit its limitations.

In my performance "Wezen—Gewording" however, I explicitly make all the mappings and sounds changeable during the performance. At times it feels for me like switching between "design" and "play" mindsets during performance, at other times I embody the instrument and feel in control of the sound while moving, but at the same time plan what changes I want to make the next time I am at the laptop. These changes are both concerning the composition and the instrument. At the same time, having the sensors coupled to sound parameters influences how I type and move my hands over the keyboard while changing the code during performance.

It may seem that the instrument could be defined as the combination of sensors, electronics and software, and that the instrument is independent of the composition it is created for. But while this may be true for artists that search for new instruments to perform their (existing) music with, others create new instruments for particular pieces and do not necessarily use the same instrument for other compositions. In this case the instrument's design is highly entangled with the composition itself—it is not used outside of the context of the composition it was created for. In certain cases, e.g.in Chikashi Miyama's work "Angry Sparrow" (Miyama 2008), the mapping of the instrument even changes over the course of the composition. There, borders between composition and instrument are hard to distinguish, as the functionality of the instrument changes over the course of the performance.

4.3 *Performers and Their Instrument*

Where human-computer-interaction (HCI) is dedicated to interfaces that allow users to control a machine, instrument builders/performers are interested in a more intimate connection with their instrument. They aim to achieve a certain state of flow, where their instrument becomes an extension of their body. Waisvisz (1999) described this as:

> During inspired performances I have experienced that a mental/physical state can emerge where a fast closed loop establishes itself between the musical intention, the muscular effort and actions, the mechanical response and the sonic feed back and the perception of this whole loop. This happens so fast that one seems to act immediately in sound and not in terms of sound and not in terms of control. Composition/performance melt into a single state of emerging, timbral, expression.

Wessel (2006) suggests the term *babbling* for the process of "non-goal-directed variation of the control parameters" as a "key to the exploration of an instruments potential for musical expression", it is the process of learning which sensory-motor actions will produce which kind of sonic output. Babbling seems to be an important step toward eventually feeling that the instrument is becoming an extension of the body.

The relationship between performers and their instrument is often described by themselves as conversations or dialogues; their instruments gain a certain amount of agency. Waisvisz (1999) is describing the relationship to his own instrument as an intimate one and his relationship to his sounds like that of a puppeteer to his puppets, dealing with gesture as a life-giving force. Similarly, Holzer (2011) describes the various relationships of the artists writing about their instruments as personal relationships, involving both intellect and emotion. Tudor comments (Kuivila 2004; Austin 1989):

> In my electronics, I work with an instrumental principle. (…) They become my friends. They have personalities, that only I see, because of my use of them. It's an act of discovery. I try to find out what's there and not to make it do what I want but to, you know, release what's there.

5 Algorithms, Software and Coding

5.1 Interactive Music Systems or Live Algorithms

Rather than taking an instrumental approach, where a new instrument is built from scratch, various artists have pursued the making of interactive systems where a computer can listen, react and surprise an improvising musician (often playing an acoustic instrument). George Lewis' *Voyager* is a well-known example (Lewis 2007):

> In my most widely performed piece, Voyager, originally programmed by me in 1987 and extensively updated since that time, improvisors are engaged in dialogue with a computer-driven, interactive improvisor. A set of algorithms analyzes aspects of a human improvisors performance in real time, using that analysis to guide another set of algorithms that blend complex responses to the musicians playing with independent musical behavior. In Voyager, the improvised musical encounter is modeled as a negotiation between improvising musicians, some of whom are people, others not; the program does not need to have real-time human input to generate music.

What is interesting is that he terms the system a "piece", suggesting that it is a composition, even though it is clearly designed for performers to improvise with it, and likely each performance will be quite different, depending on the musical vocabulary of the performers involved. The system is the musical idea, musicians enter into a dialogue with it during performance.

Young and Blackwell (2013) coined the term *Live Algorithms for Music* to describe such interactive music systems: "Live algorithms are an ideal concept: computational systems able to collaborate proactively with humans in the creation of group-based improvised music." In more detail they define the features as (Lewis 2007; Young 2005):

- a live algorithm can collaborate actively with human performers in real-time performance without a human operator
- a live algorithm can make apt and creative contributions to the musical dimensions of sound, time and structure
- live algorithms can contain a parametric representation of the aural environment which changes to reflect interaction between machine and environment.

They suggest that such live algorithms do not embody the artistic, compositional concepts of the author; they want to place the algorithm apart from the human designer of the algorithm. In contrast Drummond (2009) states that:

> Interactive music systems are of course not found objects, but rather the creation of composers, performers, artists and the like (through a combination of software, hardware and musical design). For a system to respond musically implies a system design that meets the musical aesthetic of the systems designer(s). For a system to respond conversationally, with both predictable and unpredictable responses, likewise is a process inbuilt into the system. In all of the definitions discussed, to some degree, is the notion that interactive systems require interaction to realise the compositional structures and potentials encoded in the system. To this extent interactive systems make possible a way of composing that at the same time is both performing and improvising.

Drummond (2009) further argues that traditional distinctions between composing, instrument building, systems design and performance are blurring. In his discussion of Chadabe's term *interactive composing* he writes:

> Chadabe highlights that the musical outcome from these interactive composing instruments was a result of the shared control of both the performer and the instruments programming, the interaction between the two creating the final musical response. (…) In interactive music systems the performer can influence, affect and alter the underlying compositional structures, the instrument can take on performer-like qualities, and the evolution of the instrument itself may form the basis of a composition.

and he cites Chadabe: "The instrument is the music. The composer is the performer."

By both Drummond (2009) and the LAM research network it is assumed that the algorithm (or system), once designed and written, is fixed. It is not changed during performance. The definition of live algorithms explicitly state that, once the performance has started, there is no human interference other than at the defined inputs of the system.

In both cases, the designed systems are the compositions. They describe the possibilities in which the musical output or the means of the dialogue, even if the outcome is surprising or unexpected for the performer (or even for its composer or builder).

5.2 *Music Software and Tools*

In the discussion above it is highlighted that compositional concepts are embedded into the code. The evaluation of the code is the means to perform the compositions. Clearly, coding is an integral part in the act of composition: it creates and defines the possibilities within which musical output can transform. Commercial music software embodies certain kinds of compositional or musical concepts or styles. Similarly, Tudor remarked (Kuivila 2004) on hardware that is was very hard to make certain commercial instruments do what they were not intended for. A certain musical style or way of thinking about music was embedded into that hardware. Joel Ryan[3] described musical instruments as embodiments of music theory like, for example, a piano embodies the division in octaves in a well-tempered tuning system.

McLean (2008) takes a critical stance in the discussion of software, creativity and artistic expression:

> This can lead to the bizarre situation where programmers make commercial software which practically generates music, and yet somehow the users of the software are seen as being more creative than the programmers. Here the programmers encode their musical style in the software, and the users go little beyond guiding the software to a destination pleasing to them. This can be seen in filters and plugins of music studio software as well as explicitly generative commercial applications such as Sseyo Koan Pro. The creativity of programmers is tapped into flattery of paying users.

It depends on the open-ended-ness of software in how far the artistic ideas that can be expressed with it are restricted by the concepts already embedded in the software. Audio programming environments such as Max (Cycling74) or SuperCollider (McCartney) are very open-ended and provide for many different music styles to be produced with it; yet there is a clear distinction between users of one or the other, and composers are attracted to one or the other depending on to what extent they can express their musical concepts in the language. Within SuperCollider many composers develop their own dialects or "systems within systems" and extensions to the language to "add both new possibilities and new constraints" (Rohrhuber et al. 2011). Rohrhuber et al. (2011) also discuss the "blurring of the distinction between a tool and its outcome, an application and an artwork or a model" and state that "thinking within a given language, some ideas may never occur". From this it becomes apparent that not only the programming language that is used to achieve a particular musical result is important, but also the programmer's personal capacity (or vocabulary), the dialect she can speak within that language determines what musical ideas may be expressed.

---

[3]During his talk during the Musical Organics Symposium, STEIM, Amsterdam; May 5, 2016.

5.3 Livecoding

McLean (2011) describes the process of *bricolage programming* (see Fig. 5) as "a creative feedback loop encompassing the written algorithm, its interpretation, and the programmer's perception and reaction to its output or behaviour." In his descriptive example, he shows how the original concept of the artist is changed by evaluating the code and reacting to the output it generates. The resulting composition or code is clearly shaped in the dialogue between the artist and the code she writes and the output of the machine that runs it. McLean (2011) states: "At the beginning, the programmer may have a half-formed concept, which only reaches internal consistency through the process of being expressed as an algorithm."

In the livecoding movement, bricolage programming is not only done to achieve a fixed code which can then be used as an instrument or live algorithm to play with, rather design time and play time is brought together into the live performance. Livecoding is a form of improvisation that would fit the definition of *instant composition* of the Amsterdam based "Carpet Collective" (Carpet Collective 2015):

> INSTANT COMPOSITION combines the notion of working from the moment (INSTANTaneous creation) with the intention to build something (COMPOSING a piece with an audience present). This means that for us, improvisation principles are always concerned with both the question of FREEDOM and the question of STRUCTURE.

However, whereas with most improvisation the intended structure is not apparent to the audience, until it unfolds, with livecoding—as the audience can read the code that is usually projected in a livecoding performance—the structure can be apparent before it unfolds, and the livecoder may even decide to not let it unfold,

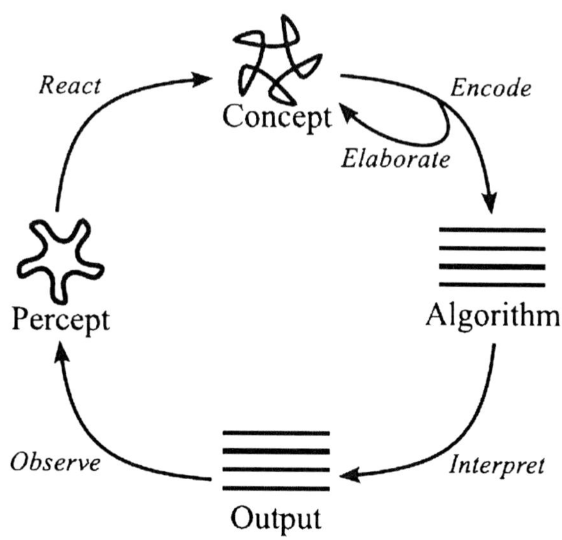

Fig. 5 The process of action and reaction in bricolage programming (image from McLean (2011), used with permission)

and change it again before it does. As DelaHunta (2001) wished for, the invisible part of the performance is made visible.

Livecoding then blurs the distinctions between composing, instrument building, and performing (and composition/instrument/performance) even further, as the computational system or code is no longer fixed once it is designed, but is adapted during the performance. A livecoder can act on all levels: changing the development of the music over time, redefining its local structure, generating events at particular times, and changing the texture of a sound while it plays.

6 Conclusion

In this article, I have reflected on my own work process, particularly in the piece "Wezen—Gewording", asking how it affects notions of composition, instrument design and performance in the context of art works using technology such as electronic circuits and software or code.

Looking at literature from artists working in the field, the boundaries between these notions have been blurring, and there are several diverging interpretations of what is what. The views on what an instrument and what a composition is seem to overlap—if we compare the statements of Behrman and Mumma (compositions existing purely in hardware), Waisvisz (calling the mapping of digital instruments the main artistic area) and the discussion on interactive music systems.

Ryan called an instrument an embodiment of musical theory; the instrument defines a set of possibilities that can be explored. A composition can then maybe be viewed as a particular path through this field—a particular exploration of these possibilities and music theory. The performance of this exploration makes it experiential both for the performer and listeners—it is the event where the exploration comes to life and is translated from being a concept to being a physical, but ephemeral, event.

With the livecoding practice, the building and conceptualising of a system is transferred into the moment of performance. The distinction between design time and play time is disappearing. As performers manipulate their instruments at the core of their functionality, they take away the notion of an instrument having a predefined behaviour. With code that can be changed and compiled on the fly, the design of an instrument, or the making of a composition, becomes a fluid process and a performance in itself.

The interplay between the processes of composing, instrument building and performing is an embodied process along the definition of Varela et al. (1991) for embodiment. This process of embodiment is very personal, and an ongoing process. This means that there are no fixed compositions or instruments, and artists each have their unique artistic expression. Nonetheless they can communicate about their processes and engage with other artists to influence each other's vocabularies.

References

Austin, L. (1989). David Tudor and Larry Austin: A conversation. http://www.emf.org/tudor/Articles/austin.html. Accessed August 16, 2016.

Baalman, M. A. J. (2016) Sense/Stage—A low cost, open source wireless sensor infrastructure for live performance and interactive, real-time environment. https://docs.sensestage.eu. Accessed August 16, 2016.

Carpet Collective. (2015). Interdisciplinary instant composition: To define, share and develop concepts of improvisation. http://www.instantcomposition.com/. Accessed August 16, 2016.

Cycling74. Max. https://cycling74.com/products/max/. Accessed August 16, 2016.

DeLaHunta, S. (2001). Invisibility/corporeality. http://www.noemalab.org/sections/ideas/ideas_articles/delahunta.html. Accessed August 16, 2016.

Drummond, J. (2009). Understanding interactive systems. *Organised Sound, 140*(2), 124–133. doi:10.1017/S1355771809000235

Hayles, N. K. (2012). *How we think: Digital media and contemporary technogenesis*. University of Chicago Press. ISBN 9780226321424.

Holzer, D. (2011). Schematic as score: Uses and abuses of the (in)deterministic possibilities of sound technology. https://web.archive.org/web/20110506065819/, http://vagueterrain.net/journal19. Accessed August 16, 2016.

Kuivila, R. (2004). Open sources: Words, circuits and the notation-realization relation in the music of david tudor. *Leonardo Music Journal, 14*, 17–23.

Lewis, G. E. (2007). Live algorithms and the future of music. *CT Watch Quarterly, 30*(2), 19–24.

Maxfield, R. (1963). *An anthology of chance*. La Monte Young and Jackson Mac Low: 965 Hoe Avenue, Bronx 59, New York, USA.

McCartney, J. Supercollider. http://supercollider.github.io/. Accessed August 16, 2016.

McLean, A. (2008). Live coding for free. In A. Mansoux & M. de Valk (Eds.), *FLOSS + Art* (pp. 224–231). Goto10 and OpenMute. ISBN 978-1-906496-18-0.

McLean, A. (2011). *Artist-programmers and programming languages for the arts*. PhD thesis, Department of Computing, Goldsmiths, University of London.

Miranda E., & Wanderley, M. (2006). *New digital musical instruments: Control and interaction beyond the keyboard* (A-R ed.).

Miyama, C. (2008). Angry sparrow. http://www.chikashi.net/. Accessed August 16, 2016.

Nilsson, P. A. (2011). *A field of possibilities: Designing and playing digital musical instruments*. PhD thesis, Göteborgs Universitet.

Oxford Dictionaries. (2016). Oxford dictionaries: Composition. http://www.oxforddictionaries.com/definition/english/composition. Accessed August 16, 2016.

Richards, J. (2011). Lead and schemas. *Institute of Contemporary Arts: Roland Magazine*, 9. http://www.dirtyelectronics.org/docs/ICA_Roland_09.pdf. Accessed August 16, 2016.

Rohrhuber, J., Hall, T., & De Campo, A. (2011). *Dialects, constraints, and systems within systems*. Cambridge, MA: MIT Press.

Schnell, N., & Battier, M. (2002). Introducing composed instruments, technical and musicological implications. In *Proceedings of the 2002 Conference on New Instruments for Musical Expression (NIME-02)*, Dublin, Ireland, May 24–26, 2002 NIME02-01.

Varela, F. J., Thompson, E., & Rosch, E. (1991) *The embodied mind: Cognitive science and human experience*. MIT Press.

Waisvisz, M. 1999. Michel Waisvisz on gestural controllers. http://crackle.org/MW's%20gestural%20round%20table.htm

Wessel, D. (2006). *An enactive approach to computer music performance* (pages 93–98). Studio Grame: Lyon, France. http://cnmat.berkeley.edu/system/files/attachments/An+enactive+approach+to+Computer+Music+Performance.pdf

Wikipedia. (2016). Musical composition: Wikipedia, the free encyclopedia. https://en.wikipedia.org/wiki/Musical_composition. Accessed May 15, 2016.

Young, M. (2005). What is a live algorithm. http://homepages.gold.ac.k/michaelyoung/LAMweb/whatis.htm. Accessed August 16, 2016.

Young, M., & Blackwell, T. (2013). Live algorithms for music: Can computers be improvisers? In *The oxford handbook of critical improvisation studies* (Vol. 2). Oxford University Press. doi:10.1093/oxfordhb/9780199892921.013.002

Instrumentality in Sonic Wild{er}ness

Antye Greie-Ripatti and Till Bovermann

Abstract In 2015, a group of six sound practitioners including the authors came together for 'Sonic Wild Code' and engaged in a series of sonic wilderness interventions with portable electronic instruments. We investigated notions of coexistence, communication and potential for interaction in the hybrid ecology surrounding the lake and settlement of Kilpisjärvi, located close to the three-nation corner of Finland, Sweden and Norway. By immersing ourselves into the vast and raw landscape of the Samiland, we researched and tested musical conversations between us players and the site which we found sounding, vibrating, and speaking for itself. This text is a collection of fragments originating in discussions between the two authors on the theme of such sonic wilderness interventions.

1 A First Day in the Field

On our first day in the field with the 'Sonic Wild Code' group, we started walking and, after an hour, reached a valley of hills. Low-pass filtered over centuries by ice waves, wind and weather, the environment felt calm yet empty. We initiated our first intervention: screaming into the landscape and listening to its response. First each person's voice alone, then together. The soft mountains echoed back at us, leaving us impressed by the power, reflection and interpretation of our own voices (Fig. 1).

A. Greie-Ripatti
Hai Art, Hailuoto, Finland
e-mail: agf@poemproducer.com

T. Bovermann (✉)
Institut für zeitbasierte Medien, Universität der Künste Berlin, Berlin, Germany
e-mail: t.bovermann@udk-berlin.de

T. Bovermann et al. (eds.), *Musical Instruments in the 21st Century*,
DOI 10.1007/978-981-10-2951-6_16

Fig. 1 The Sonic Wild Code group in the Tundra

An extract from our blog report (Greie-Ripatti et al. 2015) describes the day's progress:

> We had lunch at the Saanajärvi "päivätupa", found a gas stove, made tea and talked. A small group of reindeer came for a visit and we eventually spread out, searching for things to record and capture within our little devices. Some of us started making sounds integrating found objects: Creating audio feedback within rock formations, sampling water streams and wind. A first glimpse of an improvisational piece appeared. This was the moment Dinah Bird, one group member, referred to when she mentioned the intensity she found in being able to mix artificially induced sounds with the acoustics and soundscape of the landscape in her recordings, walking around and trying to find her own subjective listening space. Whether it was the feedback sounds, the voice improvisation, playing the blade of grass or simply our strange behaviour that again caught the attention of reindeer, we don't know for sure. But we had the feeling to have made an impression on the landscape.

2 Evolution of Outdoor Music Intervention

Musicians always performed in homes and public spaces. Over time performance stages evolved from common areas, located i.e. in the centre of settlements, to locations such as churches, concert halls and clubs, that were intentionally designed for music practice.

Fig. 2 Antye and Till performing with portable electronic instruments in Norway

But music practice is not bound to such places. Moreover, it is an integral part of, e.g., street and festival culture as well as camp-fire gatherings. Such outdoor performances enrich soundscapes of areas that are not commonly considered to be environments for making music.

Outdoor music practice exists since the emergence of human culture. Yoik, a voice practice of the indigenous people of northern Scandinavia in the stateless tribe called Sami, is an example. Inspired by the environment and common day-to-day life, Yoik is performed to the land, the self, the animal and the ethereal.[1]

The reason for people to sit at desks, heads close to computer screens has been brought onto stage in the form of playing digital instruments, practised in buildings with either seated or dancing audiences. Since then, music technology evolved and computers got smaller which opened up new possibilities for sound production and musicians. Current digitisation and amplification techniques allow not only storing but manipulating, synthesising and distributing sound in real time. At the same time, miniaturisation enables such electronic and digital instruments to be as portable as, e.g., flutes or jaw harps (Fig. 2).

[1]For a more detailed introduction see e.g. the interview by riley French (2016) with Chris Watson and Ande Somby.

3 Deconstructing Wilderness

What we call "wilderness" is hard to find or not to be found at all. It seems to be a mere romantic concept that does not exist in this pure form. Even finding a place that is "sonically wild", i.e., where one hears only natural sounds is harder than we thought. We experienced this on our journey to and around Kilpisjärvi, the northern-most settlement of Finland.

To reach it, we flew to the nearest open airport, Rovaniemi, still a 5-h bus ride away from Kilpisjärvi, going through a seemingly endless area of boreal forest until we reached the tree limit towards the open tundra. Being next to the three nation border to Sweden and Norway, Kilpisjärvi has about 100 permanent citizens, living in houses along a central street with regular border traffic of mainly trucks and tourist cars. From there, it took us another two hours to finally escape the car sounds from the local road, only to be overflown by a helicopter.

With this in mind, we state that humans colonised the planet almost completely and all notions of wilderness are romantic constructs. Even the word "wilderness" itself has a history as a political notion of colonisation (Callicott 2000).

The Ponca Native American chief Standing Bear (1998) said:

> We did not think of the great open plains, the beautiful rolling hills, and winding streams with tangled growth, as "wild". Only to the white man was nature a "wilderness" and only to him was the land "infested" with "wild" animals and "savage" people. To us it was tame. Earth was bountiful and we were surrounded with the blessings of the Great Mystery. Not until the hairy man from the east came and with brutal frenzy heaped injustices upon us and the families we loved was it "wild" for us. When the very animals of the forest began fleeing from his approach, then it was that for us the "Wild West" began.

With this background we like to talk of "sonic wilderness" as a term that describes places that appear to us sonically uncanny, alien, sublime. Their soundscapes fail our common understanding: they remain wild-in the sense of seemingly "untouched-of cultural sounds and music. It is a subjective term, one that changes with the time one spent within the surrounding of a sonic wilderness. The more one learns about the surrounding, the less it appears to be "wild" and becomes "tame" (Fig. 3).

4 Preparations and Practicalities

Feedback represents physics.

Stone-throwing represents time and the human body.

Computer represents algorithms and complex programming.

Live coding represents intellect and the human capacity to think music as a concept.

Fig. 3 Dinah Bird blending into the rock

The author's sonic wilderness interventions revealed several practical factors that greatly facilitate wilderness performances.

Lightness

Gear should be chosen based on the principles of carrying light, and maximising aptitude. Every element involved in an outdoor performance has to be brought, unless it is part of the performance site. Light yet robust material, with multiple uses is preferable over heavy, single-purpose items. Additionally, everything should be rainproof and electronic components should be self-powered and energy-efficient.

Modularity

For the instrument setup, modularity is key. A collection of building-blocks can be combined to create a variety of different sounds. The instrumental setup becomes a toolbox that helps to adapt to the unknown sonic qualities of the performance site.

Dependency on technical material

Electronic instruments often require additional technical material. E.g., if the instrument itself does not include a transducer, a loudspeaker and appropriate cables are required. While small speakers are preferable for their lightness, they mostly sound tinnier than bigger ones. Besides the inherent sonic possibilities of the instrument itself, its final sound depends on the specifics of the chosen additional materials.

Sound mixing

A matrix mixer allows to mix everything connected to its inputs to everything at its outputs. Effects can be driven into feedback, while still maintained with a mix of external material. Figures 8 and 9 show two performance set-ups, each built around a matrix mixer.

5 Sonic Ecology

Outdoor sites for sonic interventions can be differentiated roughly into *terrain* and *inhabiting life forms* (both non-human and human). Together with us visitors and our instruments, such sites form a temporal *ecology of sonic wilderness intervention*.[2]

This ecology possesses a unique soundscape and can be differentiated into *keynote sounds*, *signals* and *soundmarks* (Schaffer 1994).

> The *keynote sounds* of a landscape are those created by its geography and climate: water, wind, forests, plains, birds, insects and animals. [...]
>
> *Signals* are foreground sounds and they are listened to consciously. [...]

[2]See e.g. the highly site and time specific sound works by Grill (2014).

> The term *soundmark* is derived from landmark and refers to a community sound which is unique or possesses qualities which make it specially regarded or noticed by the people in that community.

These elements form the performance ecology's *sonic conditions* and determine both the intervention's starting point and its development. There are several ways to actively integrate the site into an intervention: Structural elements such as rocks or trees influence the sonic perspective of artists and their audience. Being in front of or behind a rock, close to a stream or on top of a tree affects the sonic experience of the intervention.

Engaging with non-human agency not only means to identify with the site and nourish the temporary ecology but also to possibly identify an audience or even playing partners. In our case those were the wind, the hills, plants, reindeer, lemmings, rocks and a waterfall.

The site shapes the timbral character of musical instruments. Particularly instruments which integrate feedback into their sound generation depend highly on their surrounding. They pick up local acoustic properties of, e.g., rock formations or the open reverberation properties of a forest. In a sense, *the site is a crucial part of the instrument itself.* For other electronic instruments, non-intrusive sensors like barometric, gas, temperature or humidity placed into (or onto) site-residing elements such as mud, water, plants or mushrooms allow the site to contribute its condition into the performance.

6 Immersing and Dissolving

> One can just go and be.
>
> Sometimes, dropping yourself to the ground is enough.
>
> Just fall and watch ants or frozen structures.
>
> It can catapult you instantly to just being.
>
> It is a personal experience.
>
> A childhood memory.
>
> When alone with the Land and prepared for a wilderness performance, you can allow yourself to reflect on your self-being.

You can dive into a conversation with the site. You have the opportunity to (re-) connect with it, to try your borders, shout. Or just, very quietly, whisper. Narrate a story meant only for you and your environment, the Land. Find yourself and get connected with the Land. Gain and immediately follow new ideas. These steps may help you to immerse and dissolve:

Meditation is a strategy to get into an attentive mind-space. Meditate to be in the moment. No purposeful listening is needed. Select a spot where you feel safe, as meditation practice relies on trusting the surroundings. Meditate in the Land, meditate the Land.

Contemplation means getting to know what is already at the place where you plan to play. Contemplate the Land. Practice passive and active listening to get an idea of your surrounding.How do you anticipate it to affect your performance?

Taking action means playing, making music. Make use of the Land as material and playing partner. Express your mood and incorporate what you found while contemplating. Improvise, recognise the Land as your playing partner.

Reflection means to consciously take the time to revisit what happened. To give afterthoughts, observations and new ideas a dedicated space to form themselves. How did the Land affect your playing? How did your playing affect the Land? How are you feeling? Are things different than before?

7 Active and Passive Listening

Observe.

Take in.

Embrace.

Passive listening means taking in everything that surrounds you. Take a walk, stop at random spots, listen closely to what you hear. This practice is closely related to *Soundwalks*, an "excursion whose main purpose is listening to the environment. It is exposing our ears to every sound around us no matter where we are" (Westerkamp 1974).

In contrast, *active listening* means mixing present sounds by positioning ourselves with respect to the sound sources. One can search for sounds, make it a task to identify as many sounds as possible, or create a dynamic "live mix" by moving from one place to another, pushing certain sounds into the foreground over time; pass them from left to right.

Amplification, headphones and a (stereo) microphone support active listening practice because they introduce an abstraction layer. That their *immediateness* is different from listening directly shifts one's experience from hearing "nature" or "the environment" to thinking and perceiving in more abstract terms such as loud, quite, harsh, soft, high-pitched, or repetitive (Fig. 4).

8 Intervention Structure

Derive all sound from the environment.

Process and interact with the space and play it.

First pre-condition: start with microphones, or other capturing devices.

Fig. 4 People listening in the snow

Use a sustainable and ecological energy source.

Incorporate digital processing.

Process sound live, code live.

Possibly add a built-in modular synthesiser.

A wilderness intervention can be composed with the help of *event scores*. They were introduced by George Brecht as collections of written instructions intended to be either followed or explicitly disobeyed (Robinson et al. 2005; Ouzounian 2011). Such a compositional approach provides a framing in which performers can move around freely. They allow a performance to be re-enacted, either at the same place or somewhere else, very likely with a completely different outcome but still identified as the same piece.

Sonic wilderness interventions also thrive on improvisation, the-possibly complex-process in which artists contribute to a piece by selecting *while playing* from an extensive repertoire of figures and phrases. Choice is based equally on subjective listening and the direction towards which the artist intends the piece to develop. Improvisation means to ground the selection of phrases and musical expressions not only on the piece itself but also on the impressions from the site.

The two concepts of event scores and improvisation complement each other. Rules introduced by an event score may be interpreted as guidelines for an otherwise improvisational performance. They offer a way to interpret the site as a playing partner to which one can act and react, listen and talk. Consequently, improvisation techniques can be interpreted as rules of an event score.

9 Digital Music, Computers Versus Nature

Connecting specifically digital music with nature is a curious starting point for these experiments. Using powered PA systems and more sensitive radio FM transmission technology resulted in the conclusion that it is important to move off the grid. The setting is too delicate to invade with powerful machines and impose digital music on natural places. The music has to be derived from the environment and developed from there. The power, the sound source and the electricity must come from the site, that seems like the ultimate goal. A collaboration of digital space and the environment.

By employing a variety of live sampling applications, where a sound is recorded and processed in realtime, a potential symbiosis is formed with the sonic ecology and the musicians playing. The listening is performed both ways. Respect to the environment is established.

Complex apps for tablets offer live sampling, processing, granular syntheses, and all possible audio manipulation. Sonic results merge into a music which is electro-acoustic by definition. A minimal simple setup to go out and work with is a tablet with live sampling apps such as Borderlands, FieldScaper, SAMPLR, AUM (complex app mixer), an attachable microphone, a battery powered speaker, a stereo field recorder and for documentation a camera with tripod (Fig. 5).

Fig. 5 Antye's field setup

10 Voice—The Embodied Instrument

There is one instrument one always carries. It embodies identity, and many consider it the most personal instrument. There will soon be eight billions of them. The voice is part of the human body and, if you stroll through "wilderness", you have a strong self-powered instrument right with you (Fig. 6).

The human voice is a source for melody, rhythm, and acoustic intervention. At the same time, it can imitate sounds like wind or dripping water. Whether used individually or in groups, it easily becomes part of the environment and a natural source to work with. Acoustic scenes immensely contribute to its appearance. For example, screaming in a valley manifests itself differently than when shouting against a powerful waterfall or humming in a cave.

Composing vocal pieces along the landscape brings us to Yoik. Yoik is a voice practice by the Sami people and, while the Sami culture has a poetic approach to language, Yoik is wordless (Wikipedia 2016):

> [T]here are no references to how and where yoik originated. According to the oral tradition, the fairies and elves of the arctic land gave yoiks to the Sámi People. Just Quigstad, who recorded the Sami oral tradition, has documented this legend in several works. According to music researchers, Yoik is one of the longest living music traditions in Europe.

Fig. 6 Shouting with and against a waterfall

Screaming, yoiking and whistling is used in mountain areas by indigenous people to communicate between villages. The whistling language “El Silbo” practiced on the Canary island La Gomera is one example which has been declared as World Cultural Heritage by the UNESCO.

Our voice is a powerful instrument, especially when use without words or language, it can contribute immensely to an outdoor intervention, communicating with the sonic environment. Using voice without words can also remind us of a more animalistic, ancient thread within us and our subconscious. It leads to intriguing and sometimes funny results.

Humans tend to control their vocal expressions closely and self-consciously train not to express themselves non-verbally. An outdoor intervention can open up this possibility.

Use your voice!

11 Live Coding

> Live coders expose and rewire the innards of software while it generates improvised music and/or visuals.

Typically, the rewiring as described in the above quote from Toplap (2011) is performed via textual interfaces, a drastic contrast to the organic environments in which sonic wilderness interventions take place. While they may seem impractical, it is rather a question of understanding the benefit of such live coding interfaces within a wilderness intervention. Their power lies in their flexibility: starting with parameters and value ranges (e.g. frequency or amplitude mapped to controllers) to the DSP algorithms themselves; nearly everything can be adjusted or changed on the fly, while performing.

The addition of generative elements such as the BetaBlocker environment make a live coding environment good for pad sounds and ambient/evolving elements (Bovermann and Griffiths 2014).

Next steps in research suggest to interpret and reflect natural processes by integrating data drawn from external sensors: in a DSP-oriented language, microphones are the easiest to integrate but one can also include environmental data such as the local temperature, humidity, light, colours, or gas concentration, captured with sensor elements in realtime.

12 Ensemble Playing

> Take X amount of people to the field.
>
> Find an instrument or select a landscape.
>
> Bring: acoustic and digital sound making devices, sound objects, portable battery powered Speakers, cameras, microphones, recorders.

Listen, start to play.

Listen to the landscape, to the non-human, and to each other.

Let it happen.

It ends when it ends.

Playing and interpreting the landscape in multiple ways by multiple players has great potential. A wilderness ensemble performance thrives particularly on the unfolding multitude of instruments and approaches. The variety of the participating concepts contributes to the narrative created. Examples for complementing practices are: Real-time sampling combines effectively with physical interactions and movement. An auditory feedback system interacts with acoustic properties of the surrounding landscape. Contributions from a DIY pocket synthesiser and the rhythm of tangible actions such as throwing stones on natural surfaces culminates in complex sonic results.

The more the environment as material informs the sound processing, the more the two will blend into each other. The more the human interaction intermixes and harmonises with the storytelling of the land, the more the entire story is resonating. Being creative in the field largely depends on non-intentional listening. One person's perception can be misleading but a group's perception based on collective listening supports the emergence of a shared sonic truth, a truth that can in fact be felt (Fig. 7).

Fig. 7 Ensemble—taking in the land, adding something as a group

13 Solitude

Solitude.

Introversion, contemplation, introspection, self-reflection, daydreaming.

Being self-absorbed, immersed into oneself and the environment.

Self-sunkenness.

Even if you are by yourself and do not notice anyone listening, there is always something or someone around you that will be affected by your actions. At the same time your surrounding has an effect on you, both subliminally and consciously. Playing in what at first seems to be solitude invites to explore possibilities and embrace the surrounding. By close observation you can find actors and inspiring elements on site from which you can choose playing counterparts: There are sonic cues such as the rustling of leaves in the wind or the humming of a distant street. There are visual cues such as the shape of the horizon or the colour variations of the moss next to you. There are dynamic cues such as the movement of water or the behaviour of visiting (wild?) animals. You can select from those sounds, shapes and movements and make them part of the same piece you are playing.

A prominent participant of your performances-one that actually takes part in every single intervention you do-are you yourself. How does it feel to recognise yourself as artist and audience at the same time? Being the only member of the (human) audience and the "solo artist" at the same time questions the "performance" as the core element of music making. The act of playing rather becomes an opportunity to reflect upon decision processes and let oneself drift without the pressure to perform for others. The absence of a critical audience can be liberating. You alone decide: What are the rules for the performance? What are the rules for listening? Will you play solely for your own pleasure, or do you, e.g., practice for a future performance? Do mistakes vanish into the void of distant remembrance, or do they stay and be subject of further interpretation? Will your performance only exist within the moment or do you record it with the aim to turn it into a lasting piece? Will you allow yourself to rethink those decisions while performing?

14 Unfolding Instrument Design

Amplification or synthesis? Feedback or re-synthesis?

Harmony or noise? Generated or sample based?

Acoustic or electronic? Haptic or code?

The gestalt of a sonic wilderness instrument is within the artists's choice. It consists of a multitude of different elements, ranging from objects found at the site over bodily elements such as the voice up to technological artefacts like samplers, microphones, transducers and computers.

Adaptability

A core feature of a sonic wilderness instrument is its adaptability: it gets re-invented constantly depending on the playing situation. Its gestalt therefore reflects the site it is played at as well as the performer's mood and emotional state. Playing a wilderness instrument is musicking in its purest, utopian form; a never-ending process of design, build, play, practice, refine, repeat (Green 2014).

Experimentation

To engage in sonic wilderness interventions means to experiment also in designing the instrumental setup. How do certain sensors behave when applied to objects found on site? How does sound get picked up from a transducer? Which instrumental parameters are musically most interesting?

Energy

Instruments with electronic components and amplification require electricity. As an alternative to providing the energy via batteries, it can also be harvested on site. Depending on the location, solar, wind, water or biochemical processes can be used. The natural fluctuation of such sources (changes in wind speed or clouds overshadowing solar panels) could even be directly used as an additional sensory element for the system by means of power starvation techniques (as known from circuit bending techniques).

Amplification

The way how signals are picked up and, after creative processing, rendered as acoustical waves has a massive influence on how the electrical signals of an instrument are perceived. Apart from the obvious variations in size and amount of loudspeakers, also their specific sound-generation technique can be altered.[3]

Modularity

A sonic wilderness instrument is often not a single object but a setup consisting of several parts as shown in Figs. 8 and 9.

The instrument could be further extended by integrating networking capabilities. Then it would not necessarily be located solely at the performance site but in part at other locations, connected wirelessly.

[3]See e.g. the alternative methods of sound generation in the Resophonic Manta and the Bass Manta by Snyder (2011).

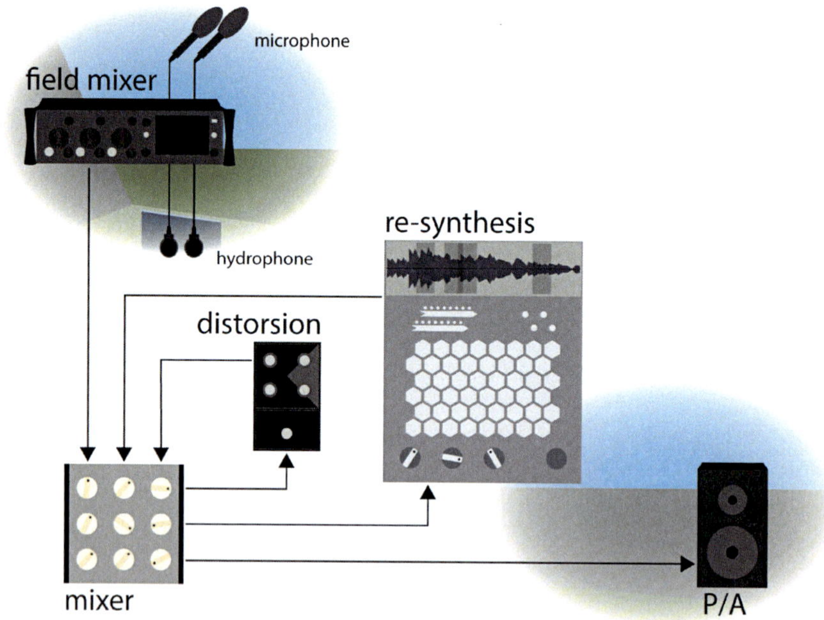

Fig. 8 The combination of an artificial feedback system (a distortion effect wired into a feedback loop) complemented by a set of microphones on a field mixer is picked up by a digital re-synthesis system that can be played percussively

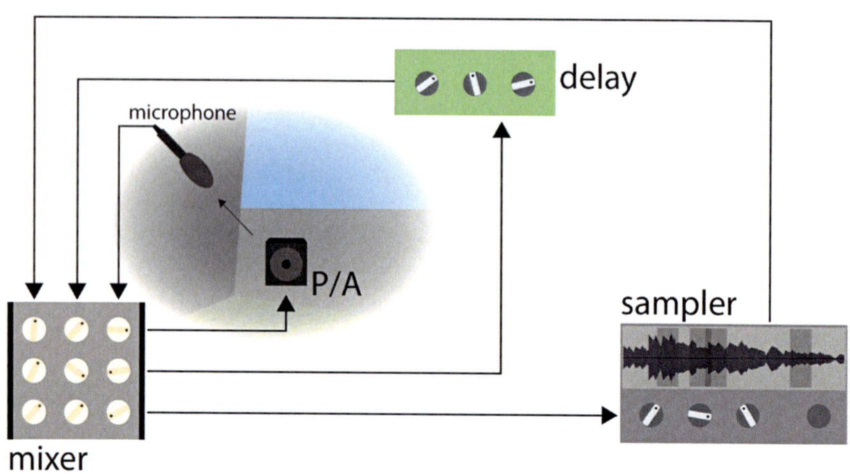

Fig. 9 An acoustic feedback loop can be directly played as well as artificially altered with a delay. The sampler can be used to preserve phrases, repeat and alter them in the ongoing session

15 Instruments to Complement Sonic Niches

Each potential site for a sonic wilderness intervention is a unique constellation of sonic elements. Making music at this site means to add something to it, to complement it and interpret the resulting soundscape as music. Since it is unclear what kind of sounds to expect from the site, it is beneficial to pack instruments by which one can create a broad variety of sounds, so one can contribute sonic elements that complement the sonic niches within the environment.

An instrument with a diverse timbral repertoire such as a digital synthesiser allows to play percussive drum-like sounds as well as sustained drones. Thus it can be played in situations that call for percussion (to impose structure) as well as when, e.g., a gurgling creek inspires to play a low-pitched pad. Instruments based on sampling technology or acoustic feedback, on the other hand, can be used to pick up and extend site-specific sonic cues. Yet again, to introduce sounds contrasting to the site's soundscape, instruments based on classic additive or subtractive synthesis are useful. All in all, it is the inclusion of both *environmental* as well as *artificial* sounds which enables the player to react musically to the ecology of sonic wilderness intervention.

Altering sounds with filters and effects adds another layer of sound shaping. Here, time-affecting effects like artificial reverberation, echo and granular re-synthesis can be differentiated from sound-shaping effects like distortion, modulation or filtering.

The combination of such electronic and digital sound making and shaping techniques with acoustic elements such as resonating bodies found at the site results in hybrid sound structures. Hybrid, in two ways: they integrate digital elements with acoustics and they allow to draw sonic characteristics from the site yet imprint it with artificially induced elements. Figures 8 and 9 show examples of such instrument setups.

16 Interfaces for Playing

An instrument's form and intended playing style thoroughly influences the character of a sonic wilderness intervention. If the instrument e.g. supports immediacy in sound generation (you press a button and a sound appears or changes) and features a simple playing interface, it makes it easy to react to the performance site. Contrastingly, an instrument that semi-automatically generates musical gestures (e.g. a drum machine that creates rhythmical elements) introduces a musical layer that can unfold without the performer's dedicated attention. This allows her to focus on other elements of the intervention (de Campo 2014). A combination of these two instrument types into one setup means that the performer can shift her attention between the two musical layers: she can either adjust parameters of the generative part, or play on top of its output. This is especially useful in wilderness interventions because it is often not clear from the beginning of the performance what level of control fits to a specific playing situation. The lack of rehearsal time at the site itself requires a setup as dynamic as possible. It is possible to integrate the two levels of interaction-immediate versus generative control-into one instrument by means of a live-coding interface that gives access to the mapping algorithms between (pre-defined) instrument components (Bovermann et al. 2014). It allows the performer to decide while playing: "I want to have fine-grained manual control over the rhythmical elements and then record them into a slowly changing a pattern", or "Let's change the scale from Dorian to Lydian".

17 Documentation

Documenting magic is an art form in itself.

The format in which a sonic wilderness intervention is documented is a significant statement on the artists' viewpoints. Within the Sonic Wild Code sessions, Dinah Bird, an experienced field recordist and experimental radio artist recorded and captured our interventions exclusively.[4]

Independent of the intended usage of such documentation, we recommend to capture every wilderness session in as varied forms as possible. From our experience the actual moments of intervention are so precious and intense that a recording is often a welcome help to re-imagine it later-on. If possible, sound and video recording should be combined with photography to capture the intervention in its context. One approach is to tightly integrate the documentation process into the set-up, possibly recreating a more subjective view of the intervention from the artist's perspective.

Documenting while playing can distract the performers from the creative process itself. Careful planning and prior preparation can help here as well as inviting a dedicated documentarist to participate. Such an external documentation adds a subjective perspective and captures the performance from a distance. In our venture it turned out to be even more interesting when the documentation is performed *actively*, i.e., the recordist moves around the site, changing focus between its sonic ecology and the sounds added by the sonic intervention. The documentation becomes a composition in itself, an interpretation of the moment that captures the soundscape, the playing, and the various sound sources (Fig. 10).

[4]The result can be listened to at archive.org (Bird 2015).

Fig. 10 Dinah Bird and Vygandas Simbelis capturing the stone field session with audio and video

18 Implications

A great number of artists of the 20th century contributed to the liberation of sound as a diverse musical material with vast creative possibilities. Sonic wilderness intervention and outdoor music are powerful examples of such contributions. Still, they are only at their beginning stage and struggle with technical circumstances as well as conceptual and philosophical questions.

Technical challenges include that commonly available electronic musical instruments are rarely suited for being played both in and with a sonic wilderness: Not only do we need to develop instruments that are more capable in incorporating aspects of their environment into their sounds, further, we should integrate mechanisms that enable them to harvest their electricity needs from their surroundings, e.g., via solar cells, wind turbines or electro-chemical reactions.

Conceptually, sonic wilderness interventions challenge the common understanding of performance not only by breaking up the dualistic approach of *performer versus audience* but also by questioning its anthropocentric viewpoint: interventions are equally intended to be both perceived and experienced by non-human agencies. This circumstance immediately raises questions that require more investigations and, most essential, a personal experience of sonic wilderness interventions:

Does a stone field listen?

What does it mean to communicate with birds and wind?

Are we still Nature?

Can we deepen our understanding of ourselves by making music in sonic wilderness?

References

Baird Callicott, J. (2000). Contemporary criticisms of the received wilderness idea.

Bird, D. (2015). Reindeer crossing concert. https://archive.org/details/ReindeerCrossingConcert_201509. Accessed August 5, 2016.

Bovermann, T., Egermann, H., Foerstel, A., Hinrichsen, A., Hildebrand Marques Lopes, D., Pysiewicz, A., et al. (2014). 3dmin-challenges and interventions in design, development and dissemination of new musical instruments. In *Proceedings of ICMC/SMC*. Athens: National and Kapodistrian University of Athens.

Bovermann, T., & Griffiths, D. (2014). Computation as material in livecoding. *Computer Music Journal*, *38*(1).

de Campo, A. (2014). Lose control, gain influence—Concepts for metacontrol. In *Proceedings of ICMC/SMC*. Athens: National and Kapodistrian University of Athens.

Green, O. (2014). Nime, musicality and practice-led methods. In *Proceedings of the International Conference on New Interfaces for Musical Expression* (pages 1–6). London, United Kingdom: Goldsmiths, University of London. URL http://www.nime.org/proceedings/2014/nime2014_434.pdf

Greie-Ripatti, A., Bovermann, T., Bird, D., Erdmann, A., Lindström, K., Simbelis, V., et al. (2015). Sonic Wild Code report. http://fieldnotes.hybridmatters.net/posts/sonic-wild-code. Accessed August 5, 2016.

Grill. T. (2014). On world construction, variation: Duoddaris. In *Proceedings of the 2014 Conference on Multimedia Modelling*.

Ouzounian, G. (2011). The uncertainty of experience: On George Brecht's event scores. *Journal of Visual Culture*, *100*(2), 198–211.

Riley French, J. (2016). Yoik. http://www.caughtbytheriver.net/yoik-jez-riley-french-chris-watson-field-recording-norway-ande-somby-nils-aslak-valkeapaa-sami-ulla-prittijarvi-johan-anders-baer/. Accessed August 5, 2016.

Robinson, J., Fischer, A., & Robinson, J. (2005). George Brecht events: A heterospective.

Schaffer, R. M. (1994). *The soundscape: Our sonic environment and the tuning of the world*. Destiny Books.

Snyder, J. (2011). Artistic and engineering research projects. http://scattershot.org/research.htm. Accessed August 5, 2016.

Standing Bear, L. (1998). *The great new wilderness debate* (pages 201–206). Athens: University of Georgia Press.

Toplap. (2011). Toplap about. http://toplap.org/about/. Accessed August 5, 2016.

Westerkamp, H. (1974). Soundwalking. *Sound Heritage, III*(4).

Wikipedia. (2016). Joik. https://en.wikipedia.org/wiki/Joik. Accessed May 20, 2016.

Instrumental Modality. On Wanting to Play Something

Bjørnar Habbestad and Jeff Carey

Abstract Composer/Performer Jeff Carey and Performer/Composer Bjørnar Habbestad have collaborated since 2002, developing a chain of works, performances and improvised events as the duo USA/USB. Parallel to their artistic work runs a development effort in SuperCollider, now branched into what is known as Modality (http://modality.bek.no)—a network of developers collaborating on the creation of a toolkit to support live electronic performance environments. In this interview Habbestad and Carey share insights to this development process whilst discussing what it means to play an instrument in a computer music context.

1 Tracing the Event

BH: As a starting point, I thought to go back to our first meeting, in Amsterdam in 2002, at Robert van Heumen's LiSa[1] class at STEIM. You were preparing for studying Sonology at the conservatory in the Hague, and I was in my first year as a postgrad student at the Amsterdam Conservatorium. And I think it is fair to say that we came from different musical backgrounds, yourself from audio engineering, punk rock and hardcore performance, and me being a classically trained flutist with a preference for contemporary music. But still we somehow shared a lot of interests and observations. Despite having radically different skill sets and technical knowledge, we identified similar problems with the whole laptop performance paradigm that was around at the time, for example. And Robert's class was an environment where these topics came up. I think it provided the opportunity to start talking about things, even if none of us ended up using LiSa in the end.

[1]http://steim.org/2012/01/lisa-x-v1-25/.

B. Habbestad (✉) · J. Carey
The Arne Nordheim Centre for Artistic Research Norwegian Academy of Music, Oslo, Norway
e-mail: bhabbestad@gmail.com

T. Bovermann et al. (eds.), *Musical Instruments in the 21st Century*,
DOI 10.1007/978-981-10-2951-6_17

JC: I like to think of LiSa as a tool that illuminated an issue, and when we realised what the problem was, it was time to leave it, dig deeper and search elsewhere for answers.

BH: And what was the issue?

JC: Gaining control, I think. Finding ways to create mapping situations that met our musical goals.

BH: This idea of influencing the tools we work with is another recurring topic of ours. Not wanting to accept the limitations by given tools and immediately available thinking. This was present already when I first saw you perform, with your big mixer feedback system.

JC: It was only four flight cases.

BH: … but I sensed that what you wanted to do was something else, or something more, something more direct. And I *definitely* wanted something more and something else than what my flute and the feedback system I was using offered. I had the distinct feeling that I had pushed the mixer feedback and guitar pedal approach to flute playing to its natural conclusion, and that now was the time to move on towards another approach.

JC: I think both our instruments were interesting to work with in terms of sound material, but also very frustrating, because they weren't as flexible to work with, you couldn't move from A to B, musically, without risking that the sound collapsed. So a lot of the music was dictated by the instrument, that is, the way of playing dictated what the music could do. And the dissatisfaction of this situation was a major motivation for developing something new.

BH: So rather than having your instrument dictating the composition …

JC: You might want to start as an architect and determine a structure, but then there is this uncooperative instrument …

BH: I remember being critical of many live electronic performances at the time, I found that they often were sounding out the immediate options of a given software. I used to say that I could "hear" people's controllers, because the way they were used was so one-dimensional and fixed. Another important factor, I think, was that we kept talking about instrumental music. You would often return to saxophonists like Brötzmann or Evan Parker.[2] Not that we didn't talk about electronic music, but there was quite often references to 'bodily practices' in our discourse.

JC: There is a strange connection between this and the dissatisfaction with the software tools that we just mentioned. With a computer instrument you are limited to the ideas put into your software. And at the time, computer music seemed to me to be a difficult strategy for a direct or fluent way to express music. Within system based approaches or controller based approaches, you are still stuck within a framework. Especially compared to the sonic palette and the immediacy of an instrumentalist. Here the ability to make a change is limited only by thought and skill, whereas with software synthesis there are

[2]http://www.paristransatlantic.com/magazine/interviews/parker.html.

often built-in limitations on the sonic palette by either the synthesis routine or the functions that control it. We wanted to build "discoverability" into our approach.

BH: I think our aims at the time were completely utopian. The systems we had in mind were, more or less, completely out of our reach. But we allowed ourselves to be ambitious enough to articulate ideas that we are still working on, almost 15 years down the road! And today it's easy to say that, well, things take time, or maybe we have not realized all we wanted. But another perspective would be to say that we thought some really good thoughts, and asked some really, really difficult questions, that we have been refining, re-answering, reframing. And I think this testifies to the benefit of the quality of the questions.

JC: And what were the questions again?

BH: Hah! How to make a live electronic instrument that allows for immediacy of sonic detail and variation, while simultaneously opening up for global control in a compositional way? Or something like that.

JC: Yes, and in that way fusing what we found most valuable from the improvised music scene with the formal thinking of composed music traditions of electronic and contemporary music. So we started looking somewhere between the laptop-paradigm and the hyper-instrument-paradigm, somewhere between Cascone and Machover.

BH But this is way before developing the idea of Modality, I don't think we saw where we were heading, at least not at the time. Where we did not want to go was much clearer.

JC: I think we quickly arrived at a point where it became apparent that the idea of a hyper-instrument was not really what we were interested in. This was definitely a first step, an opening of the idea of an interface to something outside the traditional instrument, but not enough. What became clear was that our collaboration was about pulling away from the notion of expanding an acoustic instrument. Rather, it was about moving towards the idea of an instrument not residing in the physical instrument at all, but in a computer abstraction of an instrument, or actually somewhere in between the two.

BH: This reminds me of how I perceived the role of the flute in my feedback system. I was conscious of the fact that I was playing an instrument with a lot of cultural connotations, but I felt at the time that I was not really playing 'the flute', but rather using it to influence a larger system. And so, going towards a hyper-instrumental paradigm would have been moving backwards, towards creating an extension of an instrument with a fixed sonic identity. I was more interested in creating a system where the instrument would be one of several factors that would have an impact on the final result, than having somebody else's preconceptions of the flute, or my own for that sake, define it.

JC: We were not that aware at the time, but what this led to was the notion that physicality would interface into a larger and more abstract system. Fundamentally, this was really what we discovered, but did not yet fully see the consequence of.

2 “Mapping Is Where It’s At!”

BH: The first real piece we did together was Machine Gun Etiquette (MGE), a three part composition requiring radically different kinds of processing and synthesis. An automated phrase sampler recording and transforming flute material, real-time synthesis based on analysis of the flute sounds, granular processing, and a computationally heavy time stretch procedure. All this were posing challenges just in terms of its administrative tasks: Stuff needed to be done at certain times in order for other things to happen, etc. This was in the middle of the development of MKeys,[3] no?

JC: In a way this was a step back into composition, away from the development, as the framework wasn’t really there yet. So we needed pragmatic solutions. And this was a valuable lesson: it provided a test bed for our ideas that made it clear that our musical expectations were not met with the technical solutions we had at hand. A central challenge was the transition from distinctly different sections—the moving from A to B—a musically necessary and obviously simple formal situation. But it involved mundane tasks having to be taken care of, working synchronously and asynchronously at the same time, in order to prepare for an upcoming change. I think the main lesson from MGE was that we discovered that we needed a way to offload the responsibility of moving the instrument forward onto somebody (or something) else than the performer. So preparing the software to move forward had to become part of the instrument itself.

The use of networked computers, all the logic involved in starting and stopping events, all this points towards the piece being about navigating a larger system. That the totality of our resources made out this shared instrument that we had to influence in different ways. It had some secretarial dimensions to it, administrative tasks to prepare for the computation. Or—preparing for the *real time moment*. It is composition on its head, in a way: always evaluating what happened and interpreting how that might influence where we are going.

[3]MKeys (2005)—or *modal keys*—was the first generation software written by Jeff Carey, that allowed for a layered keyboard. MKeys later became the foundation or starting point for the work with Modality.

BH: I find it interesting that we established a compositional discourse that involved an improvisational ethics and a formal structure at the same time. That we wanted both worlds present.

JC: Yes—trying to reconcile the "what happened school" to the "school of prescriptively declaring what is next". It does illuminate that there are these two worlds, and that these musics interfere somehow in our desire to make a new instrumental paradigm.

3 What Is the Instrument? How Do We Play It? and What Does It Mean?

BH: As we are tracing *why* we ended up doing what we did, why did we ask *these* questions and so forth, it seems to me that they mainly came out of practical musical situations. And post *Machine Gun Etiquette*, I actually thought: Wow, this is coming together, this is going somewhere, we will actually make a tool with the powers that we need and desire.

JC: Well, *The Respirator,*[4] following MGE was definitely a multidimensional leap forward. Where MKeys allowed us to repurpose the computer keyboard through the notion of a *layered interface*, the first respirator mirrored this *layeredness*, allowing numerous layers of instruments that you could choose to work with. That is, you could choose to focus on an aspect of the physical interface and select which aspect of the sound-producing part of the instrument this would address. I think the respirator was about interacting, a little bit at that time, with a much larger instrument. So the grand abstraction is the thing that is modified as it is going, interacted with, but simultaneously commanded. Moving from the organic instrument we talked about earlier, towards having the ability to make executive decisions, globally and immediately. And this is not really from the conductor-point-of-view, but more from the musicians point of view.

BH: So it is in a way empowering the musician's ability to make micro-decisions with global impact?

JC: Yes. It allows you to scale your level of interaction, so you can tune the system, and then take command of it. There are two modes of intervention, at least this is what became clear with the respirator.

BH: And this effectively addresses the duality that we talked about—allowing for both split-second decision making *and* architectural influence.

JC: And as such, collapsing the duality of the composer and performer. To me, this is an analogy of the management of the composition process. The respirator is about interrogating the spaces in between all these analogies: the

[4]A solo piece for flute and electronics by Bjørnar Habbestad, using the eponymously named software and hardware https://respira2or.wordpress.com/about/.

conductor versus musician, the composer versus material generator, the moment versus the architecture. I perceive the code for all the musical tasks as this arduous interface that has to be navigated. So scaling mapping systems —or making mapping systems that can be navigated as musical instruments in themselves—is the concern of our work. There are so many details to work on, adjusting parameter by parameter, but we wanted the ability to first look at a single one, then a bunch of them, and then at ALL of them, in order to easily push them around.

4 From Piece to Tool

BH: And the next piece would be *Chop-Chop (2010)*, your first live multichannel piece?

JC: Yes. 'Chop-Chop' was made leading up to the first Modality meeting in 2010. In summary the big thing here was unifying input protocols, so that there is no difference between HID,[5] MIDI,[6] OSC[7] or whatever. This became the starting point for Modality.

BH: I see a pattern evolving here: the lack of coherence between the artistic idea and the available tools, trying to do something which resides in the outskirts of what is technically available. If *Chop-Chop* was a testbed for modality, *MGE* was the same for MKeys and my Respirator-piece had the same function for the MKeys II/Respirator software. I have the feeling that the Modality team now has a more unified, global way to approach software development.

JC: I think our perspective initially was that of musical need, of production need. And we had a somewhat limited ability to envision the entirety of the language within which we were working.

BH: And at a point it became clear that the only way to achieve our goals, to collectively tackle the snags and limitations we found within SuperCollider itself, would be to open up our questions to a larger group. I remember that 'our problems' also proved to be 'other people's problems', and that the first Modality meeting opened up to solving many of these. The very act of putting a diverse group of people in a room and then asking 'what's the best way to solve this problem' has proved to be a very strong tactic.

JC: I think it is about the balance between wanting control and having control, both musically and technically. And here, to me, the composition comes as a utopian promise, equipped with a set of notions that need to be considered.

[5]Human Interface Device, a computer device and communication protocol allowing for interaction between humans and computer.

[6]Musical Instrument Digital Interface, a communication protocol describing musical information.

[7]Open Sound Control, a protocol for networking sound synthesizers, computers and other devices.

And as end users of a framework of our own imagination, as both developers and musicians, we try to get these compositions to fit within our structure of working.

BH: Yes, being able to influence *how* you work. And if the piece is the problem, and the instrument the tool to solve it, we have to understand not only how to make the instrument, but how to use it, too. This is one of the strengths of the Modality group I think. The circumstance that the developers are musicians and vice versa allows for looking at development from many angles. Because a tool is not a static object, a tool implies a practice. Something that something is being *done* to. So it is an active term…

JC: …an object with all its metadata, and its actions, and ideas of context.

BH: …well, to any tool there are actions. Hence it is not possible to reduce it to an object alone, it has actions connected to it, what Gibson called *affordances*. So the shaping of our tools is also the shaping of our actions and possibilities that we have. And if anything, this whole development process has been about negotiating how to shape these tools and our conceptions of them. By carving out new possibilities in software we are unleashing new ways to act, listen and think with technology. This pushes us to refine the software. Which *again* allows for a different action, etc. There is a strong connection here, not only a feedback, but a real dialectic between thinking and doing.

JC: I think it is true that the pieces have driven much of the development, but then the use of software in each piece has led not only to refinement, but also to a lot of reflection around the processes which, at certain points, has not really narrowed but exploded the amount of work needed. Where we potentially are at—in a bit of time—this is a place we could not have foreseen without all these plateaus, all the stops. Each subprocess has opened up certain options, some possibilities which have proven to be ways of approaching parts of this utopian situation that we imagined in 2002.

5 Preparing the Real Time Moment

JC: As Modality is approaching a finished state, as much as any software is ever finished, it allows us to imagine where to go next. And to me, Modality is a window for being able to examine physicality in a way that we yet haven't been able to. Specifically this is about examining the way that different physical gestures interact with the same sound processes or code bases. We can now start looking at how we can exploit and develop this unified data stream that we have access to through Modality, and see more clearly what are the implications of having put these things together. I think there is another 15 years of music in our hands here. Remember—I am working on a daily basis with technology that we made from respirator and MKeys. That is hundreds of concerts, thousands of hours of music making. And the Modality plateau will open up for a continued musical expression of our ideas.

BH: Could you say something about the whole need to perform? Where does it come from?

JC: For me, music is largely a social activity, both in the act of making and experiencing it. And I think that as empathetic social creatures, seeing and hearing a performer tends to have a stronger impact on us, compared to musical situations where performers are absent. And personally, as a performer, my ability to react to other musicians and the audience in ways that are not entirely anticipated or pre-determined is of critical importance.

BH: I wonder if we attach more value to expressions that we understand as linked to physical gestures, if the connection itself warrants some authenticity, some liveness or realness.

JC: When performing in improvised situations, or performing works with variability or indeterminacy, something that requires decision making on the part of the musician, I find it is imperative to be working with a tool that allows you to discover the musical landscape you are creating, whilst in the making of it. Not having to separate the planning and execution of a sound or sound event into different moments, but allowing for an actual interpretation of the context the moment you are experiencing it. If this equals a quality in itself on a general basis is hard to say, but it lies to the core of why I work as I do.

Jeff Carey and Bjørnar Habbestad continue to collaborate from two different sides of the Atlantic. Both are eagerly awaiting the next release of the modality toolkit.

Appendix

MKeys, a basic software utility that allowed for paged toiling of musical functions: Starting/stopping of synthesis processes, sample playback and recording, algorithmic tasks, transformations etc. Developed by Jeff Carey between 2005 and 2010.

Modality, a software framework bringing control input for electro-instruments into a unified structure. Various control sources such as HID, OSC and MIDI are all accessed in a uniform manner, with the goal of unifying their interface to allow for easy swapping and trading of control sources among different processes. Developed by the Modality team http://modality.bek.no/.

Machine Gun Etiquette, composed in 2005 by Jeff Carey and Bjørnar Habbestad, premiered May 13th at NuMusic, Stavanger. Later performed in Norway, Holland and Germany. The piece was comprised of three large form sections: (1) duo for notated acoustic flute with SuperCollider processing, (2) duo for amplified flute with live transformations and electro instrument (3) trio for transformed sampled flute and two live electro instruments.

On the technical side, MGE included networked control over flute processing, paged transformation instrument with foot pedal controllers for flute and paged transformation and synthesis with mouse, keyboard, foot pedals and bank of four joysticks. Compositionally, the goal was to make a piece characterised by a progression of material and interactive development, starting out with the internal sound of the acoustic flute. The exploration of the technicalities and 'administrative' responsibilities of the piece coupled with the desire for interaction between two players led to developing a single instrument from two networked computers. Computer A was used for sampling and transforming flute sounds. Computer B, via network control, triggered various transformations to the flute signal chain and administered sample recording on computer A. Samples from computer A were also copied over the network for further processing on computer B.

Respirator, composed in 2009 by Bjørnar Habbestad, premiered at Borealis 2010. A multi-layered interaction framework where each layer was associated with 'musical behaviours' or sections of a piece. A layer would have its own signal routing framework, where physical controls like foot pedals or computer keyboards would be redefined. Each layer, simultaneously running/sounding, allowed up to 4 series of transformations on any number of parallel transformations. Entry and exit of a layer would execute a collection of necessary functions: transformations to turn on or off, sample recording, muting of inputs, changes of mix settings or start/stop algorithmic functions etc.

Chop-Chop, composed 2010 by Jeff Carey at Landmark, Bergen. Later performed at DNK in 2011 and in Baltimore 2012. A 16-channel surround sound piece for electro-instrument based on the development and transformation of a small collection of source sound synthesis materials—phase modulation and feedback based voices—with direct physical control over voices, transformations, signal routing and panning.

The piece, divided into 9 subsections based on differing musical behaviour, was meant to focus on the physical development of sonic material, spatialization, and compositional logic over time. As such it wanted to explore the idea of a changeable interface. Mappings per section changed, allowing the performer to move back and forth to turn various transformers on and off. In some section the mappings changed from toggles to drum pads where a pair of pads would inform a particular algorithmic process by 'pumping' energy into a leaky integrator mapped to time stretch buffer pointers, filter tutors and amplitudes. This piece was the 'proof of concept' into/for the larger Modality project. Its experimental implementation had some limitations but the notion that collections of musical behaviours could be matched with a variety of control scenarios in a manageable instrumental setup was a limited success leading towards the full development of Modality.

Authors Biography

Bjørnar Habbestad (1976) is a flutist and PhD researcher at the Norwegian Academy of Music in Oslo. He is educated in music, art history and philosophy from Bergen, London and Amsterdam, now working as a soloist, chamber—and ensemble musician in Scandinavia, Europe, Asia and the US. He is a founding member of the N-Collective, Artistic Director of +3DB records and former curator at Lydgalleriet, a Bergen based gallery for sound art. Habbestad's artistic and research interests cover musical grounds from classical contemporary to noise, electro-acoustic and free improvised music (www.bjornarhabbestad.com).

Jeff Carey (1973) focuses on novel sound generation techniques ranging from feedback systems to non-standard synthesis. His electronic compositions are the expressions of structural possibility using composed algorithms where sound is elastic, non-linear, multi-dimensional, and part of an indivisible whole. As an electro-instrumentalist, he is dedicated to making self-built synthesis software a viable live performance instrument: no editing and no non-realtime operations. Carey studied Audio Technology at American University (1994) and computer music at the Instituut voor Sonologie in the Koningklijk Conservatorium (2002). He is a founding member of the N-Collective (www.jeffcarey.foundation-one.org).

Instruments for Spatial Sound Control in Real Time Music Performances. A Review

Andreas Pysiewicz and Stefan Weinzierl

Abstract The systematic arrangement of sound in space is widely considered as one important compositional design category of Western art music and acoustic media art in the 20th century. A lot of attention has been paid to the artistic concepts of sound in space and its reproduction through loudspeaker systems. Much less attention has been attracted by live-interactive practices and tools for *spatialisation* as performance practice. As a contribution to this topic, the current study has conducted an inventory of controllers for the real time spatialisation of sound as part of musical performances, and classified them both along different interface paradigms and according to their scope of spatial control. By means of a literature study, we were able to identify 31 different spatialisation interfaces presented to the public in context of artistic performances or at relevant conferences on the subject. Considering that only a small proportion of these interfaces combines spatialisation and sound production, it seems that in most cases the projection of sound in space is not delegated to a musical performer but regarded as a compositional problem or as a separate performative dimension. With the exception of the mixing desk and its fader board paradigm as used for the performance of acousmatic music with loudspeaker orchestras, all devices are individual design solutions developed for a specific artistic context. We conclude that, if controllers for sound spatialisation were supposed to be perceived as musical instruments in a narrow sense, meeting certain aspects of instrumentality, immediacy, liveness, and learnability, new design strategies would be required.

A. Pysiewicz (✉) · S. Weinzierl
Audio Communication Group, Technische Universität Berlin, Berlin, Germany
e-mail: andreas.pysiewicz@tu-berlin.de

S. Weinzierl
e-mail: stefan.weinzierl@tu-berlin.de

T. Bovermann et al. (eds.), *Musical Instruments in the 21st Century*,
DOI 10.1007/978-981-10-2951-6_18

1 Introduction

The arrangement of sound in space as an integral part of the musical composition can be observed as a central concern of current art music practice. In the field of electronic and electroacoustic music in particular, one will hardly find musical performances which do *not* make systematical use of multiple loudspeaker projection techniques or site-specific spatial arrangements. Along with the description of different artistic concepts related to the organization of sound in space (e.g., Stockhausen 1959; Leitner 1971; Xenakis 1992), also the techniques for sound spatialisation have been addressed with respect to a formalization and classification of their artistic and technical principles within the last decades (e.g., Roads 1996; Zvonar 2000; Baalman 2010; Sannicandro 2014).

While the critical examination of space as a musical parameter is not a general novelty, increasing attention has been paid to different aspects of real time *sound spatialisation* as performance practice. This observation is indicated by a growing number of concepts and instruments that have been developed to explore performative degrees of freedom for the spatial presentation of music, primarily introduced by communities in the context of the *New Interfaces of Musical Expression* (NIME), the *International Computer Music Conference* (ICMC) or the *Sound and Music Computing* (SMC) conferences. Understanding space as a significant musical parameter (alongside with pitch, timbre, intensity and duration), it is only consistent that spatial sound controllers are regarded as specific instruments of musical expression, too.

Beyond these general considerations, however, there is a lack of systematic approaches to identify, analyse and contextualize real time spatial sound controllers in the interplay of musical expression, artistic performance practice, audio technologies, and the field of human-computer interaction (HCI), as represented by the theory of digital musical instruments (DMI).

In the following, we will use this broader perspective and attempt to provide a structured review of instruments and interface concepts introduced to enable performers to control spatial parameters from the early days of electronic music performances to date. After a brief introduction to the terminological and methodological framework used to categorize a collection of spatial controllers, we develop a systematic taxonomy, by which we obtain a clearer, theory-based perspective on current trends of interface design in the domain of musical expression and provide an outlook on conceptual consequences for potential future developments.

2 Defining the Contexts: Space and Spatialisation of Sound

Along with the use of technology to expand musical boundaries, the notion of space as a musical parameter is considered as one constitutive element of electroacoustic music. Unlike with other parameters such as timbre or pitch, it is challenging to define the notion of *space* precisely, since it can refer to a multitude of phenomena ranging from architecture, room acoustics and spatial hearing to conceptual and metaphorical uses, such as the tonal space or structural concepts of musical foreground and background.

To avoid any terminological confusion, we will here refer to *sound spatialisation* to indicate a group of techniques for organizing and manipulating the spatial projection and movement of sound in a physical or virtual listening environment (Valiquet 2011).[1] Similarly, the term (electroacoustic) *sound diffusion*, frequently used synonymously for sound spatialisation, refers to the real time distribution of sound throughout space by controlling the relative levels, equalization and localization of sound during performance. While often related to the performance practice of fixed media compositions (stereo or multi-channel), the source material may also be generated live.[2]

Roads (1996, 451) emphasizes the duality of physical and perceptual layers within the act of sound spatialisation and states the dramaturgic and structural importance of spatial movements. Thus, not only the technical developments, but also the knowledge of mechanisms and cues for spatial perception, as it was acquired throughout the 20th century, is crucial for the application of spatial sound projection techniques. Refer to Brech (2015) for a historical review on the examination and musical operationalization of spatial perception, with Roads (1996) and Blauert (1997) providing comprehensive insights on the relevant psychoacoustic phenomena.

When investigating on the role of spatialisation in compositional and performance practice of Western art music, one will find space-related techniques throughout musical genres and eras such as early antiphonic choral writings or specific orchestral techniques reaching back to the 16th century. However, spatialisation was not generally considered a crucial parameter of musical expression before the advent of electroacoustic music being obviously related to the availability of appropriate technical resources. With new technical means, composers started to think about spatial organization in their pieces in a very different way, involving new forms of spatial aesthetics. Many composers quickly adapted their musical concepts to the new (spatial) techniques resulting in a demand for new and better technologies to realize their refined spatialisation conceptions. As a consequence, technological advance in the last two decades of the 20th century, the rise of digital production technologies and the increasing efficiency of spatial rendering algorithms, especially methods of sound field synthesis (*wave field synthesis* or

[1]Accordingly, the term *spatial music* was coined to highlight electroacoustic compositions in which the dynamic projection of sound sources is an integral part of compositional process. While the practice of spatialisation can be applied to any kind of spatial sound projection, it mainly refers to the field of electroacoustic music.

[2]Sound diffusion is originally used for the live presentation of *acousmatic music*, a form of electroacoustic music composed for (multiples of) loudspeakers using recorded sound material out of their original context. Interestingly, sound diffusion as performance practice is conceptually related to one specific control interface: the fader board of mixing desks (see our taxonomy).

higher order ambisonics, to mention only two important techniques) has greatly affected (and still affects) performance practice of electroacoustic music.

2.1 Sound Spatialisation as Performance Practice

With respect to the concepts and tools for sound spatialisation, different systematic approaches have been developed. Malham (1998) outlines three basic techniques of sound spatialisation: binaural reproduction of the sound field (by providing the signals directly at the ears), stereophonic sound projections by means of loudspeaker orchestras (a.k.a. sound diffusion) and sound field synthesis techniques (e.g., by means of ambisonic systems).[3] Whereas binaural techniques do not play a noteworthy role for performance practice, the latter two can be considered as well established paradigms of realtime spatialisation.[4]

Following Lynch and Sazdov (2011) who differentiate between three artistic concepts of spatialisation in electroacoustic music we can categorize three main approaches of sound spatialisation considering all possible means:

- Sound spatialisation based on properties of the fixed audio material (mainly related to the spectral[5] or temporal features).
- Sound spatialisation based on algorithmic or stochastic[6] processes (not related to the analysis of the audio material), controlling the spatial presentation without direct access to spatial parameters.
- Sound spatialisation based on the direct access to the spatial projection of the sounds and/or manipulation of their properties in real time (by means of decorrelation, panning, or more complex methods of sound field reproduction).[7]

The major distinction between these categories lies in the degree of active control, the performer can exercise over the spatialisation process during the performance. While the spatialisation within the first two categories is highly determined by either the texture of the material itself or the algorithm in use, the last

[3]Since we cannot address the technical principles of sound field synthesis here, the reader can refer to Geier et al. (2010) for further details on wave field synthesis, ambisonics techniques and recent stereophonic panning methods.

[4]The transition from amplitude panning techniques to methods of sound field synthesis represents a paradigm shift of sound spatialisation (Geier et al. 2010): from a channel-based approach (controlling a single channel assigned to one loudspeaker) to an object-based approach (controlling a sound object in space).

[5]For a comprehensive review of *spectral spatialisation* techniques, see Jaroszewicz (2015).

[6]It might seem paradox to include stochastic processes to a category mainly defined by determined characteristics, however they are grouped here due to their decreased realtime controllability in terms of exact spatial deployment.

[7]This category may also include mapping strategies in which the synthesis process of the sound material directly affects its spatialisation, in contrast to the static spatialisation process of fixed audio material in the first category.

category offers a direct mapping of controller data to spatial parameters. All these types of spatialisation can generally be applied during performance, yet the latter paradigm represents the most common approach to sound spatialisation to date. Correspondingly, Baalman (2010) describes exclusively spatial techniques that are based on the direct manipulation of spatial parameters ranging from control over location and trajectories of sound sources to more complex parameters such as enhanced acoustic characteristics of the space. Having identified some major concerns of spatialisation practice, in the following we take a look at the distinctions different concepts and implementations of common sound *spatialisation systems*. It is important to note that the terms *spatialisation controller*, *interface, instrument* or *system* are used inconsistently and interchangeably in relevant literature, most commonly denoting an (electronic) apparatus to control the spatial behavior of sound. While *controller*, *interface* or *instrument* frequently refer to the specific device the performer is operating, the *spatialisation system* often means the set of (digital) components to render audio streams for the spatialisation process.

2.2 Sound Spatialisation Controllers in Context of Digital Musical Instruments

As outlined above, spatialisation can be considered as established artistic practice in the broader field of electroacoustic music and live electronic music. We have discussed that prevalent spatial sound techniques can generally be applied both to the production process (mainly in the studio) and to the real time presentation of music in the respective performance space. Zvonar (2000) formally differentiates between the live performance approach to sound spatialisation and techniques for pre-composed spatial arrangements of sound, such as environmental multichannel soundscape, classic studio-based multi-track composition and automated spatial control. Accordingly, only some available implementations of sound spatialisation systems are suited for use during performance. While several spatialisation controllers have been designed explicitly as studio production means, other system designs have simply met the limits of contemporary technologies, be it in terms of computational power for spatial rendering or the lack of suitable control interfaces.[8]

Furthermore, the complexity of the control task can be considered as another substantial obstacle for real time spatialisation. Such control and mapping related issues are well-known and dealt with in the field of human-computer interaction (HCI) and especially in the interaction design for interfaces of musical expression, an applied subfield of HCI. Therefore, it seems reasonable to regard the means for

---

[8]One can consider Stockhausen's *Rotationstisch* (a loudspeaker mounted to a rotating turntable system) as typical tool for spatial studio composition (Brech 2015). The spatialisation system used by Chowning to realize his simulation of moving sound sources (Chowning 1971) represents a typical studio approach. Simultaneously, it was clearly limited by processing performance of the 1970s (Zvonar 2000).

spatial sound control used in live musical performance from the perspective of design practice of digital musical instruments (DMIs). This potential link has already been roughly explored in previous research (Wanderley and Orio 2002; Marshall et al. 2007; Schacher 2007; Perez-Lopez 2015), with a particular focus on the gestural control paradigm. At the core of the DMI metaphor, as introduced by Miranda and Wanderley (2006), stands the decoupling of the physical interface (input or control device) from the sound generating system (contrasting to the integral concept of acoustic musical instruments). Both instances are connected via a mapping layer assigning outputs of the controller to the inputs of the sound rendering engine. This modularization offers new degrees of freedom for the instrument design, however, the alleged decorrelation between the physical action of the performer and the produced sound, raises new issues related to the appreciation of the artistic performance (cf. Emerson and Egermann, this volume).

Considerations on both the control interface and the mapping structure are crucial for the instrument design in order to minimize control complexity without limiting its functionality. For a systematic outline of mapping strategies refer to Miranda and Wanderley (2006). Marshall et al. (2007) discuss common control issues and introduce three levels of spatial sound control parameters which are related to (1) the position, orientation and movement of the sound source and sink, respectively, (2) characteristics of the sound source (and sink), and (3) environmental and room model parameters (Marshall et al. 2007, 229). For a list of typical parameters related to all three levels see Table 1.

Beyond the aforementioned control aspects, the interaction interface includes the feedback side—be it visual, auditory, or tactile-kinaesthetic feedback—primarily experienced through the physical device itself and secondarily as an intended (auditory) result of the sound generation process (Miranda and Wanderley 2006, 11).

In order to compare and analyse musical interfaces appropriately, different classification systems have been developed, the most common one going back to Miranda and Wanderley (2006). Based on the resemblance to existing musical

Table 1 Spatialisation system control parameters (based on Marshall et al. 2007; Perez-Lopez 2015)

Sound source[a] position and orientation	Sound source[a] characteristics	Environmental/room parameters
Position (X, Y, Z)	Size	Size
Elevation	Directivity	Presence
(Trajectories)	Presence/distance	Early reflections
	Brilliance/warmth	Reverberation
		Reverb. Cut-off Freq.
		Doppler effect
		Air absorption
		Equalization
		Geometry

[a]Parameters refer to sound source and sink respectively

instruments, the authors distinguish between augmented musical instruments, instrument-like or instrument-inspired controllers, and alternate controllers. Especially the category of *alternate controllers*—subsuming various different interface concepts beyond the physical-mechanical interaction paradigm of acoustical instruments—can be broken down into sub-categories related to their sensing functionality relative to the human (Paradiso 1997; Mulder 2000): *touch controllers* react on direct physical manipulation (like a button or knob); *non-contact* or *expanded-range controllers* provide a limited sensing range for control gestures without physical contact (e.g. by using an infrared sensor system). *Wearable* or *immersive controllers* capture the control gestures with few or no restrictions to the movement since the performer is always in the sensing field (either by using, e.g., a sensor glove, suit or wide-range camera tracking system).

A special form of wearable controller can be found in biofeedback interfaces allowing for the acquisition of electrical signals generated by the human's muscles, eyes, heart or brain. Although present for over 50 years now in the field of music and interactive media art, these interfaces have played no significant role as spatial performance instruments, most likely due to the limited controllability and bandwidth of some of the captured parameters (such as brain waves).[9]

For a larger subgroup of alternate controllers, Overholt (2011) uses the term *borrowed controller* in order to emphasize that these have not originally been designed as a musical interface, such as video game controllers, camera tracking systems, etc. Interestingly, most spatialisation controllers can be assigned to this category.

Related to the control paradigm, further criteria to distinguish between different realizations of spatialisation controllers can be addressed. With respect to DMIs, Pressing (1990, 14) and Birnbaum et al. (2005, 193–94) propose multidimensional description spaces dealing with different aspects related to the controller and its relation to both the performance and the performer. Perez-Lopez (2015) derives a set of dimensions relevant for the analysis of spatialisation systems, including:

- *Role of the performer*—the performer exclusively controls spatial parameters in contrast to a performer who controls both spatialisation and sound synthesis.
- *Required user competency*—casual untrained users in contrast to trained expert users aiming at expressivity and virtuosity.
- *Number of performers*—most spatialisation instruments have been designed for a single performer; however, the control task could also be (functionally) shared by a group of performers.

[9]There is consensus that *Music for Solo Performer* (1965) by Alvin Lucier, scored for "enormously amplified brainwaves and percussion", was the first composition to make use of a biofeedback interface to control percussion instruments by the resonance of the performers brain activity (Miranda and Wanderley 2006). Several further artistic experiments have followed using biofeedback interfaces. Refer to Miranda and Castet (2014) for a comprehensive review on brain related interfaces.

- *Multiplicity of control*—denotes the relationship between the quantity of simultaneous control streams available and the requirement to control these parameters continuously (as opposed to a default state when no control signal is present).
- *Control Monitoring*—related to the real time feedback modalities provided by the system on the executed control (e.g. by using a graphical user interface).

Having discussed the premises of spatialisation as performance practice in the field of electroacoustic music and contextualized real time spatialisation controllers within the discourse of DMIs and HCIs, we will provide a systematic inventory of spatialisation controllers presented to the public from the 1950s till today in the following.

3 A Systematic Inventory of Spatial Sound Controllers for Real Time Performance

There have been a few recent attempts to review the evolution of spatialisation controllers from a historical and musicological perspective (e.g., Brech 2015; Brech and Paland 2015). Some authors have explicitly focused on spatialisation interfaces for real time performances of music (Mooney 2005; Johnson et al. 2013, 2014a, b), others have discussed more recent developments of sound spatialisation systems and spatial rendering frameworks (Marshall et al. 2007; Perez-Lopez 2015; Peters 2011; Peters et al. 2009; Schacher 2007) as the core component of common software solutions for sound spatialisation.

By providing a classification system and a first systematic inventory of spatialisation controllers, our contribution aims at providing deeper insight into design and performance practice of spatial sound controller. In order to guide future design efforts, we intend to gain a better understanding about the concepts that led to the specific developments for sound spatialisation practice.

3.1 Study Design and Methodology

Having outlined common categories and dimension spaces related to DMIs and their adaptions to spatialisation instruments, we suggest three dimensions for a taxonomy of spatialisation controllers (Fig. 1).

The first dimension is derived from the extended DMI taxonomy adopted by Miranda and Wanderley (2006) and Mulder (2000), which has been discussed above. As outlined before, due to conceptual similarities of spatialisation interfaces for real-time performances and digital musical instruments, we consider it reasonable to classify the controllers under the terms of musical instruments. However, since spatialisation instruments are rarely directly derived from traditional musical

CONTROLLER TYPE / INTERFACE

- **Instrument-like and augmented controllers**
  simulating, inspired, or augmented with traditional/extended techniques
- **Touch controllers**
  haptic /tactile interface
- **Non-contact, extended range controllers**
  free gestures in a limited sense range
- **Wearable or immersive controllers**
  gloves, suits, camera tracking; performer always in sensing range
- **Mixed controllers**

CONTROLLED SPATIAL PARAMETERS

- **Diffusion parameters**
  spatial position, spread, timbre etc.
- **Sound source related parameters**
  incl. orientation, trajectories and characteristics, sink respectively
- **Room parameters**
  acoustical parameters, physical models, or algorithmic/stochastic behaviour

SCOPE OF CONTROL

- **Exclusive spatial control**
- **Including sound generating/synthesis control**

Fig. 1 Dimension space for the classification of spatialisation controllers

instruments,[10] we have customized the set by combining some categories and adding another category (a *mixed set* of multiple sensors combining different categories) common for the practice of spatialisation. Since the category of *alternate controllers* tend to be the rule rather than the exception, we will not use the term *alternate* and instead directly refer to the control interface paradigm these instruments follow.

The second dimension refers to the scope of spatial parameters, which are controlled by the performer in real time using the spatialisation interface (see

[10]It remains a matter of ongoing discourse, whether certain kinds of production or reproduction devices (the record player or a mixing desk, for instance) can be considered as musical instruments. See Hardjowirogo (this volume), for a thorough discussion of musical instrument identity issues.

previous section). Similar to the first dimension above, we have adapted the categories to embrace all relevant parameters ordered according to their complexity. Here, we decided to distinguish explicitly between diffusion parameters, i.e. parameters controlled by a channel-based diffusion system such as sound presence, position, or spread of the sound image, in contrast to spatial parameters of a sound source related to an object-based spatialisation approach. Since we could identify only a small number of complex spatial parameters in our explored data (such as room or environmental parameters, extended physical models or spatialisation algorithms), we created a shared category for all remaining parameters.

Referring to the role of the performer and the scope of integrated control options, the last dimension differentiates between instruments to control the spatial sound projection exclusively and instruments designed to handle the sound synthesis process as well. This category is closely related to the general concept of the instrument being either explicitly a spatialisation instrument or a self-contained music instrument with additional means for the spatial sound projection. In the latter case, we would expect a high correlation with the augmented instrument category in the first dimension of our taxonomy.

3.1.1 Research Focus, Limitations and Resources

As outlined before, we aimed at gathering data on all kinds of real time spatialisation instruments used in the history of western art music practice ranging from early developments in the middle of the 20th century to present implementations. The most important prerequisite for the spatialisation controller to be included in our survey was their potential to be used in real time as part of an artistic performance. Thereby, it was irrelevant whether the spatialisation instrument has been presented only once in a single performance or artistic demonstration or whether it has gained a certain popularity for spatialisation practice.

Another conceptual requirement was the existence (or at least specification) of a defined control interface for the human-computer interaction. Most software-based spatialisation systems provide open interfaces to connect to any sensor or control interface of choice. While this aspect of modularity can provide advantages under certain conditions, we only considered the controller as a fully developed musical instrument, if it incorporates a specific control interface.

Our inventory is based on different sources, which have been analysed systematically in relation to relevant content. These include secondary sources, i.e. textbooks or survey articles on spatialisation practice that have been published since the 1990s. All relevant texts evaluated in our study can be found in Roads (1996), Manning (2013), Brech (2015), Brech and Paland (2015), and Johnson et al. (2014a), along with the primary resources cited therein.

Moreover, we searched the proceedings of relevant international conferences as well as the major journals related to computer music and technology for musical expression, including all years of the *New Interfaces for Musical Expression (NIME, 2001–2015)*, the *International Computer Music Conference (ICMC, 1975–2015)*,

Table 2 Inventory of spatialisation controllers (in chronological order)

#	Name of controller/project	Year
1	*Potentiometre relief portico* (1951), *Pupitre d' space* (1952), P. Schaeffer	1951
2	Rotation Mill (*Tonmühle*), 1959, 1970 for Stockhausen, Expo 1970 in Osaka	1960
3	*Photocell mixers* (1967, 1968), F. Rzewski, D. Behrman	1966
4	Spherical sound controller for German Pavilion, World Expo 1970 (TU Berlin)	1970
5	*SAL Mar construction* (S. Martirano)	1971
6	Circular relay switch (B. Leitner)	1971
7	*HaLaPhon* (different versions, 1971–1985) (H.P. Haller, P. Laszlo)	1971
8	Loudspeaker orchestras (*Gmebaphone, Acousmonium, BEAST* et al.)	1973
9	*Hybrid IV* (Kobrin 1975)	1977
10	*SSSP*—a computer-controlled sound distribution system (Federkow et al. 1978)	1978
11	*Trails*: an interactive system for sound location (Bemardini and Otto 1989)	1989
12	*EIS*—the expanded instrument system (Oliveros 1991)	1991
13	*Data Glove* real time control of 3D sound by Gesture (Harada et al. 1992)	1992
14	*MusicSpace*: a midi-file spatialisation tool (Pachet and Delerue 1999)	1998
15	*M2 diffusion*—the live diffusion of sound in space, (Moore et al. 2004)	2004
16	*Orb3*—adaptive interface for real time diffusion (Livingstone et al. 2005)	2005
17	light-emitting pen controllers (Brown et al. 2005)	2005
18	*NAISA spatialization system* (Copeland 2014)	2006
19	*DJ Spat*: spatialized interactions for DJs (Marentakis et al. 2007)	2007
20	multi-touch soundscape renderer (Bredies et al. 2008)	2008
21	*Pointing-At Glove* and 3D-DJ App, (Torre et al. 2009)	2009
22	*Grainsticks*, collaborative sound installation (Leslie et al. 2010)	2010
23	*Bodycoder system (for V'OCT* ritual) (Bokowiec 2011)	2011
24	*GAVIP* (*Gestural auditory visual interactive platform*) (Caramiaux et al. 2011)	2011
25	*The sound flinger* (Carlson et al. 2011)	2011
26	The radiodrum for real time sound spatialization (Ness et al. 2011)	2011
27	*WFS gesture control* (Fohl and Nogalski 2013)	2013
28	*tactile.space* (Johnson and Kapur 2013)	2013
29	*SSN*—sound surfing network (Park et al. 2013)	2013
30	*tactile.motion*: an iPad-based performance interface (Johnson et al. 2014a, b)	2014
31	Holistic 3D sound controller (Diatkine et al. 2015)	2015

and the *Sound and Music Computing (SMC, 2004–2015)*, and the *Computer Music Journal* (1977–2015). In total, our search yielded around three dozen spatialisation instruments (Table 2).[11]

[11]The exact figure varies between 31 and 38 depending on the way of counting different versions or parallel developments of basically the same spatialisation instrument. In the following, we will consider the minimal size of the sample for the sake of simplicity.

It should be noted that a couple controllers have been excluded from the sample for various reasons: Some research papers lack a transparent concept of the controller in focus, its mappings or used components; others do not explicitly include a defined controller device, but provide an open interface to connect an arbitrary controller or sensing system.

3.2 A Classification of Real Time Spatialisation Controllers

Firstly, we will describe the found manifestations and the general distribution of spatialisation instruments in our classification space. Secondly, we will take a closer look at the clusters and present the individual controllers briefly highlighting their most important specifications. Finally, we will recapitulate and contextualise our findings.

Figure 2 presents the distribution of the collected controllers over the categories defined above. About one third of the controllers include sound generation means, while the majority has been designed as exclusive sound spatialisation device. As expected, the category of *instrument-like and augmented controllers* is hardly represented in the sample: we could identify only two instruments of this kind. Moreover, controllers providing no sound generating means are not represented in the group of *alternate mixed controllers*. Most of the controllers for both sound synthesis and spatialisation synthesis can be found in the group of *touch controllers* that are used to diffuse sound in space. The group of *touch controllers* with control of sound source position and characteristics generally constitutes the largest portion of the sample (around one quarter of all observed instruments), closely followed by the adjacent group of controllers for sound diffusion.

In the following, we will briefly address the single controllers observed in our study. For more comprehensive information on the respective instruments refer to the given sources.

3.2.1 Augmented Controllers

Control of Sound Source Related Parameters One instrument could be classified as augmented controller for sound synthesis and spatial control. ***DJ Spat*** was presented by Marentakis et al. (2007) as a spatialisation system to augment the DJ interaction metaphor related to the use of a turntable[12] during performance. Using motion-tracking sensors and further haptic control elements, the performer was enabled to control the spatial organization of the played sounds as a "bi-product" of

[12]Again, the question might arise if this gestural interface can be considered as an augmented instrument linked to the discourse of whether a DJ-turntable represents a musical instrument or not. At this point, we avoid to comment on this topic by using the term *augmented controller* in reference to a well-established control interface for musical performances.

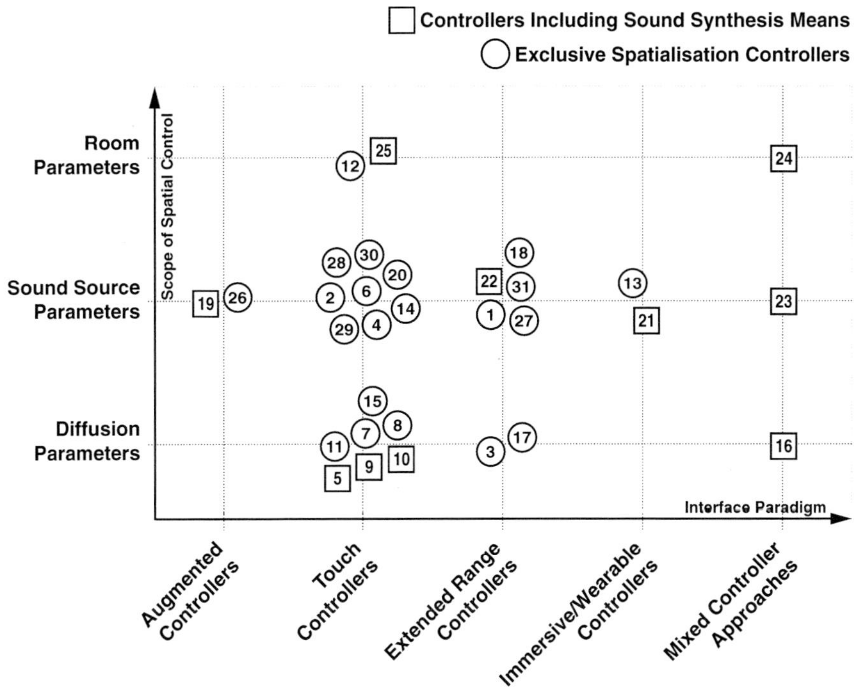

Fig. 2 Classification of real time spatialisation controllers (the numbers in the figure refer to the numbering in Table 2)

his musical performance. The researchers mapped the angular displacement of the performer's hand on the record to the sound source position reproduced through a circular loudspeaker array.

As a counterpart, one augmented controller exclusively for spatialisation was presented by Ness et al. (2011). They developed an approach to use the ***Radiodrum***, a gestural control system from the late 1980s inspired by the playing of a drum with sticks, as a spatialisation system. The authors mapped, among others, controller's position data to the positions of sound sources in space.

3.2.2 Touch Controllers

Control of Sound Diffusion Parameters The spatialisation systems from this cluster differ in terms of the used technologies, but show several similarities in relation to the control paradigm. The first three instruments in our list explicitly include control elements for sound synthesis.

The ***Sal Mar Construction*** was designed in the early 1970s by composer Salavator Martirano as a real time system for composition and performance of electronic music (Franco 1974). The large electronic instrument consists of analogue circuits and digital modules, which allowed the performer to use different sound synthesis techniques. It incorporates a multi-channel matrix system with up to 24 discrete outputs for operations in real time performance. By means of numerous touch-sensitive switches located at the front panel, the sounds could be distributed throughout the performance space.

A similar concept of spatialisation was realized with the ***Hybrid IV*** system developed by Edward Kobrin in 1975. It consists, analogously to the hybrid approach by Martirano, of several analogue sound generating and processing components which are controlled by a computer system for composition and real time use (Fedorkow et al. 1978). The multi-channel matrix provides 16 outputs to be routed to a loudspeaker array, likewise by using switches and buttons to assign the signals to the loudspeakers.

The ***SSSP Sound Distribution System*** was developed by Fedorkow et al. (1978) in the late 1970s. It was inspired by the general design approach of the two systems described above, but combines a compact and modular design with 16 output-channels and the use of various input devices such as a keyboard and a digitizing tablet. The polyphonic sounds are synthesized by another module of the SSSP system and can be controlled using the same interface.

The following controllers are mainly related to the practice of sound diffusion. While some of them offer sound manipulation means, they are considered as exclusive spatialisation controllers which have been created for the purpose of projecting sound in space.

The ***HaLaPhon*** was invented in the late 1960s by Hans-Peter Haller and Peter Laszlo as an instrument to control diffusion and spatialisation in real time performance (Brech and von Coler 2015). The system uses an amplitude panning mechanism to realize virtual (mainly circular) sound movements controlled by switches or automation. Several versions followed making increasingly use of digital technology while maintaining the general concept of the device.

The most important sub-division of instruments in this cluster embraces different kinds of ***Loudspeaker Orchestras***. The ***Gmebaphone*** (later known as *Cybernephone*) was introduced in 1973 (Clozier 2001), one year later the ***Acousmonium*** (Brech 2015). Several further developments followed, the ***BEAST System*** (Harrison 1999) being one of the most recent and important approaches. While some those systems differ significantly, they all share some fundamental features: They are—to a certain extent—modular in design (for a portable use) and they comprise a fader board based controller assigning the sounds to a multitude of included (and very specific) loudspeakers using amplitude panning methods. These spatialisation controllers can be considered as one crucial aspect of common diffusion practice in electroacoustic music.

Another example for a diffusion system following a similar concept is ***TRAILS*** (*Tempo Reale Audio Interactive Location System*), an interactive controller for sound localization (Bernardini 1989). The matrix-based system was presented in 1988. In contrast to the loudspeaker orchestras mentioned above, it did not explicitly define a particular loudspeaker configuration.

The ***M2 system*** was presented by Mooney et al. (2004) as another modular diffusion system consisting of a sound rendering engine (software running on a computer system) and a specifically designed fader board. The system allows for some specific assignment and grouping functionalities.

Control of Sound Source Related Parameters This cluster of dedicated spatialisation controllers comprises the largest portion of spatialisation instruments in our sample. One of the first spatialisation controllers recorded in literature is the ***rotation mill (Tonmühle)*** which was conceptualized already around 1960 at the Technical University in Berlin and later designed for Karlheinz Stockhausen to be used for his spatialisation approach at World Exposition 1970 in Osaka, Japan (Gertich et al. 1996). The functionality of device was very straightforward using a kind of rotational resistance patched to a circle of loudspeakers. By manually turning the crank of the mill, the sound source could be shifted circularly through the audience surrounded by 50 loudspeakers.

A similar control device was realized by Leitner in 1971 as a ***Circular Relay Switch*** with a manual crank (Leitner 2016). The sound sources could be distributed throughout the space by using up to 20 loudspeakers dynamically controlled by means of additional encoder knobs.

Another innovative approach was a ***spherical sound controller***, which was also designed for the West German pavilion at the World Expo 1970 by the Electronic Music Studio at the Technical University of Berlin (Brech 2015; Gertich et al. 1996). The spherical controller consisted of 50 sensor buttons, each representing a loudspeaker group in the spherical concert hall. By this means, sound sources could be projected and moved in space.

An example for a different sound spatialisation metaphor was presented with ***MusicSpace*** by Pachet and Delerue in 1999. The authors presented a typical approach for considering sound sources as objects, which can be freely localized and moved in the projection space. A standard mixer and object-based sound-rendering engine is used to define positions of sounds or groups of sounds in the performance space.

A different concept of haptic interface was presented by Bredies et al. (2008) referred to as ***Multi-Touch Soundscape Renderer***. The authors use a tabletop device with multi-touch sensing as direct manipulation interface which can be used by multiple users simultaneously. Sound objects represented in the graphical user interface can be manipulated through touch gestures. The object-based sound reproduction is achieved by using wave field synthesis and a circular speaker array.

Tactile.space, showed by Johnson and Kapur (2013), represents a similar interface design approach, which also makes use of a surface-based tabletop interface. Analogously, sound objects' positions can be changed by means of touch gestures, although an amplitude panning method is used to project the sound in space. By presenting ***tactile.motion***, the authors provide an application for mobile tablets to be used as a controller instead of the tabletop interface (Johnson et al. 2014b).

Another innovative mobile spatialisation system was presented by Park et al. (2013). The ***Sound Surfing Network (SSN)*** is a system that can be used for smart phone-based sound spatialisation. The application is divided into two entities: the performer-app is used to control the object-based sound sources in the performance space, the audience-app turns each smartphone into an element of the loudspeaker array on which the sound is spatialised.

Control of Room Related Parameters We found one device that could be classified as hybrid controller of sound synthesis and spatialisation. The ***Sound Flinger***, presented by Carlson et al. (2011), is an instrument for haptic spatialisation within a quadraphonic sound system. Using four motorized faders, the sounds can be moved around the circle, using intuitive gestures mapped to physical models, which affect both the spatialisation and the sound processing.

Another single spatialisation instrument could be assigned to the category of dedicated spatial controllers: the ***Expanded Instrument System (EIS)***, continuously developed since 1963 by Oliveros (1991), is a performance environment that was intended to give the performer control over the acoustic space. This is mainly achieved by means of delay and reverb enhancing the sound image created through amplitude panning. The control interface consists of several foot switches. Multi channel speaker configurations are used to reproduce the extended acoustical scene. Figure 3 illustrates a selection of different controllers for the spatialisation of sound in real time.

3.2.3 Extended Range Controllers

Control of Diffusion Parameters The ***Photocell Mixer*** (also *Photoresistor Mixer*) is an early example for an extended range controller exclusively for sound diffusion. It was created in the late 1960s by Frederic Rzewski (1968) and David Behrman (Holmes 2012, 430). Although both composers have developed their mixers independently, the mixers are very similar in design. They consist of panels with several groups of photocells integrated in the signal circuits. Illuminating the photocell of one signal path with a penlight assigns the signal to the one of four outputs and the respective loudspeaker. Thus, the performer diffuses the sound by moving the penlight over the photocells.

Brown et al. (2005) developed an approach to sound diffusion which similarly makes use of hand-held ***light emitting pen controllers*** which are tracked by means of a camera system. Referring to the spatial conductor metaphor (Marshall et al. 2007, 232) to decouple diffusion practice from the fader board, gestures performed with the hand-held pen torches are mapped to diffusion parameters of the sound image reproduced by means of the loudspeaker array.

Control of Sound Source Related Parameters ***Grainstick*** is the name of a hybrid controller system developed by Leslie and colleagues in 2010. It demonstrates a further approach of multimodal interfaces for a sound field reproduction system by using a combination of infrared motion tracking and accelerometer data of hand-held controllers. Controller data is used for sound synthesis and spatialisation within an elaborated mapping structure. The process can be controlled collaboratively by a group of performers.

Fig. 3 A selection of controllers for the real-time spatialisation of sound. *Top left* Pierre Schaeffer playing the *Pupitre d'Espace*, *photo* Maurice Lecardent, 1955 © INA; *top right* the *spherical spatialisation controller* for the World Expo 1970 in Osaka, developed at the Technische Universität Berlin, used by Fritz Winckel, Manfred Krause in the background, photo: TU Archive, 1970 © TU Berlin; *lower left* the controller board of the *Gmebaphone2*, developed at the GMEB in Bourges, France, photo by MIMO - Musical Instrument Museums Online (CC BY-NC-SA); *lower right* The *Pointing-At Glove* developed by Giuseppe Torre, photo by courtesy of Cillian O'Sullivan and John McCall

All remaining extended range controllers in this cluster are exclusive sound spatialisation systems. One of the first recorded and most frequently cited spatialisation instruments is the ***Pupitre d'Espace*** (space console) presented by Pierre Schaeffer in 1952 (Battier 2015).[13] The controller worked with four induction coils

[13]The *pupitre d'espace* is a further development of a controller introduced in 1951 as *pupitre potentiométrique de relief*. The device had the same functionality but worked with controlling three wires which are linked to potentiometers to adjust the signal level send to each loudspeaker (Battier 2015, 127).

mounted around the performer as receiver rings. A further coil was held in the hand of the performer. By moving the coil between the receiver rings, four currents were induced. These controlled the amplifiers sending the signals to the four loudspeakers surrounding the audience space. The speakers were arranged according to the positions of the coils in three spatial axis, thus achieving quasi-periphonic sound spatialisation.

The ***NAISA Spatialisation system*** has been developed since 2006 by Copeland (2014) as an interactive performance system for spatial sound projection to up to 24 surrounding loudspeakers. At the core of the gestural spatialisation control is a six degrees-of-freedom motion tracking system with a magnetic sensor connected to the hand of the performer. Gestural movements of the hand are then mapped to the sound sources' positions and orientations. The performer can move quite freely in a certain range around the receiving sensor. The software running on a computer system allows for a very variable use of the system.

Fohl and Nogalski (2013) present another spatialsation system that uses a infrared-camera system for performer's gestural tracking. Their ***Gesture Controller for a WFS System*** approach makes use of markers mounted to the hand of the performer for gesture recognition by means of the camera system. The control over the wave field synthesis renderer is achieved through the mapping of predefined gestures to certain functionality such as positioning, movement patterns or switching of sound sources.

A ***holistic spatialisation system for multiple sound sources*** is presented by Diatkine et al. (2015). The researchers use a short-range infrared sensor to track hand gestures and map this data to position data of sound sources, which are reproduced via a higher-order ambisonics and dynamic binaural rendering process. Consequently, the system uses headphones instead of loudspeakers to reproduce the sounds limiting its use to rehearsal contexts rather than to a concert performance.

3.2.4 Immersive/Wearable Controllers

Control of Sound Source Related Parameters The ***Pointing-At Glove*** was developed by Torre (2013) as a gestural spatialisation controller system incorporating sound generation and manipulation capabilities. The controller glove was realized using a six degrees-of-freedom sensor allowing for the control of periphonic 3D sound spatialisation, including the vertical dimension. Here, the use of the *3D-DJ* (Torre 2013) application enables the performer to render three-dimensional audio scenes by means of the gestural control data tracked by the glove.

In their conference contribution from 1992 Harada, Sato, Hashimoto and Ohteru present an early design for an immersive, wearable controller system exclusively for spatial sound projection. The authors make use of a ***Data Glove*** for the recognition of a performer's gesture. An amplitude panning method controlled by means of midi control data is used to route the sound signals to a 3-dimensional loudspeaker array.

3.2.5 Mixed Controller Approaches

All controllers within this category are hybrid controllers to be used for spatialisation and sound generation.

Control of Diffusion Parameters With ***Orb3*** Livingston and Miranda (2005) present an adaptive sound synthesis and diffusion system. Three mobile sensing globes, which comprise a set of eight different sensors each, collect data within the performance environment. The authors demonstrate an approach of indirect and subconscious user interaction where control data is obtained from several sensors and adaptively mapped to processes controlling the sound diffusion.

Control of Sound Source Related Parameters Another controller design using a set of different controllers was demonstrated by Bokowiec (2011) with his version of a ***Bodycoder System***. The approach is based on a sensor array integrated in a wearable controller system. It mainly consists of two sensing gloves. The system provides motion detection, 12 haptic switches and four bend sensors to control the different functions during performance including sound and video manipulation and the spatialisation through a multichannel loudspeaker array.

Control of Room Related Parameters The last controller in our inventory which is not exclusively designed for sound spatialisation was introduces as ***GAVIP*** (Gestural Auditory and Visual Interactive Platform) by Caramiaux and colleagues in 2011. The authors created a virtual space architecture with different means for gestural tracking (camara based and with gyroscope sensors). To achieve inter-modal interaction coherency, the sensor data was mapped to a complex physical model, which then was used to render the 3D audio-visual scene including the synthesis of a virtual dynamic sound field by means of wave field synthesis.

4 Discussion

The main objective of this work has been to conduct an inventory of controllers for the real time spatialisation as part of musical performances, and to classify them both along different interface paradigms and along their scope of spatial control. By means of a thorough literature study, we were able to identify 31 different spatialisation interfaces presented in the context of the most relevant conferences and/or mentioned in a selection of important monographs on the subject.

Considering the significance attributed to *space* as a musical parameter throughout the discourse on electroacoustic music, this seems only a modest number, also compared to the abundance of musical interfaces for sound synthesis, which were presented in the same sources to the same communities. Hence, despite the great interest in concepts and techniques for live-interaction and music, the spatialisation of sound still seems to be regarded as an aspect of musical composition rather than of musical performance. Whether intentionally or not, the classical concept of the performer as *sound generator*, with the spatial organisation of sound remaining part of the compositional process, is still prevalent.

The reason for this allocation of roles can hardly be assigned to a lack of technical resources, as it is demonstrated by the 31, quite diverse, approaches described above. Most of them are controllers for spatialisation only, while only a minority (10 out of 31) combines sound synthesis and spatial presentation with one, hybrid interface. Whether this again reflects a traditional *role model* with respect to the control of sound and space, or whether it reflects practical challenges in interface design and user interaction, is difficult to decide.

Almost all instruments for spatialisation are dedicated to control the position or the movement of individual sound source or of sound sources as a group (diffusion approach), while the control of the (virtual) spatial environment itself is rarely (only with 3 out of 31 tools) addressed.

Concerning the interface paradigm used, the vast majority of existing interfaces are *touch controlled*, whereas there are only few examples for immersive or wearable spatial controller systems, which are not restricted by operating an immobile apparatus or by a limited spatial range of operation. Taken both aspects together, there seems to be quite a lot of potential for further developments.

The artistic context which most spatialisation controllers were developed for, is quite specific in most cases. Already the early devices presented in the course of the 20th century were custom-built according to the requirements of individual composers and compositions, such as the *rotation mill* for Stockhausen, Martirano's *SalMar Construction*, Schaeffer's *Pupitre d'Espace*, or the *loudspeaker orchestras* designed by Bayle or the *Groupe de Musique Expérimentale des Bourges*. Also more recent developments of the last two decades have often been applied to unique artistic projects, and we could not identify a generic control paradigm or a generic interface used for a larger repertoire of music.

The only exception seems to be the *fader board paradigm* which has become an integral and constitutive part of the sound diffusion practice in the context of acousmatic music and performance practice with loudspeaker orchestras. Although the mixing desk is not only used for the spatialisation there, but also for the assignment of different loudspeaker groups with their individual sound character, this seems the only example where a spatialisation interface is re-used for the performance of different music by different performers. It is no coincidence that this is the only musical area where a notion of *performative virtuosity* with respect to sound spatialisation could develop.

With this exception, the spatial enactment of sound as performance practice is still characterised by rather individual design solutions. Many of these are modular systems consist of existing interfaces and universal components, such as *borrowed* sensing and control devices (e.g., gamepads, infrared tracking systems, gyroscopic sensors) which allow performers "(…) *to map their performative gestures to any number of musical parameters*" (Johnson et al. 2013, 271). The spatial rendering tends to be handled by common spatialisation applications such as *Spat*, *SSR*, or *Zirkonium*, or customised solutions based on common frameworks such as *SuperCollider*, *Max* or *PureData.* Even some *vintage* spatialisation controllers, formerly built as analogue electronic circuits, have been redeveloped using these software environments, such as Behrman's *Photocell Mixer* (Behrman 2016;

Holmes 2012, 430), Oliveros' *Expanded Instrument System* (Oliveros 2008), or the *HaLaPhon* recreated as a *Max*-patch to re-enact pieces by Luigi Nono or Pierre Boulez (Ferguson 2010).

Although such a modular approach offers the shortest and most flexible access to spatial control, the related interfaces will probably not be identified as *musical instruments* by performers (other than those who developed it) and by the audience. As pointed out by Hardjowirogo (this volume), the establishment and the cultural embeddedness of an interface within a certain aesthetical practice are crucial aspects of *instrumentality*, which can only be reached by repeated use—not to mention other criteria such as the immediate connection between the performer's actions and the sonic result or the perceived *liveness* on the side of the audience.

Whether it is desirable to devise a spatial controller as a musical performance instrument, is, of course, a matter of the artistic premises, on which its development is based. Given the structural importance of sound projection in space in current musical practice, however, the creation of tools for its real-time control seems only consequent. If these are supposed to be recognized as musical instruments in a narrow sense, the related technical challenges are still awaiting successful solutions.

References

Baalman, Marije A. J. (2010). Spatial composition techniques and sound spatialisation technologies. *Organised Sound, 15*(3), 209–218. doi:10.1017/S1355771810000245

Battier, M. (2015). Recent discoveries in the spatial thought of early musique concrète. In M. Brech & R. Paland (Eds.), *Compositions for audible space. The early electroacoustic music and its contexts. Music and sound culture* (pp. 123–36). Columbia: Transcript Verlag.

Behrman, D. (2016). *Personal interview with David Behrman in Berlin*, May 14, 2016.

Bernardini, N. (1989). *Trails: An interactive system for sound location*. Ann Arbor, MI: Michigan Publishing, University of Michigan Library.

Birnbaum, D., Fiebrink, R., Malloch, J., & Wanderley, M. M. (2005). Towards a dimension space for musical devices. In *Proceedings of the 2005 Conference on New Interfaces for Musical Expression* (pp. 192–95). Singapore: National University of Singapore.

Blauert, J. (1997). *Spatial hearing: The psychophysics of human sound localization*. Cambridge: MIT press.

Bokowiec, M. A. (2011). VOCT (Ritual): An interactive vocal work for bodycoder system and 8 channel spatialization. In *Proceedings of the International Conference on New Interfaces for Musical Expression*.

Brech, M. (2015). *Der Hörbare Raum Entdeckung, Erforschung Und Musikalische Gestaltung Mit Analoger Technologie*. Columbia: Transcript Verlag.

Brech, M., & Paland, R. (Eds.). (2015). *Compositions for audible space. The early electroacoustic music and its contexts. Music and sound culture*. Columbia: Transcript Verlag.

Brech, M., & von Coler, H. (2015). Aspects of space in luigi nono's prometeo and the use of the Halaphon. In M. Brech & R. Paland (Eds.), *Compositions for audible space. The early electroacoustic music and its contexts. Music and sound culture* (pp. 193–204). Columbia: Transcript Verlag.

Bredies, K., Alexander Mann, N., Ahrens, J., Geier, M., Spors, S., & Nischt, M. (2008). The multi-touch sound scape renderer. In *Proceedings of the working conference on advanced visual interfaces* (pp. 466–469). AVI '08. New York, NY, USA: ACM. doi:10.1145/1385569.1385660

Brown, K., Alcorn, M., & Rebelo, P. (2005). Sound diffusion using hand-held light-emitting pen controllers. In *proceedings of international computer music conference*.

Caramiaux, B., Fdili Alaoui, S., Bouchara, T., Parseihian, G., & Rébillat, M. (2011). Gestural auditory and visual interactive platform. In *14th International conference on digital audio effects (DAFx-11)*, 69.

Carlson, C., Marschner, E., & Mccurry, H. (2011). The sound flinger: A haptic spatializer. In *Proceedings ICMC 2011*.

Chowning, J. M. (1971). The simulation of moving sound sources. *Journal of the Audio Engineering Society, 19*(1), 2–6.

Clozier, C. (2001). The gmebaphone concept and the cybernéphone instrument. *Computer Music Journal, 25*(4), 81–90.

Copeland, D. (2014). *The NAISA spatialization system*. April. http://www.darrencopeland.net/web2/?page_id=400

Diatkine, C., Bertet, S., & Ortiz, M. (2015). *Towards the holistic spatialization of multiple sound sources in 3d, implementation using ambisonics to binaural technique*. In *Proceedings*.

Fedorkow, G., Buxton, W., & Smith, K. C. (1978). A computer-controlled sound distribution system for the performance of electroacoustic music. *Computer Music Journal*, 33–42.

Ferguson, P. (2010). Development of a 3D audio panning and realtime visualisation toolset using emerging technologies. Ph.d. thesis, Edinburgh Napier.

Fohl, W., & Nogalski, M. (2013). A gesture control interface for a wave field synthesis system. In *Proceedings of NIME 2013*, 341–346.

Franco, S. (1974). *Hardware design of a real-time musical system*. Champaign: University of Illinois at Urbana-Champaign.

Geier, M., Spors, S., & Weinzierl, S. (2010). The future of audio reproduction: Technology—formats—applications. In M. Detyniecki, U. Leiner, & A. Nürnberger (Eds.), *Adaptive multimedia retrieval. Identifying, summarizing, and recommending image and music* (pp. 1–17). Springer.

Gertich, F., Gerlach, J., & Föllmer, G. (1996). *Musik, Verwandelt: Das Elektronische Studio Der TU Berlin 1953–1995*. Wolke.

Harada, T., Sato, A., Hashimoto, S, & Ohteru, S. (1992). Real time control of 3D sound space by gesture. In *International computer music conference proceedings*.

Harrison, J. (1999). Diffusion: Theories and practices, with particular reference to the BEAST system. *eContact*, 2.

Holmes, T. (2012). *Electronic and experimental music: Technology, music, and culture*. Routledge.

Jaroszewicz, M. (2015). Compositional strategies in spectral spatialization. California: University of California Riverside.

Johnson, B., & Kapur, A. (2013). Multi-touch interfaces for phantom source positioning in live sound diffusion. In W. S. Yeo, K. Lee, Alexander Sigman, (Hyunkyung) H. Ji, & G. Wakefield (Eds.,) *13th international conference on new interfaces for musical expression, NIME 2013, Daejeon, Republic of Korea, May 27–30, 2013* (pp. 213–216). nime.org.

Johnson, B., Murphy, J. W., & Kapur A. (2013). Designing gestural interfaces for live sound diffusion. In *Proceedings of the 39th international computer music conference, ICMC 2013, Perth, Australia, August 12–16, 2013*. Michigan Publishing.

Johnson, B., Norris, M., & Kapur, A. (2014a). Diffusing diffusion: A history of the technological advances in spatial performance. In *International computer music conference proceedings* 2014.

Johnson, B., Norris, M., & Kapur, A. (2014b). Tactile.motion: An iPad based performance interface for increased expressivity in diffusion performance. In *Music technology meets philosophy—from digital echos to virtual ethos: Joint proceedings of the 40th international computer music conference, ICMC 2014, and the 11th sound and music computing conference, SMC 2014, Athens, Greece, September 14–20, 2014*. Michigan Publishing.

Leitner, B. (1971). *Sound architecture—space created through traveling sound*. New York: ARTFORUM.

Leitner, B. (2016). *Atelier Bernhard Leitner*. In Amt der Niederösterreichischen Regierung. Bielefeld: Kerber Verlag.

Leslie, G., Zamborlin, B., Jodlowski, P., & Schnell, N. (2010). Grainstick: A collaborative, interactive sound installation.

Livingstone, D., & Miranda, E. (2005). *Orb3: Adaptive interface design for real time sound synthesis and diffusion within socially mediated spaces* (pp. 65–69). Singapore: National University of Singapore.

Lynch, H., & Sazdov, R. (2011). An ecologically valid experiment for the comparison of established spatial techniques. In *International computer music conference proceedings* 2011.

Malham, D. G. (1998). Approaches to spatialisation. *Organised Sound, 3*(2), 167–177.

Manning, P. (2013). *Electronic and computer music*. Oxford: Oxford University Press.

Marentakis, G., Peters, N., & McAdams, S. (2007). DJ spat: Spatialized interactions for DJs. In *International computer music conference proceedings* 2007.

Marshall, M. T., Malloch, J., & Wanderley, M. M. (2007). Gesture control of sound spatialization for live musical performance. In M. S. Dias, S. Gibet, M. M. Wanderley, & R. Bastos (Eds.), *Gesture-based human-computer interaction and simulation. Lecture notes in computer science 5085* (pp. 227–38). Berlin: Springer.

Miranda, E., & Castet, J. (Eds.). (2014). *Guide to brain-computer music interfacing*. Berlin: Springer.

Miranda, E. R., & Wanderley, M. M. (2006). *New digital musical instruments: Control and interaction beyond the keyboard* (Vol. 21). AR Editions, Inc.

Mooney, J. (2005). *Sound diffusion systems for the live performance of electroacoustic music*. Sheffield: University of Sheffield.

Mooney, J. R., Moore, A., & Moore, D. (2004). M2 diffusion: The live diffusion of sound in space. In *International computer music association*.

Mulder, A. (2000). Towards a choice of gestural constraints for instrumental performers. In M. Wanderlay & M. Battier (Eds.), *Trends in gestural control of music* (pp. 315–35).

Ness, S., Odowichuk, G., Driessen, P., Tzanetakis, G. (2011). Controlling Real time sound spatialization using the radiodrum. In *International computer music conference proceedings* 2011.

Oliveros, P. (1991). The expanded instrument system (EIS). In *International Computer Music Conference Proceedings* 1991.

Oliveros, P. (2008). The expanded instrument system (EIS): An introduction and brief history. *Future of Creative Technologies, Journal of the Institute of Creative Technologies, no., 1*, 21–24.

Overholt, D. (2011). Violin-related HCI: A taxonomy elicited by the musical interface technology design space. In A. L Brooks (Ed.), *Arts and technology—second international conference, ArtsIT 2011* (pp. 80–89). Berlin: Springer.

Pachet, F., & Delerue, O. (1999). Music space: A constraint-based control system for music spatialization. In *International Computer Music Conference Proceedings* 1999.

Paradiso, J. A. (1997). Electronic music: New ways to play. *IEEE Spectrum, 34*(12), 18–30. doi:10.1109/6.642965.

Park, S., Ban, S., Hong, D. R., & Yeo, W. S. (2013). Sound surfing network (SSN): Mobile phone-based sound spatialization with audience collaboration. In *Proceedings* (pp. 111–14). Korea.

Perez-Lopez, A. (2015). 3DJ: A supercollider framework for real-time sound spatialization. In *Proceedings*. Graz, Austria.

Peters, N. (2011). *Sweet [Re] production: Developing sound spatialization tools for musical applications with emphasis on sweet spot and off-center perception*. McGill University.

Peters, N., Lossius, T., Schacher, J., Baltazar, P., Bascou, C., & Place, T. (2009). *A stratified approach for sound spatialization* (pp. 219–24). Citeseer.

Pressing, J. (1990). Cybernetic issues in interactive performance systems. *Computer Music Journal, 14*(1), 12–25.

Roads, C. (1996). *The computer music tutorial*. MIT press.

Rzewski, F. (1968). A photoresistor mixer for live performance. *Electronic Music Review, 4*(4).

Sannicandro, V. (2014). *Space and spatialization as discrete parameter in music composition: A space-oriented approach to écriture; from acoustic through informatics to musical notation*. Berlin: epubli GmbH, Berlin.

Schacher, J. C. (2007). Gesture control of sounds in 3D space. In *Proceedings of the 7th International Conference on New Interfaces for Musical Expression* (pp. 358–362). NIME '07. New York, NY, USA: ACM. doi:10.1145/1279740.1279819

Stockhausen, K. (1959). Musik Im Raum. *Die Reihe, Berichte—Analyse, 5*, 67–72.

Torre, G., Sazdov, R., & Konczewska, D. (2009). MOLITVA–Composition for Voice, Live Electronics, Pointing-At Glove Device and 3-D Setup of Speakers. *Proceedings of the International Conference on New Interfaces for Musical Expression (NIME)*, 330.

Torre, G. (2013). The design of a new musical glove: A live performance approach. Limerick: University of Limerick.

Valiquet, P. (2011). The spatialisation of stereophony: Taking positions in post-war electroacoustic music. In *International computer music conference proceedings 2011*.

Wanderley, M. M., & Orio, N. (2002). Evaluation of input devices for musical expression: Borrowing tools from Hci. *Computer Music Journal, 26*(3), 62–76.

Xenakis, I. (1992). *Formalized music: Thought and mathematics in music, revised edition*. Stuyvesant.

Zvonar, R. (2000). An extremely brief history of spatial music in the 20th century. *Surround Professional Magazine*.

Lucille Meets GuitarBot: Instrumentality, Agency, and Technology in Musical Performance

Philip Auslander

Abstract The relationship between musicians and their instruments in performance has been characterized in a variety of ways that tend to describe the instrument either as an entity inseparable from the musician or as an entity with relative autonomy. Through the trope of ventriloquism, Philip Auslander looks at how two musicians working in very different genre contexts construct their respective relationships to instruments in performance. Both blues guitarist and singer B.B. King and classical violinist Mari Kimura treat instruments as entities separate from themselves and performers in their own right: King by naming his guitar Lucille and constructing a narrative around his relationship with her, and Kimura through her interaction with GuitarBot, a digital musical instrument. By dramatizing the ventriloquial relationship between player and instrument and creating the impression that an instrument possesses an identity and agency, both King and Kimura enact the fantasy of instrumental autonomy that underlies the ventriloquial relationship between performer and instrument. But because the digital technology Kimura employs allows GuitarBot a greater degree of (apparent) autonomy than Lucille, who is always under King's visible, physical control, it enables her to push the enactment of this fantasy further toward the uncanny to show us what it might look like for a performer to interact with a genuinely autonomous musical instrument.

Authorized reprint of the article "Lucille Meets GuitarBot: Instrumentality, Agency, and Technology in Musical Performance" (*Theatre Journal* 61, 603–616. Johns Hopkins University Press 2010).

P. Auslander (✉)
School of Literature, Media, and Communication,
Georgia Institute of Technology, Atlanta, GA, USA
e-mail: auslander@gatech.edu

T. Bovermann et al. (eds.), *Musical Instruments in the 21st Century*,
DOI 10.1007/978-981-10-2951-6_19

1 Introduction

Although the relationship between instrumentalist and instrument in musical performance has been imagined in a number of different ways, there is a basic distinction between those who posit the instrument as an extension of the performer and those who see the two entities as separate. Edward T. Cone presents a strong version of the former position. Using primarily the performance of classical music as his point of reference, Cone argues that whereas a singer is akin to an actor in that the singer "enacts a role, portrays a character" through the lyrics of the song, and thus maintains a distinction between the performer's voice and that of the thing being performed, the instrumentalist's voice melds with that of the instrument to become "a compound creature, the musician-*cum*-instrument" (Cone 1974, pp. 105f). Matthew Gelbart extends Cone's position into the realm of rock music by making a parallel argument regarding the conventions surrounding the performance of rock during its formative era (roughly 1956–1970) and describing Chuck Berry's guitar as a phallic "extension of [his] own physical body" (Gelbart 2003, p. 208).

By contrast, Susan Fast describes the performed relationship between another rock guitarist, Jimmy Page of Led Zeppelin, and his guitar in terms that suggest they are two separate entities. Fast describes Page as continually moving his guitar around his body, which "gives the impression that this instrument has life in and of itself—that it is a force, a presence. Moving it points up its materiality—it is not just a static piece of technology but also a body with which the player has a physical relationship…" (Fast 2001, pp. 151f). This characterization suggests that the instrument, as an entity separate from the musician, enjoys a degree of agency—it is something with which the musician must engage in order to produce the music. Derek Miller takes this idea of instrumental autonomy and agency a step further, albeit in a somewhat different direction, by defining "musical performance [as] a double performance: a technological performance by an instrument and a technical performance by a musician" (Miller 2011, p. 262). In Miller's view, it is from the collaborative relationship of these two ontologically distinct entities, each engaged in its own kind of performance, that musical sound arises.

In an essay titled "Instrumentalities," David Burrows pursues the latter direction of positing instrumentalist and instrument as separate entities and proposes ventriloquism as a metaphor for understanding their relationship. Although this is not Burrows's point, this metaphor suggests that one can see the musician as making the instrument sing by "throwing" his or her musical voice into it. Burrows actually uses the figure of ventriloquism to suggest another aspect of the relationship between player and instrument, describing ventriloquism as "not simply [an act] of concealment and transformation but [one that] involves splitting the performer's personality and displacing part of it onto an alter ego that acts as a foil, not a clone" (Burrows 1987, p. 123). With this metaphor, Burrows importantly posits the musical instrument not as a McLuhanesque technological prosthetic that extends the capacities of the human body, but as an entity perceived as distinct from, and in tension with, the musician. Like the ventriloquist's dummy, this entity is made to appear to have its own agency with which the musician must negotiate in order to make it sing. In reality, of course, the instrument is subject to the musician's agency in the sense that, like the ventriloquist's dummy, it is mute without human intervention, but the illusion of the instrument's (semi-) autonomy is fundamental to instrumental performance in most Western musical genres.

A possible reason for the cultivation of this illusion is that conventional Western musical performance is a demonstration of skill undertaken, as Stan Godlovitch has pointed out, under "accepted artificial constraints." Godlovitch means that instrumental performance is not simply about producing particular sounds, but rather about producing them by means that reflect the traditional values of a community of musicians. These traditional values, which forbid such tactics as redesigning instruments to make them easier to play or playing a violin part on a synthesizer (Godlovitch's examples), demand not only that certain sounds be made, but that they be made under circumstances that make them difficult to produce so that musical performance "becomes a ritual requiring skill":

> In these respects, [musical] performance shares much with exploring and athletics. Specialized gear notwithstanding, ardent mountain climbers do not typically solve their challenge by blasting and bulldozing so as to furnish level terrain where once there were cliffs; nor do they hasten ascent to the peak in helicopters. Being an accomplished guitarist is in part being able to subdue confidently the treacheries of the guitar. (Godlovitch 1985, p. 71)[1]

In other words, musical performance is not just about achieving certain sonic ends—it is crucially also about perceptibly overcoming challenges presented by the means used to achieve those ends.

Without in any way discounting the real and hard-earned skill of musicians, I insist, Godlovitch notwithstanding, that what counts, ultimately, is audience perception, not actual degree of difficulty. Here, it is necessary to make a distinction between audiences. Godlovitch's argument pertains, in part, to the idea of skilled musicians as constituting something like a guild; in order to claim membership in this "exclusive community," one has to demonstrate the necessary level of technical accomplishment (ibid.). Presumably, one's fellow guild members are in a position to judge one's technical ability in and for itself by assessing one's instrumental technique, handling of the conventional repertoire, and so on. Most audience members, however, are not also guild members and therefore do not possess the requisite knowledge to make an informed judgment of instrumental skill. As Theodore Gracyk puts it, "If one does not know the demands of the particular instrument, one cannot judge the virtuosity displayed. And this may be the situation more often than not" (Gracyk 1997, p. 145). It becomes necessary, then, for the musician to perform instrumental skill in a way that will make it apparent to a more general audience.

[1]It is important to note that Godlovitch takes the classical solo recitalist as the model of musical performance on which to base his philosophical inquiry.

This dimension of musical performance is related to the process Erving Goffman calls "dramatization." For Goffman, dramatization is the making-visible of aspects of one's work routine that are not readily visible to the audience so as to make a desired impression. He identifies violinists, along with prizefighters and surgeons, as not needing to indulge in dramatization, because "some of the acts which are instrumentally essential for the completion of the core task of the status are at the same time wonderfully adapted, from the point of view of communication, as means of vividly conveying the qualities and attributes claimed by the performer" (Goffman 1959, pp. 30f). I differ with Goffman's inclusion of violinists among those who do not need to dramatize, for the acts that are "instrumentally essential" for the performance of music (e.g., fingering or striking keys, bowing or plucking strings, blowing into apertures or vibrating reeds, and so on) are not in themselves expressive of "the attributes claimed by the performer." Musicians therefore must engage in additional actions to communicate those attributes to the audience, especially the nonspecialist audience. To be an accomplished guitarist in the eyes of this audience therefore is *to appear to be able* to subdue the guitar, which means that the guitar has to be constructed in performance as something that presents obstacles for the player to surmount, something that resists the player in some way and is not simply a tool that yields readily to his or her use. Investing the instrument with agency, constructing it as an entity with a will of its own, as the ventriloquist constructs the dummy, is a way to achieve this effect in performance.

Here, I will discuss the questions of instrumentality and agency, and their relationship to technology, in musical performance using two contemporary performers as case studies: blues guitarist and songwriter B.B. King and classical violinist and composer Mari Kimura. Both may seem somewhat idiosyncratic in their respective engagements with instruments: King is famous for having named his guitar Lucille and treating it as a person unto itself, and *GuitarBotana*, the composition and performance of Kimura's that I will examine, involves a digitally programmed robotic musical instrument as well as a standard violin. I will use the trope of ventriloquism as a means of unpacking the layers of agency, both overt and covert, in their performances to suggest that both performers, while innovative, do not so much challenge the ways that agency and instrumentality play out in conventional musical performance as dramatize them.

2 Lucille

As Burrows suggests, musicians can displace their own agency onto the instruments they play in ways that constitute those instruments as (semi-)autonomous entities to which they relate as performing partners rather than just tools. I once saw Judd Hughes, a virtuosic country guitarist who played lead in Patty Loveless's band, hold and manipulate an acoustic guitar as if it were an unruly alter ego, like a barely trained Great Dane over which he had temporary control but that could get away from him at any moment. A more celebrated example is B.B. King, who in naming his guitar Lucille encourages his audience to perceive it as a separate being, and implies that his relationship with it is fraught with the complexities attending heterosexual relationships between men and women.[2]

King's relationship with Lucille is indeed complex, and I cannot hope to do it justice here. The guitar is said to be named for a woman over whom two men brawled at a juke joint in Arkansas where King played in the late 1940s, early in his career; the fight led to the immolation of the place, a story that itself could have been taken from a blues ballad. King consistently treats Lucille as an entity separate from himself, both discursively and in the way he performs with her. He frequently gives Lucille instructions, saying "One more, Lucille" when he wants to play another chorus, or "Take it easy" when he plays pianissimo. He confirms his ventriloquial relationship with Lucille in the way he seeks to make her sing in his displaced voice: "The one thing that I'm concerned about today, to make Lucille sound even more like singing, more in the style of my singing."[3] King defers to Lucille at moments when he claims to find himself unable to speak, suggesting that his voice and Lucille's are expressively interchangeable. In his recording of the song "Lucille," one of the places where he has recounted the story of how the guitar got its name, King says at one point: "Sometimes I get to a place where I can't even say nothing." This remark is followed immediately by guitar playing, to which King responds appreciatively, "Look out!" as if addressing the actions of another. At a different moment in the same song, he says, "Sorta hard to talk to you myself. I guess I'll let Lucille say a few words, and then…." His voice trails off as the guitar takes over; when he resumes speaking at the end of the instrumental passage, he does not pick up where he left off—it is as if Lucille had completed the thought for him.[4]

He also describes Lucille as a distinct individual, with her own sensibility, from whom he must coax musical sound:

> It seems that it loves to be petted and played with. There's also a certain way you hold it, the certain noises it makes, the way it excites me… and Lucille don't want to play anything but the blues…. Lucille is real, when I play her it's almost like hearing words, and of course, naturally I hear cries. I'd be playing sometimes and as I'd play, it seems like it almost has a conversation with me.[5]

King's rhetoric here is worth attending to. There is ambivalence in the way he refers to the guitar sometimes as "it" and sometimes as "her," alternately

---

[2]A number of celebrated guitarists in the blues/rock tradition have named their guitars: Eric Clapton had a guitar called Blackie; Roy Buchanan had one called Nancy; Keith Richards has called a guitar "Micawber"; George Harrison played Rocky and Lucy; Steve Vai has guitars named Evo and Flo; and the list goes on. The *Bad Dog Café* section of *The Telecaster Guitar Forum*, the online bulletin board that is my source for this information, also features entries by many lesser-known musicians listing the names they have given their instruments. This thread, which began on 27 March 2009, is available at the tdpri forum (2009).

[3]B.B. King, quoted by Kerekes and O'Neill (1996).

[4]B. B. King, "Lucille," in *Lucille*, MCA Records, 1968.

[5]King, quoted by Kerekes and O'Neill (1996).

personifying the instrument and acknowledging its status as an object. When he discusses his actions on Lucille (petting, playing, holding), he refers to the instrument as "it." But when he discusses Lucille's own musical contribution, he refers to the guitar either by name or using feminine pronouns, thus clearly positioning Lucille as an active, gendered entity separate from himself.[6] This entity has human characteristics: she speaks, cries, engages in conversation. He implies that Lucille is autonomous: she is "real" and has specific ideas about what music she will perform. However, King does not characterize Lucille as "treacherous," the word Godlovitch uses to describe the resistance the guitar offers its player. Lucille is King's indispensable creative partner and alter ego, but it is clear that her cooperation is not guaranteed: she must be cajoled. King must do what she wants ("it loves to be petted and played with") if she is to work willingly with him in playing the blues.

King dramatizes this aspect of his relationship with Lucille in performance. Like many other guitarists who are also vocalists, King often does not play when he is singing. When he sings, his guitar simply hangs against his torso on its strap while he uses his arms and hands to gesticulate in ways that underline the emotional states expressed in his songs' lyrics (Fig. 1). While singing, he stands erect, his face toward his audience or directed slightly heavenward, his eyes often closed. When he plays Lucille, however, his posture changes. He hunches over the fretboard in his left hand, his head tilted downward toward the instrument. Even if his eyes are closed, his head is positioned as if he were looking at Lucille, giving her his full attention (Fig. 2). While he is playing, every movement of his body and every facial expression is a direct response to the sounds emanating from Lucille, often on a note-by-note basis. In conjunction with what he says about Lucille, this way of performing with her suggests that when King is singing, he is free to express his own feelings as conveyed through the lyrics. If he wants Lucille to participate, however, the focus must be entirely on her and what she has to say.[7] In Burrows's description, quoted earlier, ventriloquism "involves splitting the performer's personality and displacing part of it onto an alter ego." In King's case, it is arguably not just his personality that is split and displaced, but also the two musical functions he performs: he sings as himself, but his guitar playing is displaced onto Lucille as

[6]It is important to stipulate, however, that Lucille is not a specific instrument; there have been many Lucilles over the course of King's career, though they have all been of the same model, the Gibson ES-355. But the fact that Lucille is not a particular guitar reinforces the distinction between object (it) and persona (she) that King implies in talking about her: Lucille's identity persists across multiple physical incarnations.

[7]These observations are based on King's performance of several songs on Ralph Gleason's *Jazz Casual* television show in May 1968 on the National Educational Television network. Clips of this program are available on YouTube (Jazz Casual 1968). It was also published as a DVD by Rhino/WEA in 2002.

Fig. 1 B.B. King sings on Ralph Gleason's *Jazz Casual* (National Educational Television 1968)

Fig. 2 B.B. King communes with Lucille on Ralph Gleason's *Jazz Casual* (National Educational Television 1968)

his alter ego. In this respect, King may be said to be dramatizing the relationship between singer and instrumentalist posited by Cone. Cone describes the character portrayed by the singer as the music's "vocal persona" (or protagonist) and the accompanying music as the "instrumental (or virtual) persona." Cone treats these two personae as distinct voices in the performed composition and considers the dialogue between them to express the composer's intentions. He also suggests that the relationship between them can take many forms (Cone 1974, pp. 18, 29). Gelbart proposes that in the performance of rock music, vocal and instrumental personae are fused into a single entity. By contrast, King, the blues singer, performs as the vocal protagonist while King, the blues guitarist, anthropomorphizes the instrumental persona in the "person" of Lucille.

In a discussion of a series of experiments intended to show the connections between the auditory and the visual in musical perception, a group of research psychologists describes King's typical gestures and facial expressions, noting that

> King frequently adopts an introspective demeanor, with eyes closed and a pained expression, yet stubbornly shaking his head. This *affective display* conveys an impression of stoically reflecting upon but not surrendering to difficult emotions. Periodically he stares open-eyed at the audience with an open mouth. The expression appears to convey a sense of wonder.... Judge A [one of the experimental subjects] observed that King's facial expressions often functioned to signal that certain passages were difficult but satisfying to play. (Thompson et al. 2005, pp. 207f, emphasis in original)

These authors also observe the direct relationship between King's behavior and the music he plays:

> It is notable that B.B. King's facial expressions closely track his guitar sounds.... In some cases his rapid head shaking movement mirrors vibrato on individual notes. This gesture has the effect of drawing the listeners' attention to local aspects of music, specifically to B. B. King's nuanced treatment of individual notes. (ibid., p. 208)

I suggest that the ventriloquial paradigm for instrumental performance points toward a different reading of King's performance, though not one that excludes the psychologists' analysis. Whereas the psychologists take it as given that King's behaviors express his feelings about his own playing and the music he is producing, it seems to me that the same gestures and expressions can equally well be read as his reactions to *Lucille*'s behavior. Perhaps the sounds Lucille produces arouse difficult emotions within him, and perhaps it is her ability to move him that stirs his sense of wonder. Perhaps it is Lucille's prowess at rendering difficult passages rather than his own that he signals for the audience, and perhaps he is following "her" playing with his head movements. Constituting the guitar as a separate "person" (or persona) and acting toward it as such allows King to dramatize the ventriloquial relationship between instrumentalist and instrument, a relationship that is always enacted, though not usually foregrounded, in conventional musical performance.

3 GuitarBot

Mari Kimura's *GuitarBotana* (2004) is a work for violinist and GuitarBot, a robotic musical instrument designed by Eric Singer and based on the slide guitar. The GuitarBot consists of four independently controlled strings, each of which is "fretted" by a mechanical slide and plucked by a plectrum. It cannot be played directly by human hands, but only by using a computer and MIDI (Musical Instrument Digital Interface); the computer can be programmed to play it automatically.[8]

GuitarBot is a rather large and imposing sculptural object, over five feet tall, including its base. Each string is stretched over its own vertical metal strip; metal

---

[8]Some of my description of *GuitarBotana* here repeats material that appeared originally in Auslander 2008, where I discuss Kimura and her performance in a different context.

Fig. 3 Mari Kimura and GuitarBot perform *GuitarBotana* at the Chelsea Art Museum, New York City, in 2004. (*Source* Performance video directed by Liubo Borrisov)

braces behind them hold the four strips parallel to one another. Attached to the braces near the bottom of the assembly is a metal rod that curves down to a supporting base. Although the robot is not humanoid, it is generally anthropomorphic in its size and verticality. As GuitarBot plays, it bobs and shakes on its vertical axis. In a video of *GuitarBotana* shot in an art museum, GuitarBot is perched on a white gallery pedestal, making it noticeably taller than the diminutive Kimura, who stands a few feet away from it while playing. She faces GuitarBot throughout the performance (Fig. 3).[9]

Kimura composed the music and wrote the software for *GuitarBotana*; when performing the piece, she both plays from a score and improvises. The GuitarBot's part is also both scored and improvised. Its software enables it to respond to the violinist's playing in various ways. In some cases, it follows the violinist closely and produces tones to fill out the harmony of the piece; in others, it is programmed to disregard random pitches played by the violinist, producing more open-ended situations in which its responses are relatively unpredictable. It is therefore possible for the violinist and the robot to enter into an improvisational dialogue in which the

[9]These and subsequent observations about Kimura's performance with GuitarBot are based on video by Liubo Borrisov of Kimura and Guitarbot performing GuitarBotana (Borrisov 2004).

robot responds to the violinist's playing and the violinist responds improvisationally to the robot, and so on, all within the structural constraints of Kimura's composition.

In a careful parsing of different kinds and degrees of interactivity in performance, David Saltz makes the point that in a piece such as *GuitarBotana,* the computer functions as a musical instrument (Saltz 1997, pp. 123f). Given that GuitarBot itself is a physical and mechanical object and not a computer, it is more precise to say that it is the instrument and the computer "playing" it is a musician, but I will offer a slightly refined version of this analysis below. Although King's and Kimura's respective instrumental performances are different in important ways, not least of which are the differences between the genres of the music they perform, the audiences for those genres, and the association of the blues with folk culture and art music with high culture, they both foreground musicians' ventriloquial relationships to their instruments. In *GuitarBotana*, Kimura is, in effect, playing two instruments at once. As a violinist she behaves conventionally, without drawing attention to her ventriloquial relationship to the instrument. But through her interaction with GuitarBot, which she treats as an entity separate from herself in a fashion quite comparable with King's establishment of Lucille as an autonomous agent, she engages in a meta-discourse around questions of musical identity and agency similar to those implicitly raised in King's performance, but further complicated by her use of digital technology.

When Kimura plays *GuitarBotana*, she interacts physically with GuitarBot very much as she might with a fellow human performer: as Kimura plays her violin, she faces GuitarBot, leans and gestures toward it, and watches its movements. Her facial expressions and body language look as if she were taking and giving the kinds of performance cues that musicians exchange (Fig. 4). Her proximity to GuitarBot makes these gestures seem quite intimate. At other times, Kimura does not look directly at GuitarBot, but closes her eyes or looks down. She focuses her gaze on her violin and her own playing of it, looking over at GuitarBot only every so often, as if to check in with a fellow player. Kimura uses her gaze within the performance to construct her violin and GuitarBot differently: the way Kimura looks at her violin while playing it establishes it as her instrument, while the ways she looks at GuitarBot suggest that she perceives it as another musician. The fact that GuitarBot moves as if it were leaning toward and away from Kimura as its strings sound enhances this effect by making it seem to move in response to her playing, gestures, and looks. Like King, Kimura separates the two musical functions she performs: just as King sings in his own voice, Kimura retains the identity of violinist for herself. And just as King displaces his identity as guitarist onto Lucille, Kimura displaces her agency as GuitarBot's programmer onto the instrument itself. (In saying this, I mean only that King and Kimura both perform two musical functions simultaneously and use one to foreground the ventriloquial aspect of the musician's relationship to the instrument, though not the other. The fascinating question of whether a singer has a similar relationship to the voice as an instrumentalist does to the instrument lies outside the purview of this essay.)

Fig. 4 Mari Kimura and GuitarBot perform *GuitarBotana* at the Chelsea Art Museum, New York City, in 2004. (*Source* Performance video directed by Liubo Borrisov)

Kimura's stated goal in this performance is for the audience to perceive GuitarBot as a musical partner akin to another human musician, not an instrument.[10] She furthers this goal by using her own performance as violinist to ensure that everything that happens in the performance adheres to her vision of the composition:

> My compensating for the robot's or computer's lack of musical "integrity" as the performance goes along should be hidden from, or unnoticeable to, the audience. In short, my aim is that the performance as a whole come across to the audience as if the robot or computer is thinking, feeling, and being sensitive; that it possesses the "rights and responsibilities" of a true musician. (quoted in Auslander 2008, p. 114)

[10]In his article "Live Media: Interactive Technology and Theatre," David Z. Saltz makes a useful taxonomic distinction between instrumental media, in which "interactive technology is used to create new kinds of instruments," and virtual puppetry: "The difference is that while an instrument is an extension of the performer, a kind of expressive prosthesis, a virtual puppet functions as the performer's double. In other words, instruments are something performers use to express themselves …; a puppet is a virtual performer in its own right." See *Theatre Topics* 11, no. 2 (2001), p. 126. Kimura uses GuitarBot as a virtual puppet that is ultimately under her control, but appears to the audience as a "performer in its own right".

The purpose of this benign deception is rhetorical. If King's dramatic relationship with Lucille seems to echo the complex gender politics of the blues, Kimura's presentation of GuitarBot as a separate entity addresses cultural issues pertinent to her genre—that of art music. She wishes to persuade her audience that contemporary, experimental, and electronic music belong in the same canon as the classical repertoire: "I consciously try to convey to the audience the fact that Bach, Brahms, Cage, Berio, and Robots belong together in the same evening's program" (ibid., p. 116).[11]

4 Lucille Meets GuitarBot

Although I have emphasized up to this point the similar ways King and Kimura construct Lucille and GuitarBot as autonomous musical agents through their respective performances, there is an obvious and important difference between these performances: Kimura's use of digital technology allows her to remain physically independent of her instrument, enhancing the illusion of GuitarBot's autonomy. I say "illusion" because, while it is true that GuitarBot is more autonomous than a conventional instrument, since it produces sound without being manipulated directly by a human being and the programming for *GuitarBotana* allows it to make some "decisions" on its own, it is permitted only relative autonomy. Although Kimura does not touch GuitarBot, she nevertheless determines what it plays during the scored sections of the piece through her programming of the computer that controls it. The score contains special notations that allow the performer to anticipate what GuitarBot will do during specific passages in the piece. For example, (*ii) means that "GuitarBot follows and plays the violin pitches in unison," while (*iii) means that "GuitarBot follows and plays the violin pitches, adding 4th chords in parallel motion."[12] The sections in which GuitarBot "improvises"—that is, produces relatively unpredictable output—are also determined by Kimura and marked in the score. In other words, even though Kimura does not know exactly what sounds GuitarBot will produce during those passages, she knows when those passages will occur and the basis on which GuitarBot will respond. Furthermore, as the quotation above suggests, she seems to be primarily concerned with the integrity of her composition, not with the creation of a genuinely autonomous technological musical agent. Since GuitarBot cannot think, and therefore cannot actually make musical decisions, Kimura must make it appear to think by compensating for what she calls its "lack of musical 'integrity'" through her own

[11]An index to the differences between the cultural contexts in which Kimura and King operate is that whereas it is possible that the more experimentally inclined part of the audience for art music might be open to the idea of a robotic musician, it is unimaginable that the blues audience, which subscribes to an ideology of folk authenticity, would be equally accepting.

[12]Mari Kimura, *GuitarBotana*, © 2005. The score was provided to me by the composer.

playing. Writing of the characters in MOOs,[13] Michele White observes that "characters can be programmed to … participate in events when the spectator is not engaged with the host computer. It is difficult to describe characters as subjects, even though the character 'acts' while outside the spectator's control because of the continued conflation of spectator and character" (White 2006, p. 43). Much the same is true of GuitarBot: even though it acts on its own to a certain extent and is never physically controlled by Kimura, it does not qualify as a subject apart from her. Kimura's programming of GuitarBot does not grant it true autonomy; its apparent autonomy is an effect created through the way Kimura performs with it, just as Lucille's autonomy is an effect of King's performance.

Another important difference between King's instrumental ventriloquism and Kimura's is also a direct consequence of her use of digital technology: we can always *see* King manipulate Lucille and there is ultimately no ambiguity as to who is truly vested with agency in the performance. In this regard, King's and Kimura's respective performances parallel two different moments in the history of ventriloquism. King's performance is akin to the more recent, and most familiar, paradigm of the vaudeville ventriloquist who has a dummy for an interlocutor, which originated in the mid-nineteenth century. As Steven Connor notes, our ability to make a visual connection between ventriloquist and dummy is crucial to this kind of theatrical ventriloquism: our delight at the act derives precisely from knowing, despite appearances, where the voice comes from (Connor 2000, pp. 20f). Indeed, our delight in King's facial expressions and physical gestures derives from the way they can be read as his reactions to the sounds Lucille produces, as if autonomously, juxtaposed with the self-evident fact that it is King who is playing.

As I have noted, the technology Kimura uses enables her to sever the physical connection: GuitarBot is her instrument, her dummy, her alter ego, her foil; it is controlled by the software she wrote, her violin playing, and the parameters of her composition. But there is not the evidence of direct physical control conveyed by the proximity of dummy to ventriloquist, not to mention the presence of the ventriloquist's hand on the doll. I suggested earlier that the computer could be seen as a musician "playing" GuitarBot. I will now refine that characterization by suggesting that the computer is not like a musician in itself; it serves, rather, as Kimura's "hand" that controls both GuitarBot's movements and its musical actions, in the way the ventriloquist's hand manipulates the dummy or King's hands play Lucille. Kimura's physical agency as GuitarBot's controller has been displaced onto the computer. In her performance, we see Kimura's enactment of traditional physical agency on her violin contrasted with the electronically mediated agency through which she controls GuitarBot via the computer.

---

[13]John Daintith defines MOO, which is an "acronym for multiuser object oriented," as "a system that has been developed from the early text-based multiuser adventure games, and offers a purely text-based environment allowing multiple users to … interact with other users and with end-user systems"; see *A Dictionary of Computing* (2004). In MOOs, users create characters, spaces, and objects and perform actions by typing commands.

Whereas we see King control Lucille even as we enjoy the (fictional) idea that he has to negotiate with her, there is greater ambiguity as to who has agency in Kimura's performance. This ambiguity stems largely from the probability that her listeners do not know exactly what is going on much of the time, even if they are aware of the basic situation unfolding before them. Absent such knowledge, it is not clear just from watching and listening whether, for example, GuitarBot is a playback device or is interactive, or, when it is improvising, exactly what it means for such a device to "improvise." Even if one knows the technological set-up and how Kimura uses it, what is happening on a moment-by-moment basis is still not necessarily clear unless one happens to be reading the score. This ambiguity relates directly to an issue that is emerging around performances involving human-machine interaction: how to enable the audience to understand the operation of cause and effect in such performances. Whereas some argue that performers should do things that allow the audience to understand how they trigger their technological devices and how those devices respond to performers' actions, Kimura, at least in this piece, intentionally obfuscates the precise nature of the human-machine interaction in the interest of promoting the illusion that GuitarBot is her equal partner in the performance.[14]

If King's play with Lucille parallels the interaction of the vaudeville ventriloquist and the dummy, Kimura's performance relates more closely to an earlier, less familiar practice of ventriloquism: early nineteenth-century acts in which ventriloquists threw their voices into the bodies of automata that stood apart from them and seemed to move on their own (Connor 2000, pp. 335ff). If King dramatizes the instrument's status as dummy to the musician's ventriloquist, Kimura takes that representation a step further (ironically by taking a step backward in the history of ventriloquism) by using, in addition to her violin, a robotic instrument that is physically distinct and seemingly autonomous even as it sings, like Lucille, in the ventriloquial voices that Kimura "throws" as composer, coder, and player.

Like a theatrical ventriloquist, King' deflects his audience's attention from his control over Lucille through his reactions to her, and by interacting with her as if she were a separate entity. Kimura seeks to deflect her audience's attention from GuitarBot's lack of autonomy and musicianly intelligence, but also goes beyond that. The masking of agency is central to her ventriloquial musical performance in a way that it is not to King's. John Deighton, a professor of marketing, provides a taxonomy of strategies for masking agency in performance, including his concept of "objectification." In Deighton's terminology, objectification means the masking of human agency and its apparent transfer to an object:

[14]For an essay discussing the need for transparency in human–machine interactions in musical performance, see Schloss (2003). For a discussion of the issue in more general terms, see Wechsler 2006, p. 72. See also Stuart 2003. Stuart argues that listeners to laptop music should surrender the desire for a visually verifiable relationship between the performative in itself and therefore as the proper object of their attention.

> [T]he marketer stages a performance, but the consumer perceives it as merely an objectively good performance by the product. The dramatistics are overlooked…. Hoch and Deighton (1989) describe several examples of framing events to show the product in its most attractive light while the marketer stays offstage. (Deighton 1992, p. 365)

While it may seem unconventional to describe artists' work as analogous to that of marketers, I find Deighton's account of objectification to be particularly pragmatic and direct. Both King and Kimura are akin to marketers in one respect: each seeks to sell to an audience the idea of a musical instrument as an autonomous entity in equally vivid, though different, ways. King presents his guitar as a separate, named being that enjoys a degree of independent expression and with which he is in dialogue. Kimura also interacts with GuitarBot as if it were her musical partner and engages in both scored and improvisational musical dialogue with it. But whereas King does not in any way hide his physical manipulation of Lucille even as he rhetorically constructs her as a separate entity, Kimura seeks to mask the "dramatistics" involved in her staging of GuitarBot by encouraging her audience to perceive her relationship to it as a relationship between equals, and hiding the work she does to maintain the integrity of her composition in the face of GuitarBot's unpredictable behavior.

Objectification in Deighton's sense is thus the opposite of Goffmanian dramatization: whereas dramatization makes visible the work involved in a routine, objectification masks and displaces that work. But both techniques serve the end of making a specific impression on an audience. Deighton's observation that "the marketer stays offstage" points to the fact that both King and Kimura wish their audiences to perceive them as performers engaging with other performers—Lucille and GuitarBot, respectively—rather than as manipulators of instruments. King and Kimura appear as performers in the scenarios they create as "marketers". The versions of themselves responsible for setting up the mise-en-scènes that make these scenarios possible and plausible are not exposed to the audience.

Comparing and contrasting the means that King and Kimura use in pursuit of the common goal of constructing the instrument as an entity unto itself, it is clear that while Kimura gains much through her use of digital technology, as contrasted with King's use of more conventional instrumentation, she also loses something. To understand what is lost, we must return to Godlovitch's characterization of musical performance, quoted earlier, as a demonstration of skill in which the musician "subdue[s] confidently the treacheries" of the instrument. Kimura hints at the nature of GuitarBot's potential treachery when she describes working with the machine. Referring to the instrument's four strings, she states that "I started to imagine GuitarBot as actually four individuals…. I would come in for a rehearsal and ask, 'So, how is Mr. Two today?' because he is the most temperamental of the four strings."[15] (It is noteworthy that Kimura characterizes her technological partner as male, especially in relation to King's feminization of his guitar. This leads to

[15]Quoted in Popper (2007).

speculation that, within the matrix of heterosexuality, the instrument that is at once the musician's creative partner and foil is likely to be assigned the opposite gender.)

Because Kimura plays two instruments in *GuitarBotana*, one directly and one through displaced agency, there is twice the possibility of her being betrayed by them, as Godlovitch would have it, and therefore twice the opportunity for her to show her prowess by overcoming the obstacles they present. Kimura also created another "artificial constraint" (Godlovitch's term) through her decision to program GuitarBot not just to follow her playing, but also to deviate unpredictably from it at some points, making the piece that much more difficult to perform, because it forces her to think and respond, simultaneously and very quickly, as both composer and player, in order to maintain the integrity of the work. But Kimura's desire for her audience to perceive GuitarBot as a legitimate musician actually causes her to forego full credit for her own skill as a performer. She presents herself only as a virtuoso violinist overcoming the normal technical challenges offered by that instrument; her role as GuitarBot's ventriloquist remains intentionally offstage. She masks the challenges presented by GuitarBot as an instrument in favor of constructing it as a fellow performer and collaborator.

On the other side of the ledger, GuitarBot's greater apparent autonomy in the eyes of the audience makes Kimura's performance that much more effective as a staging of the ventriloquial relationship between musician and instrument. Indeed, GuitarBot's seeming independence borders on the uncanny, which

> occurs when animate and inanimate objects become confused, when objects behave in a way which imitate life, and thus blur the cultural, psychological and material boundaries between life and death, leading to what [Ernst] Jentsch called "Intellectual Uncertainty"—that things appear not to be what they are, and as such our reasoning may need re-structuring to make sense of the phenomenon. (Hollington and Kyprianou 2007)

Animate objects like ventriloquists' dummies and automata evoke the uncanny, as does GuitarBot. At its best, *GuitarBotana* induces Jentsch's intellectual uncertainty, since GuitarBot, a machine that is not supposed to be capable of creativity, appears to act as a sentient being making music. Arguably, the version of the uncanny in performance enacted by Kimura and GuitarBot differs significantly from the version described by Matthew Causey in "The Screen Test of the Double." Discussing "the simple moment when a live actor confronts her mediated other through the technologies of reproduction," Causey posits "that the experience of the self as other in the space of technology can be read as an uncanny experience, a making material of split subjectivity" (Causey 1999, p. 385). Even though GuitarBot is not literally Kimura's double (that is, her reproduced and mediated self), it is plausible to suggest that GuitarBot acts as Kimura's Other in this performance. Since this Other's performance is a manifestation of Kimura's musical sensibility, one might stretch the point slightly and claim that *GuitarBotana* makes split subjectivity material, as Causey describes. But the technological uncanniness of GuitarBot lies in its alterity—its difference from, and apparent independence of, Kimura in the way it serves her "as a foil, not a clone" (to return to Burrows's characterization of musical ventriloquism). The fact that GuitarBot's performance is actually a displacement of Kimura's agency is suppressed in favor of foregrounding the machine's ostensible autonomy.

On one level, Kimura dramatizes the ventriloquial relationship between player and instrument in the same way King does, albeit in a very different musical context, by creating the impression that an instrument possesses an identity and agency; both King and Kimura enact the fantasy of instrumental autonomy that underlies the ventriloquial relationship between performer and instrument. But because the digital technology Kimura employs allows GuitarBot a greater degree of (apparent) autonomy than Lucille, who is always under King's visible, physical control, it enables her to push the enactment of this fantasy further toward the uncanny to show us what it might look like for a performer to interact with a genuinely autonomous musical instrument.

This essay began life as a conference paper presented at the Performance Studies Focus Group of the Association for Theatre in Higher Education's 2008 pre-conference in Denver. I would like to thank my fellow panelists Nicholas Cook and Elizabeth Patterson and the others who read versions of this paper and shared their reactions: Kevin Brown, Mark Marrington, and John Richardson. Additional thanks to Mari Kimura and Liubo Borrisov, and a very special thank you to David Saltz and Catherine Schuler.

References

A Dictionary of Computing. (2004). Accessed October 28, 2009. http://www.encyclopedia.com/doc/1O11-MOO.html

Auslander, P. (2008). Live and technologically mediated performance. In T. C. Davis (Ed.), *The cambridge companion to performance studies* (pp. 107–119). Cambridge: Cambridge University Press.

Borrisov, L. (2004). Performance video of Mari Kimura and GuitarBot performing *GuitarBotana* at the Chelsea Art Museum, New York 2004. Accessed May 4, 2009. http://www.music.columbia.edu/~liubo/guitarbotana/

Burrows, D. (1987). Instrumentalities. *Journal of Musicology, 5*(1), 117–125.

Causey, M. (1999). The screen test of the double: The uncanny performer in the space of technology. *Theatre Journal, 51*(4), 383–394.

Cone, E. T. (1974). *The composer's voice*. Berkeley: University of California Press.

Connor, S. (2000). *Dumbstruck: A cultural history of ventriloquism*. Oxford: Oxford University Press.

Deighton, J. (1992). The consumption of performance. *Journal of Consumer Research, 19*(3), 362–372.

Fast, S. (2001). *In the houses of the holy: Led Zeppelin and the power of rock music*. Oxford: Oxford University Press.

Gelbart, M. (2003). Persona and voice in the Kinks' songs of the late 1960s. *Journal of the Royal Musical Association, 128*(2), 200–241.

Godlovitch, S. (1985). *Musical performance: A philosophical study*. London: Routledge.

Goffman, E. (1959). *The presentation of self in everyday life*. New York: Anchor.

Gracyk, T. (1997). Listening to music: Performances and recordings. *Journal of Aesthetics and Art Criticism, 55*(2), 139–150.

Hollington, S., & Kyprianou, K. (2007). Technology and the Uncanny. In *Paper presented at the EVA London Conference, July 2007*. http://www.eva-conferences.com/eva_london/2007/papers. Accessed 4 May 2009.

Hoch, S. J., & Deighton, J. (1989). Managing what consumers learn from experience. *Journal of Marketing, 53*, 1–20.

Jazz Casual. (1968). Accessed May 4, 2009. http://www.youtube.com/watch?v=tBWcSc3nPow and http://www.youtube.com/watch?v=O5URVbh3KX8&feature=related

Kerekes, J., & O'Neill, D. (1996). Lucille speaks. In *The king of the blues*. Accessed July 11, 2008. http://www.worldblues.com/bbking/prairie/lucille.html

Miller, D. (2011). On Piano performance—technology and technique. *Contemporary Theatre Review, 21*(3), 261–275.

Popper, B. (2007). Robot rock. In *Brooklyn paper*, November 10, 2007. Accessed May 4, 2009. http://www.brooklynpaper.com/stories/30/44/30_44robotrock.html

Saltz, D. Z. (1997). The art of interaction: Interactivity, performativity, and computers. *Journal of Aesthetics and Art Criticism, 55*(2), 117–127.

Schloss, W. A. (2003). Using contemporary technology in live performance: The dilemma of the performer. *Journal of New Music Research, 32*(3), 239–242.

Stuart, C. (2003). The object of performance: Aural performativity in contemporary laptop music. *Contemporary Music Review, 22*(4), 59–65.

tdpri forum. (2009). Accessed April 27, 2009. http://www.tdpri.com/forum/bad-dog-cafe/155962-clapton-had-blackie-roy-buchanan-had-nancybb-king-has-lucille.html

Thompson, W. F., Graham, P., & Russo, F. A. (2005). Seeing music performance: Visual influences on perception and experience. *Semiotica, 156*(1–4), 177–201.

Wechsler, R. (2006). Artistic considerations in the use of motion tracking with live performers: A practical guide. In Susan Broadhurst & Josephine Machon (Eds.), *Performance and technology: Practices of virtual embodiment and interactivity* (pp. 60–77). Houndmills, UK: Palgrave Macmillan.

White, M. (2006). *The body and the screen: Theories of internet spectatorship*. Cambridge, MA: MIT Press.

No Flute Is an Island, Entire of Itself. Transgressing Performers, Instruments and Instrumentality in Contemporary Music

Bjørnar Habbestad

Abstract What does an instrument offer us, and how does playing it, change this? Oscillating between personal experiences and select theoretical positions, the author discusses relationships between instrument and performer. Through a questioning of the validity of subject/object positions, the dilemma of instrumentality is introduced and the relevance of transgression used as an entry to a rethinking of instrumentality.

"*I want to build a trumpet!*", exclaimed Joel Ryan, to a room full of Super Collider apprentices, gathered at STEIM sometime in early 2002.[1] He had once more stopped talking about the computer code for his newly developed performance instrument and ventured into philosophy and aesthetics. Much to my immediate disappointment, he continued: "*... a trumpet, not a flute*". But the argument behind such a (to me) disheartening statement soon became apparent. Joel's utopian 'trumpet' was a digital software instrument capable of sustaining multiple types of use and abuse: A simple but solid design allowing the performer a wide range of performance options, from a thin airy shadow of a sound through warbling pitch clouds to snarly brass rants. And for him, the idea of a trumpet housed a much wider spectrum of sounds and performance strategies than that of a flute. So the excellence of the trumpet metaphor was established, and I found myself reluctantly agreeing. Compared to the sound of a Chet Baker, Axel Dörner or Dizzie Gillespie, my flute seemed to offer me little. Their freedom, their influence on their instrumental practice seemed without borders. Did I not approach my instrument of choice in a manner that could satisfy listeners like Ryan?

[1]STEIM SuperCollider workshop, oct 2002. See http://www.steim.nl and http://v2.nl/archive/people/joel-ryan.

B. Habbestad (✉)
The Arne Nordheim Centre for Artistic Research,
Norwegian Academy of Music, Oslo, Norway
e-mail: b@bjornarhabbestad.com

T. Bovermann et al. (eds.), *Musical Instruments in the 21st Century*,
DOI 10.1007/978-981-10-2951-6_20

15 years later, I ask myself whether it is possible to separate an instrument from its instrumentalists in this manner. Can we distinguish one from the other and find the point where the instrument-object ends and the instrumentalist-subject starts? Can we ask 'how is a performer or a performance defined by an instrument?' Or even the other way around: 'how is an instrument defined by its performance?' All these questions supersede the primary question of this text and the book it is presented in; the desire to understand instrumentality, and to investigate the role it plays in the development of musical performance technologies. So, before approaching the idea of instrumentality, new or old, we need to clarify the connection between instrument and performer.

Many discourses on this relationship inherit modes of thinking emanating from the platonic division of subject and object, such as the musical discourses of analytic philosophy (Davies 2001). Despite the rigour and thoroughness of many such accounts, a potential risk in the application of these theories is the reduction involved in perceiving the instrument as a physical object to be operated by a human subject. We will return to this issue throughout the text. Reviewing the introductory anecdote, it is hard to say whether Ryan's antipathy for the flute was aimed at the instrument or its instrumentalists. Clarifying that question is of little importance in itself, but pursuing the questions that surround such a statement allows us to inspect the fruitfulness of the subject/object separation itself. We will start by investigating two opposite entries to our dilemma: Gibson's concept of *affordances* and Giddens' concept of *agency*.

1 What Does the Instrument Offer Us?

"*The psychologists assume that objects are composed of their qualities. But I now suggest that what we perceive when we look at objects are their affordances*", writes James Gibson (Gibson 1979, p. 134). His archetypical example is the affordance of a door knob, and its inherent invitation to be turned. Affordances for Gibson comprise the *action possibilities* that an object offers to an organism in an environment. Let's note that the idea implies a systemic, or ecological, approach.[2]

The principal affordance of a flute could be said to be its ability to be blown and keyed—a set of affordances it happens to share with the trumpet, the second instrument of our little tale. Actually, it takes some degree of familiarity with these instruments and their instrument families to acknowledge the different approaches needed in order to make the air resonate and cause sound to be heard. There seems to be little in the flute's design that in itself informs us about the shaping of lips, the soft straining of different facial muscles while modulating the oral cavity and

[2]In my summary of Gibson's view there are components (*the object, its actions possibilities, an organism*), all residing in an environment. A central tenet in ecological thinking is that any change to a component influences the entire eco-system. We shall later see how this relates to a performance-paradigm.

relaxing the lower throat. Most people would find it probable that a flute works differently than a trumpet, but besides their obvious differences in construction, a certain muteness of the object on the topic of its practice becomes apparent when we inspect it closely enough. Let's think of it as the gap between performer and instrument. If we ascertain that the instrument alone does not offer us much information on its use, it seems logical to continue that to a certain extent the instrument's affordances do not present themselves until they are engaged, until there is an effort to bridge the gap, so to speak.

The element of mystery, or even awe, that can resound in the performer as she starts uncovering the ins and outs of her chosen instrument can function as an indication of the width and the depth of the knowledge and experience needed to truly *know* it. This *knowing,* the competence which is part experience, part instinct, part knowledge, part taste, part absolute and insisting, part relative and suggestive, is what performers do with instruments.[3]

The apparent muteness of an instrument is broken by its use, by listening, learning, imitating and innovating within historic and social contexts. This suggests that if we want to understand the affordances of a flute, we need to interrogate flutists on their practice. This notion is easily supported by commonsense logic: an instrument itself is unthinkable, separate from the very operations or actions that are to be performed on it. As the instrument-as-made is an active concept, the making of a *tool*—something which comes attached with actions—we must understand the *instrument-as-played*, not *the-instrument-as-constructed* or *-observed*. This echoes an important distinction in Gibson's writing: rather than thinking of affordance as a "*release*" of qualities residing in an instrument, we must bear in mind that "Affordances are properties taken with reference to the observer" (Gibson 1979). So affordances reflect not only the object perceived, but also the organism that perceives it.

2 Playing, Participating, Influencing

Giddens' concept of agency belongs to his theory of '*structuration'*, an account for the relationship between societal forces (structure) and individual freedom (agency). The concept of structuration (…) "*expresses the mutual dependence of structure and agency*" (Giddens 1984), emphasising their interdependence as they take place in time. Agency is subsequently described as an agent's "*capacity to have acted differently*" in a given situation, and structure as "*rules and resources, organized as properties of a social system*" (ibid. p. 66).

[3]And this doing, this act, has been targeted as an epistemological category by a long line of thinkers, from Wittgenstein to Merleau-Ponty, Bourdieu to Rheinberger. In recent musicology, Christopher Small's concept of "musicking" targets this. Older sources includes Vladimir Jankelevitch, who stresses the ineffable aspect of music as a live event.

Both the instrument and its performance culture could be understood as structure ('rules and resources') in such a perspective. The same could be said of genre conventions, musical form, technology, software, composer/performer relationships or instrument building traditions. A performer's agency, the conceptual counterpoint to structure, could then be read as her ability to negotiate, intervene or oppose with such structures, her ability to choose to *act differently* than what is immediately defined by the structures. This metaphoric application of the structure/agency dichotomy to musical performance ties agency to notions of power and influence rather than to tradition, skill and aesthetic value. To what degree does the performer interact with structuring forces? To what extent does the artistic practice of instrumentalist X highlight something about instrument Y that we were previously unaware of? Asking questions that might identify the scope of action available to performers, their changes in "agentic orientation" and the "different ways in which [their] agency interacts or interpenetrates with different forms of structure" (Emirbayer and Mische 1998) could be approaches to understanding more of the relationship between instruments and their performers. This way we can aim to better articulate what takes place in the gap between them.

A risk apparent in a direct transferral of Giddens' thinking to musical practice would be the potential of projections of genre-based stereotypes. A typical example would be to connect improvisation to agency and scored performance to structure, without asking what other factors might influence their different musical situations, leading to a thin reading of surface observations.

Let us, as a test, put performer agency on a scale, as a quality ranging from low to high, and try to imagine some different performative situations. Our test case will be a comparison of a DJ performing with two CD-players, a cross-fader and a volume knob and a pianist performing a Beethoven Sonata on a Steinway A model grand piano. Which agent is bestowed with the higher degree of agency? On the one hand, the pianist has a well functioning musical system available for minute manipulations, offering a wealth of sonic and rhythmic opportunities. The DJ seems to be more limited, with the ability to start and stop the sound, make cross fades and turn up or down the volume. Easily, we give the piano player the upper hand. But, the pianist is bound to realise the work represented by the notation in front of her, the pitches available are set, many indications are given also on her tempo, dynamics and sound. Whereas the DJ gets to choose which songs to play according to his taste or to his reading of the audience as the event is taking place. In this perspective the higher degree of agency seems to belong to the DJ. As the example shows, agency is not a transparent entity, but dependent on the context in which it resides. It is therefore vital that we conceive of agency in musical performance as a multidimensional space engaging with equally multidimensional notions of structure.

Are the two above perspectives alone sufficient to inform our understanding of the relationship between instrument and performer? It seems that both concepts only partially succeed when applied as analogies to musical practice. Invoking both concepts simultaneously though, producing a certain dissonance, creates a tension field that seems more productive than relying on either of the two. We precede this investigation with a description of a concrete musical situation that might illuminate the distinction between the-instrument-as-made and the-instrument-as-played.

3 Transforming Sound and Skill

In 1975, Robert Dick released a volume on contemporary flute techniques[4] entitled "The other Flute" (Dick 1975), a gesture that seems to indicate and acknowledge both that the performance culture of the flute was limited at the time, and that there was indeed "another" instrument available somewhere.[5] 20 years later, as a young music student, I was impatient to discover and master new music, but a period of tendonitis stood in my way. I was instructed to refrain from practicing, so to fight a growing sense of desolation, I decided to take on a musical situation where I to a lesser degree would be faced with the shortcomings of my under-rehearsed technique. So I started taking lessons with improviser and noise artist John Hegre.[6] Through our weekly sessions I discovered free improvisation as a method to generate other types of interactions between players, sounds and instruments than those accustomed to me in classical and contemporary scored music. I experimented with hyper-microphonic amplification and soon adapted known flute techniques towards an amplified performance paradigm. After some time, I started developing novel flute techniques, almost inaudible on an un-amplified instrument. The most radical example would be that I generally refrained from blowing the instrument at all, treating it as a resonating percussive tube, with keys and tongue providing different ways to hit, slap and stroke it, producing filtered resonances as a result. This work transformed into an autonomous search for a different perspective on my instrument. A recalibration of my understanding of the sonic possibility space of the flute. As there was indeed another flute, behind the image of the instrument I was taught in music conservatory, different also from the one described by Dick, I started thinking of it as a "non-flute", escaping pitch, tonality and the lush, harmonious timbre as the principal affordances of my instrument.[7]

---

[4]'Extended technique' is the nomenclature used to describe any performance technique that goes beyond that of traditional performance practice. For the flute this normally includes various types of percussive techniques using keys and or lips, vocal techniques, multiphonics, timbral transformations, air sounds, noise spectra and microtonal intonation.

[5]Dick is but one of many flutists behind such publications: Bartolozzi (1967), Artaud (1980), Levine and Mitropoulos-Bott (2002).

[6]See https://en.wikipedia.org/wiki/John_Hegre.

[7]One could argue that the inclusion of microphones to my practice constituted a fundamental change of my *dispositif*, to the extent that it was no longer 'a flute'. I disagree with this position, as I see the use of microphones as a transformation of the auditory perspective applied to the instrument, rather than the invention of a new one.

Some concrete changes in competence took place as a function of this transformation. My use of the flute as a percussive instrument led to the development of an independent right and left hand fingering technique, a parametric separation of motor control to obtain different parallel rhythmic structures, a skill with no use in traditional flute playing.[8] Later this was extended to include separation of articulation and breath tempi, as well as foot movements for controlling effect pedals, resulting in six independent rhythmic layers or temporal streams.[9] Who or what changed in this situation, the instrument or me? Does the instrumentality of the flute change through such a process?

4 The Dilemma of Instrumentality

Both Gibson's ecological and Giddens' systemic models offer holistic views on our topic, with an emphasis on the reciprocal relationship between instrument and performer, a fundamental understanding of their interdependence. Still, the proximity to the chicken/egg-dilemma seems inevitable: What came first, the flute or the flutist? The impossibility of answering this question holds the key to understanding the degree of ontological interconnectedness between the two.

This is the dilemma of instrumentality: how can we articulate traits of an instrument without referring to its practice, and hence its performers? As a blunt descriptor, instrumentality seems to avoid the practice surrounding it altogether, almost suggesting an ahistoric, essentialist definition of distinct qualities in an object. The very question '*what are the defining qualities of an object's instrumentality*' could be considered reductive, minimising its usefulness when we want to understand performance as a practice. Following what seems to be an entanglement of the performer-instrument relationship, we must search for a wider conception of what instrumentality is, and where it is located.

5 Addressing Instrument and Performer as One

Recent developments in what has been coined "Critical organology" (Tresch and Dolan 2013) present a rethinking of the study of instruments, moving away from the idea of taxonomic discipline of inventories and collections and moving towards a broader field of historical interpretation. This opens up for a more ecological reading:

[8]A technique that later proved valuable when performing works by Brian Ferneyhough.

[9]These experiences were fundamental to the later development of Modality, a process described elsewhere in this volume.

> We want to think about instruments as actors or tools with variable ranges of activity, with changing constructions and definitions, and with different locations in both technical and social formations. We want to ask, What aspects of instruments have been variable (or have been seen to be), and what were the consequences of that Variation. (ibid.)

Addressing the instrument as both actor and tool, within several parameters and layers, in order to identify the variation, or the *difference*, (Derrida 1982) opens up the field in a manner echoing how critical musicology influenced the understanding of performance in the 1990s (Abbate 2004). This thinking also echoes the development of object-oriented ontology which has made a substantial impact on aesthetic practice and philosophy made evident for example by OCTOBER devoting a large section of a recent issue to this topic (Apter et al. 2016).

6 Reintroducing the Object

For the sake of brevity, the potential of understanding performance within the framework of new materiality needs to be dealt with at another time, but for our purpose it seems worth noting how Donna Haraway's writing has touched upon similar issues for many years already:

> [The] boundaries of [bodies as objects of knowledge]… materialize in social interaction among humans and non-humans, including the machines and other instruments (…) that functions as delegates for other actor's functions and purposes. "Objects" like bodies do not pre-exist as such. (Haraway 1992, p. 298)

Haraway's well-known preoccupation with the idea of *situated* and *embodied* knowledge (Haraway 1988) resonates well within our discussion, although the above quote emanates from a description of the relationship between researcher and his lab instruments. She connects this thinking to that of Bruno Latour, and especially to his description of how machines are part of social relations "through which actants shift competences" (Latour 1990, quoted in Haraway 1992). This description seems specially relevant to our situation, where we lack a way to articulate what takes place between performer and instruments. Haraway's thinking also removes us from having to address the dilemma of instrumentality, as she so eloquently describes the interaction of humans and non-humans.

In an account of his actor-network theory, Latour turns to a definition of agency as "doing something, that is, making some difference (…) transforming some As into Bs through trials with Cs" (Latour 2007). This seems to be a much more active description of agency than the one we found in Giddens' theory. He continues:

> Without accounts, without trials, without differences, without transformation in some state of affairs, there is no meaningful argument to be made about a given agency, no detectable frame of reference. An invisible agency that makes no difference, produces no transformation, leaves no trace, and enters no account is not an agency. (ibid.)

This emphasises the connection between agency and influence. An agent that does not leave a trace is not an agent according to Latour, and although he most certainly did not have musicians in mind per se, it is an interesting experiment to transfer this to a musical situation. It is a well known fact within the circles of festival organisers, curators and musicians in the contemporary music circuit that a composer such as Phil Niblock performs on stage playing sound files from his laptop, listening to his own music from the stage more than performing it. Within Latour's thinking, this event might not be awarded agency at all. The question then becomes: is agency in the Latourian sense a prerequisite for performance? What then about the countless performances of classical music where the "invisibility" of the performer has been cultured and canonised in order to set "the work", "the composer" or "music itself" at the centre? Let us pursue musical acts with Latour's notion of the transformational character of agency in mind.

7 The Role of Transgression

Transgression has a complex genealogy in many different scholarly disciplines but central to many of them is the notion of its doubleness—it "violates or infringes" but simultaneously can "announce or laudate", making it "a deeply reflexive act of denial and affirmation" (Jenks 2003). In the 20th century, transgression of established instrumental practice is evident in the artistic projects of many of the most prominent performers. Harnoncourt' turn to reconstructions of historical instruments in order to redefine the sound of early music and Hendrix's complete plasticity between idiomatic guitar technique, amplifier control and the use of effect pedals are two equally obvious but contrasting examples of transgression of established instrumental technique and sound. Peter Brötzmann's redefinition of the performance practice of the saxophone in the 1968 recording of "Machine Gun" and Salvatore Sciarrino and Roberto Fabbricciani's similar redefining of the flute sound in the cycle "Fabbrica degli incantesimi"[10] in the decade between 1976 and 86 are other clear instances. By following the transformation of sound of the female pop vocalist in the music of Björk and Cher, we could trace similar situations, and continue probably endlessly by turning to other genres and conventions. The contrast to the competition-driven market of mainstream classical music is evident, where "jurors, agents, and programmers will all tell you they are looking for a performer who has something unique to say, while (…) all their values in relation to composer, score and performance tradition, tend towards enforcing conformity", to quote a recent presentation given by Daniel Leech-Wilkinson in a conference on the topic of virtuosity.[11] The emancipatory potential of describing a hitherto believed fixed state (gender identity) as a negotiable space of individual construction has

[10]See the interview "Working together. Roberto Fabbriciani in conversation with Bjørnar Habbestad" by this author in Music + Practice. To be published late 2016: http://www.musicandpractice.org.

[11]"Virtuosity—an interdisciplinary Symposium" held at the Liszt Academy, Budapest, 3–6 March 2016. See https://virtuosity2016.wordpress.com.

been proven to be extensive (Butler 1988, 2002). It is an intriguing thought to consider whether a similar effect could be observed if we would approach instrumental practice itself as performative, "instituted through a stylized repetition of acts" (Butler 1988), and thus refuting tradition's normative hold over musical agency. It could be argued that this rebuttal of the fear of transgression is one of the most important qualities from the much discredited early music scene of the 1960s. We find this mirrored in many experimental musics[12] and in much of the undocumented instrumental work that underlies compositional musics of the post-war era. As we continue, let us note that transgression is valued differently in different performance cultures.

What Latour's concept of agency awards us is the opportunity to release the instrument from fixed preconceptions about its use. Hence, the relevance of my experiences from transforming my flute technique is not dependant on whether or not I actually asserted a high degree of agency or not. Rather, their potential relevance lies in the description of the transformation of competence involved in the process. I believe that similar accounts of instrumental practices can give further insights both to future instrumental design processes and to scholarly and artistic works that seeks to scrutinise the performer-instrument relationship. As such, I see my case not as exemplary, but rather as carrying a symbolic function in order to advocate a practice-centred perspective when we approach and understand new instruments and performance environments.

8 Rethinking Instrumentality

If previous notions of instrumentality tended to favour the instrument over the performer, our current understanding of instrumentality should reflect the interdependence of the two, drawing on both Giddens and Latour. Looking as much towards what is being done to and with instruments as towards describing them as objects would be a natural first step. This means that we need to learn and understand how instruments are used and abused, created, re-contextualized and extended, not just designed, as the intentions of instrument makers are equally unobtainable as those of composers. This includes aiming to understand instrumentality as a complex function of differing practices and agents, whereby the instrument maker, the virtuoso and the maker-instrumentalist as well as the instrument itself bring different but relevant skills and knowledge sets to the table. In this perspective we could locate instrumentality as much in the performer as in the instrument.

[12]Described thoroughly by Benjamin Piekut in 'Indeterminacy, Free Improvisation, and the Mixed Avant-Garde: Experimental Music in London, 1965–1975' (Piekut 2014), in *Journal of the American Musicological Society* Vol. 67, No. 3 (Fall 2014), pp. 769-824.

In its plural form—*instrumentalities*—we can grasp the existence of several instances of the term. Acknowledging the unstable and fleeting character of performance cultures, means allowing for parallel conceptions of instrumentality co-existing in time, without being mutually exclusive. On such a basis, we can approach ideas of agency and affordances as multidimensional formations rather than as the separation of instrumental subject and object, looking towards personal skills in a performer (limitations or extensions), external objects (scores or instructions), to other performers or performance traditions (orchestra cultures, guilds, societies, genres, sub-cultures) a well as overarching shared beliefs (capitalism, anarchism).

At the outset, I asked rhetorically whether one could distinguish instrument from performer in order to locate the point where the instrument-object ends and the instrumentalist-subject starts. Neither Gibson's ecological perspectives, Giddens' theory of structuration nor Haraway and Latour's inclusion of non-human partners as actants in the development of knowledge support such a division. Rather, a nested understanding, an entanglement of instrument and performer has been proposed.

9 Approaching New Instruments

Although the relevance of the concept of instrumentality for the development of new performance technologies motivate the current volume, the alert reader will have noticed that our discussion has mainly refrained from references to any such examples. Why? Partially because the complexity found in (digital) instrument construction and development often lead to an under-development on an instrument's performance culture.[13] And partially because this lack of focus on performance often is downplayed within the discourse of technology-driven musical innovation. By using established instrumental practices as examples I have hoped to avoid questions of categorisation or definitions, in favour of principal concerns.

So—turning to our new creations—as much as asking whether or not new instruments are 'new', and what their inherent qualities might be, we should ask ourselves to what degree the performance of/with these instruments allow for new or different modes of agency, transgression or transformation. Not only the directly observable transparency of 'liveness' which Philip Auslander asks for (2009), but fundamentally interrogating the qualities available at the intersection between performer and instrument. Does this agency leave a trace? This very intersection is one of those points where musicology and organology alike needs the efforts of artistic research, to attempt to articulate "the bridging of the gap". Carolyn Abbate mentions musical performance's "*strangeness*, its unearthly as well as its earthly

[13]Cr. the many NIME-instruments that are developed, performed at conferences, never to be seen again.

qualities, and its resemblance to magic shows and circuses" (2004, p. 534), a description of mystery I would like to extend into the realm of the musical instrument. Because an instrument without secrets, without vast scopes of indecipherable opportunity, accessible only through dedicated practice and experimentation, might not be worth calling an instrument at all.

Although Joel Ryan's utopian trumpet did not manifest its transgressive powers in the performances I attended around that time, his concept made a lasting impression: To hold agency, sonic flexibility, and directness of expression as an ideal state for musical instruments. That it spawned a process of reflections around instrumentality and my musical practice proved equally important. I hereby extend my gratitude: After all, No flute is an island, entire of itself.

References

Abbate, C. (2004). Music—Drastic or gnostic? *Critical Inquiry, 30*(3), 505–536.

Apter, A. A. et al. (2016). A questionnaire on materialisms. *October, Winter 2016, 155*, 3–110.

Auslander, P. (2009). Lucille meets GuitarBot: Instrumentality, agency and technology in musical performance. *Theatre Journal.*

Butler, J. (1988). Performative acts and gender constitution: An essay in phenomenology and feminist theory. *Theatre Journal, 40*(4), 519.

Butler, J. (2002). *Gender Trouble.* Routledge.

Davies, S. (2001). *Musical works and performances: A philosophical exploration.* Oxford: Clarendon Press.

Derrida, J. (1982). *Margins of philosophy* (1972), trans. A. Bass.

Dick, R. (1975). *The other flute* (pp. 1–162).

Emirbayer, M., & Mische, A. (1998). What Is agency? *American Journal of Sociology, 103*(4), 962–1023.

Gibson, J. J. (1979). *The ecological approach to visual perception.* Psychology Press.

Giddens, A. (1984). *The constitution of society.* John Wiley & Sons.

Haraway, D. (1988). Situated knowledges: The science question in feminism and the privilege of partial perspective. *Feminist Studies, 14*(3), 575.

Haraway, D. (1992). The promises of monsters: Reproductive politics for inappropriate/d others. In Larry Grossberg, Cary Nelson, & Paula Treichler (Eds.), *Cultural studies* (pp. 295–337). New York: Routledge.

Jenks, C. (2003). *Transgression.* Psychology Press.

Latour, B. (2007). *Reassembling the social.* Oxford: OUP Oxford.

Tresch, J., & Dolan, E. (2013). Toward a new organology (pp. 1–21).

LiveCodeNet Ensamble: A Network for Improvising Music with Code

Hernani Villaseñor Ramírez

Abstract LiveCodeNet Ensamble is a laptop band from Mexico City that approaches computer music improvisation through live coding and a local network. This text discusses aspects of the ensemble's practice related to some elements of the SuperCollider program such as broadcast and history, and the concepts synchrony and collective listening in order to reflect on a collective instrument, group practice, and improvisation in the context of network music and live coding.

1 Introduction

In a group practice of computer music where on-the-fly programming and network connection are part of the infrastructure and strategy for improvising collectively with code, instrumentation is shaped by the medium and what it produces, and thus networked computers and code constitute the instrumentation of some ensembles that improvise with computers. Such group-operated instruments originate from the individual work of those who participate; a network is constructed based on programming languages, protocols, routers, laptops, personal codes, extensions, communication and synchronization systems, and the interaction between people. Code is developed both individually and collectively for the performance and the interconnection system; in a context of a culture of sharing, openness, collaboration and do-it-yourself, tools developed by other musicians and programmers are used too.

But, what is this instrument? And, how exactly does it operate? In this text, I will try to answer these questions by describing as a case study the instrumentation of *LiveCodeNet Ensamble,*[1] a group I am a part of and where I have done observations based on the ensemble improvisation and interviews with the other members.

[1]See website for details: LiveCodeNet Ensamble (n.d.).

H. Villaseñor Ramírez (✉)
Universidad Nacional Autónoma de México, Mexico city, Mexico
e-mail: hernani.vr@gmail.com

T. Bovermann et al. (eds.), *Musical Instruments in the 21st Century*,
DOI 10.1007/978-981-10-2951-6_21

The ensemble's musical improvisation practice is based on code, where a collective modular instrument is constructed and formed through a combination of software functions and group interactions. In order to describe this instrumentation, I will discuss some elements of the program SuperCollider such as *broadcast* and *history*, and as well address the concepts of *synchrony* and *collective listening*. This serves to reflect on a collective instrument that is constrained to the action of placing events in time by listening, in a collaborative practice within the context of networked computer music and live coding in group.

2 LiveCodeNet Ensamble and its Context

LiveCodeNet Ensamble (LCNE) is a group of people who improvise music on their computers using code while being interconnected to a "local network" (Barceló et al. 2004). The group was founded in October 2013 in the context of the local live coding scene of Mexico City, which was brought into being by the *Taller de Audio del Centro Multimedia*[2] and active during the period covered from 2011 to 2014. Current members of the ensemble are Emilio Ocelotl, Libertad Figueroa, José Carlos Hasbun, Hernani Villaseñor, and Eduardo H. Obieta; former member was Katya Álvarez.

Networked live coding in group in Mexico most probably has its origin with the bands *mU*[3] and *Colectivo Radiador*[4]; the former was an experimental audiovisual trio founded in 2004 based at Taller de Audio, who organized the first live coding concert in Mexico in 2006[5]; the latter is an improvisation group who has been active since 2008, realizing, among other improvisation practices, networked live coding in group, both locally and remotely. Later, during a series of live coding sessions and workshops conducted at the Centro Multimedia, other groups and ensembles were founded, such as *LiveCodeNet Ensamble* and *otú*.[6]

An important factor that motivated the foundation of *LCNE* was the encounter with different network music and live coding actors during the series of Symposiums /*vivo*/,[7] and especially the contact with band members of *The Hub*, *PowerBooks_UnPlugged*, *Benoît and the Mandelbrots*, *BiLE* and *Glitch Lich*.

---

[2]The *Taller de Audio* (until 2014, now *Laboratorio de Audio*) is a part of the *Centro Multimedia del Centro Nacional de las Artes* in Mexico City. The scope of the Audio Lab is research about sound and its relation with different artistic practices using new technologies.

[3]See MuElectro (n.d.).

[4]See Radiador (n.d.).

[5]See Toplap (2012) events page.

[6]See NMF (n.d.).

[7]The series of International Symposia of Music and Code/*vivo*/ were organized by the *Taller de Audio de Centro Multimedia* from 2012 to 2014. See Vivo (n.d.).

3 The Practice of LiveCodeNet Ensamble

LCNE is primarily an improvisation group, and they approach this practice through live coding using the common programming language SuperCollider; as each member has a personal style of live coding within the group, naturally there are differences, and sometimes even tensions arise caused by the relations and interactions that occur once the improvisation is underway. The practice of this ensemble can be defined as networked live coding in group, which contains three aspects: improvisation by code, group practice, and interconnection by local network.

The working method of *LCNE* is decentralized, that is to say, there is no director, but each member assumes the "role" (Lisiecka 2013) of an active improviser. Usually, the group starts improvising from macro-structures, that means, they determine rough structures based on changes throughout the performance, for example, changes in the rhythmic intention, or the moving from an abstract form based on sound textures to a more rhythmic part.

Improvising with code denotes a programming practice in process, which is different from programming a structured instrumentation beforehand; although in its performances *LCNE* does not start programming codes from the scratch, there is still much space left for programming and modifying the codes used on-the-fly. Each member prepares his or her codes in a different way, so that they remain open to modification during the performance; some prefer more elaborated structures, whereas others start from small pieces of code and develop them during the concert; structure or process, the differences between these two approaches provoke tensions. Winberg (2005) proposes "two main driving forces for designing and participating in interconnected musical activities: *process*-centred forces and *structure*-centred forces" (p. 256); to the first one he attributes a focus on the player's experience, and to the second one a focus on the musical outcome.

4 The Instrument

The networked live coding infrastructure of *LCNE* is a combination of informatics, telecommunication and audio technology in which the elements broadcast and history, as well as synchrony and listening, play an important role with regard to the group dynamics at the moment of improvising music. To perform this practice, the members of the ensemble use programming code written in the SuperCollider language, thus enabling them both to live code and connect themselves to a local network through a router, which Rohrhuber et al. (2007) describe as "programming languages that integrate network models and sound synthesis" (para. 9); this way the code that is written in a programming language describes sound sources and routines in the form of a text which activates sound at the moment of being declared.

The interconnection of the ensemble is a combination of hardware-software, in which the computers are connected to a router by a code[8] that allows the transmission of data using broadcast addresses, and thereby forming a local network with a star topology. At the same time, this connection code enables using the *History* class and its GUI, from which two communication channels are created between the members; text messages and declared code are mixed in this window which works in a similar way a chat system does. To maintain synchrony, the ensemble uses the SuperCollider extension BenoitLib and its synchronized clock; furthermore, each computer sends its stereo output to a sound mixer where the signals are mixed before finally being transmitted to the sound system. This is the interconnection core of *LCNE;* playing together involves being synchronized, seeing each other on the network and listening to each other during the improvisation.

But, how can the instrument of *LCNE* be defined? We can say that it is a distributed techno-social instrument composed of the network infrastructure, the individual codes and the network of relations between the members of the ensemble, in other words, it is an instrument of social interaction and a platform built by musicians and technical devices at the moment of improvising.

As Bazzichelli (2008) argues, artistic network practices create platforms on which "To network means to create relationship networks, to share experiences and ideas. It also means to create contexts in which people can feel free to communicate and to create artistically in a "horizontal" manner" (p. 26). On that subject, José Carlos (LCNE member-interview-, Mexico City, September 2015) points out that LCNE "has been a good method, a good tool to benefit from two kinds of learning: an individual learning and sharing of my knowledge with you, and absorbing your knowledge at the same time".

Brown and Bischoff (2005) refer to a "network instrument" which initially has been explored by groups such as *The League of Automatic Music Composers* and *The Hub* who "approached the computer network as a large, interactive musical instrument" (p. 375). The network of *LCNE* is an infrastructure that is activated by the relations between their members at the moment of improvisation, it is a platform to make music in a collective way which derives from individuality, where instrumentality is contained in the network, the computer and the code.

Since the instruments of each member are contained in a computer and designed with code, computer and code can be considered part of the collective instrument that falls within the context of the net. The band *PowerBooks_UnPlugged* states that "The laptop is their only instrument. Being synthesizer, realtime sound processor, networking device, PB_UP understands it as a fully autonomous instrument" (PB_UP, n.d.), whilst Wang and Cook (2004) say that code is "a real-time, expressive instrument" (para. 6).

---

[8]This code is written in SuperCollider and based on the broadcast connection method and History class. The code was written by Alberto de Campo and Julian Rohrhuber during a workshop of the Symposium/*vivo*/ 2012, and later adapted, by the ensemble, to the necessities of interconnection. See Villaseñor, H. (2014) for the code.

We can say that the instrumentation of *LCNE* is a modular instrument formed by software-hardware which is interrelated with the members of the ensemble, that is to say, people and infrastructure form a "socio-material assemblage" (Sorensen 2007) the moment the instrument is activated by the group. Likewise, this interconnected instrumentation facilitates an interaction between people who improvise in the network, based on the elements broadcast and History which display the personal programming to the rest of the group every time a line of code is declared, thus, the act of programming together transforms the individual activity in an act of collective creation.

The instrumentation of *LCNE* points at a decentralized practice, the processual activity of writing and shaping code on-the-fly allows the collective development of ideas, in the same time and space, which result rather from the collective listening than the written code. Regarding the design of these instruments, part of its success lies in this aspect of socialization, in allowing that the instrument is activated according to the network of relations, which detonates the process of the collective interconnected improvisation.

Each member has a different approach to build his or her instruments; one approach is to program instruments previously with an input in order to manipulate them during the live performance, while another approach is to program the instruments directly at the moment of the performance, sometimes even from the scratch. In both cases, the instrument is present in the mind of the live coder who modifies or develops it during the improvisation, so, this is about instruments changing over the time.

Other approaches deal with the use of synthesis or with sounds that are recorded on buffers, organizing them in routines. Usually, these codes are simple and allow both manipulation and rewriting at the moment of playing; in that regard, Emilio (*LCNE* member-interview-, Mexico City, September 2015) comments on the instrumentation he uses, "I always try to make my code and instruments as economical as possible, first, to not overload my computer […] and, on the other hand, to have them present in my head, so I can always rebuild them." For her part, Libertad (*LCNE* member-interview, Mexico City, November 2015) says, "I try to use simple structures that I have prepared previously, with variables of which I know how they work, which I know to a certain point, and which can be modified in many ways […] I am also looking for them to be very versatile". The design of each individual instrument requires to remember it exactly in order to access its programming at the moment of improvisation. Each member prepares his or her code to shape a modular instrument, which finally adopts its form in a collective way.

While I have argued so far that the collective instrument of *LCNE* is built by individual instrumentation, this construction is not simply summing up all the individual instruments; it is a constant exchange between the individual and the group at the moment of sharing information. The key moment is when we declare our code, as there two things happen simultaneously: the individual code is transmitted to all the computers and, at the same time, it is translated into sound. The moment a player presses the keys to evaluate a new fragment of programming, the code is distributed to the computers of all members of the ensemble by the broadcast method; in this way, the declared code bursts onto the screens of the other players and becomes visible for everybody in a temporal line which links our code sequence in the History window. In turn, the code activates the procedures it describes to produce sound through the interpretation and conversion processes performed by the computer, thus, the individual code, converted into sound, is mixed collectively within the sound system in order to be transmitted within the space where the improvisation takes place; in that moment we are listening to the collective result and seeing the other players' code on our screens. The collective listening is activated and the players interact based on what they hear, on the individual sound mixed with the group sound. In that regard, Emilio comments: "I think we are already listening to ourselves on an individual and collective level; I think the collective listening is the most difficult".

Regarding synchrony, when code is declared it is placed in a precise time interval; although synchrony allows to solve a simultaneity problem of the rhythmic events, at the same time it disables the internal pulse of the player to follow the ensemble. In any case, the pulse is programmed and quantized, and permits that each sound event or change of code starts on time in the next time interval available; consequently, we have a structure where it does not matter in which moment we launch a new change, it will be integrated to the synchrony pulse provided by the software.

Listening is probably the most important level of interaction in a collective improvisation practice with computers; while we are listening we make decisions about the modifications of our code in order to establish an interaction with the rest of the group throughout the improvisation. We listen and declare new fragments of code in the desired moment, and it does not matter whether this action—hitting the enter key to evaluate the code—is carried out precisely or not, the computer is in charge to place each event in a quantized temporality. This can actually be seen as problematic, because it reduces our practice to simply evaluate code fragments that will be placed in time by listening to the emerging sound and determining the moment in which we want to evaluate new code fragments.

In that way, networked live coding in group becomes a practice of individual interventions in which each player releases new code fragments in time and, in which the only active interaction occurs on the basis of the listening; so we are cast adrift to synchrony, and other potentially possible forms of interaction inherent in the dynamics of interconnectedness can be inhibited. But, it is precisely these types of collective instruments which have the potential to inspire collaborative work based on interchange and sharing. Nonetheless, other forms of group participation and interaction should be made explicitly available while improvising so as not to reduce this practice to a collective instrumentation with strict synchrony.

5 Conclusions

The instrument of *LCNE* creates a condition of possibility to share code to improvise collectively through data transmission based on the broadcast method, which becomes visible to the ensemble through the History class. During improvisation, the collective listening guides the players in their sound dialogue, and synchrony enables them to organize simultaneous events in time. However, these characteristics alone do not make the instrument collaborative, but the network of relations being established at the moment of improvising together.

By any means, it is necessary to be explicit about interactions occurring in the realm of listening in computer music improvisation. Finally, there are still some questions remaining: Is it possible to assign a collaborative role to the instrumentation? Are synchrony and listening the only ways of interaction in this kind of ensembles?

References

Barceló Ordinas, J. M., Íñigo Griera, J., Martí Escalé, R., Peig Olivé, E. Y., & Perramon Tornil, X. (2004). *Redes de computadores*. Barcelona: Universitat Oberta de Catalunya.

Bazzichelli, T. (2008). *Networking: The Net as Artwork*. Aarhus: Digital Aesthetics Research Center.

Brown, Ch., & Bischoff, J. (2005). Computer Network Music Bands: A History of The League of Automatic Music Composers and The Hub. In A. Chandler & N. Neumark (Eds.), *At a Distance: Precursors to Art and Activist on the Internet*. Cambridge: The MIT Press.

Lisiecka, K. (2013). Group as a Unit of Analysis. In A. Nowak, et al. (Eds.), *Complex Human Dynamics: Understanding Complex Systems*. Berlin: Springer-Verlag.

LiveCodeNet Ensamble. (n.d.). Retrieved from https://livecodenetensamble.wordpress.com/

MuElectro [Youtube channel]. (n.d.). Retrieved from https://www.youtube.com/user/muElectro

NMF. (n.d.). Retrieved from http://networkmusicfestival.org/nmf2014/programme/performances/otu/

PB_UP. (n.d.). Retrieved from http://networkmusicfestival.org/nmf2013/programme/performances/pb_up/

Radiador. (n.d.). Retrieved from http://www.radiador.cc/en/

Rohrhuber, J., de Campo, A., Wieser, R., van Kampen, J. K., Ho, E., & Hölzl, H. (2007). Purloined Letters and Distributed Persons. In *Music in the Global Village Conference*. Budapest.

Sorensen, E. (2007). The Time of Materiality. *Forum: Qualitative Social Research*, 8(1). Retrieved from http://www.qualitative-research.net/index.php/fqs/article/view/207/457

Toplap (2012). Retrieved from http://toplap.org/wiki/ToplapEvents/

Villaseñor, H. [Code project]. (2014). Retrieved from https://github.com/hvillase/lcne

Vivo. (n.d.). Retrieved from http://vivo.cmm.cenart.gob.mx/

Wang, G., & Cook, P. R. (2004). On-the-fly Programming: Using Code as an Expressive Musical Instrument. In *Proceedings of the 2004 International Conference on New Interfaces for Musical Expression* (pp. 138–143). Hamamatsu.

Winberg, G. (2005). Local Performance Networks: musical interdependency through gestures and controllers. *Organised Sound, 10*(3), 255–266. doi:10.1017/S1355771805000993

Three Flavors of Post-Instrumentalities: The Musical Practices of, and a Many-Festo by Trio Brachiale

Dominik Hildebrand Marques Lopes, Hannes Hoelzl and Alberto de Campo

Abstract This article offers a shared account of one out of a pluriverse of possible musics in the 21st century, with three personal perspectives. It is written by a trio of musicians/artists/researchers, who have been co-inventing an idiosyncratic style of music, including its instruments, compositional strategies, and performance systems. We articulate our artistic aims in a *many-festo*, discuss the background that informs our thinking, give examples of related artistic instrument design, and explain aspects of our own work that exemplify our essential insights and resulting tenets.

1 Introduction

Trio Brachiale exists since 2010, when Dominik Hildebrand Marques Lopes, Hannes Hoelzl and Alberto de Campo first played together as a group at an evening with the Society for Nontrivial Pursuits in Berlin. We share much common ground, being performers, composers, coders, luthiers; and we are all inspired by *second order cybernetics* (von Foerster), observation of processes with nontrivial behavior, the possibilities arising from working with code, and the roles medieval and baroque combinatorics (A. Kircher, R. Llull) play for modern media-technological society (S. Zielinski). This makes the group an ideal platform for experimenting

D. Hildebrand Marques Lopes (✉) · H. Hoelzl · A. de Campo
UdK Berlin, Grunewaldstr 2-5, 10823 Berlin, Germany
e-mail: d.himalo@udk-berlin.de

T. Bovermann et al. (eds.), *Musical Instruments in the 21st Century*,
DOI 10.1007/978-981-10-2951-6_22

with ideas about musical practices for which we share intuitive understanding. The music we make is a dense, high energy flow of improvised structures, reminiscent of free jazz improvisation, but deploying purely electronic sound sources. The trio constantly strives to extend this vocabulary by creating processes in electronics and computers, and learning to become fully aware of their immense flexibility and possibility spaces.

It is common knowledge today that the combinatorial power of modern equipment, open source software and shared technical knowledge has led to an amalgamation of the formerly separate fields of instrument building, composition, live music performance, recording and mixing into one unified musical practice. Many of the more preparatory tasks involved can now be made accessible in realtime, and thus can be played with in a live setting. This obviously changes received notions of instrumentality.

When working with processes with some degree of autonomy, a musician's decisions affect more than just the next musical moment—actions may have consequences across multiple time scales (cf. Roads 2001: pp. 1–42). Thus developing processes, activating them, observing them and influencing their behavior becomes a central part of live performance practice. Delegating the details of musical tasks to complex algorithmic machinery frees the musician(s) to invent nontrivial ways of influencing these processes, which can then be shared among the musicians and machines of the ensemble. We consider such systems *instances of MetaControl*.

This article begins with a many-festo[1] declaring our artistic aims. Then we discuss the conceptual background that informs our activities: Cybernetic thinking and its relation to music; the notion of possibility spaces and evolving technology; the roles of decision making on different time scales; and the ways limitations and constraints open up new possibilities and close others. All of these inform a central concept we propose, *Second Order Virtuosity*, which we expand upon by giving examples of artistic instrument design in this spirit, and by discussing aspects of our own work to exemplify our essential tenets.

[1]The many-festo exists in many animated versions; for some net-based ones, see: Trio Brachiale (2016).

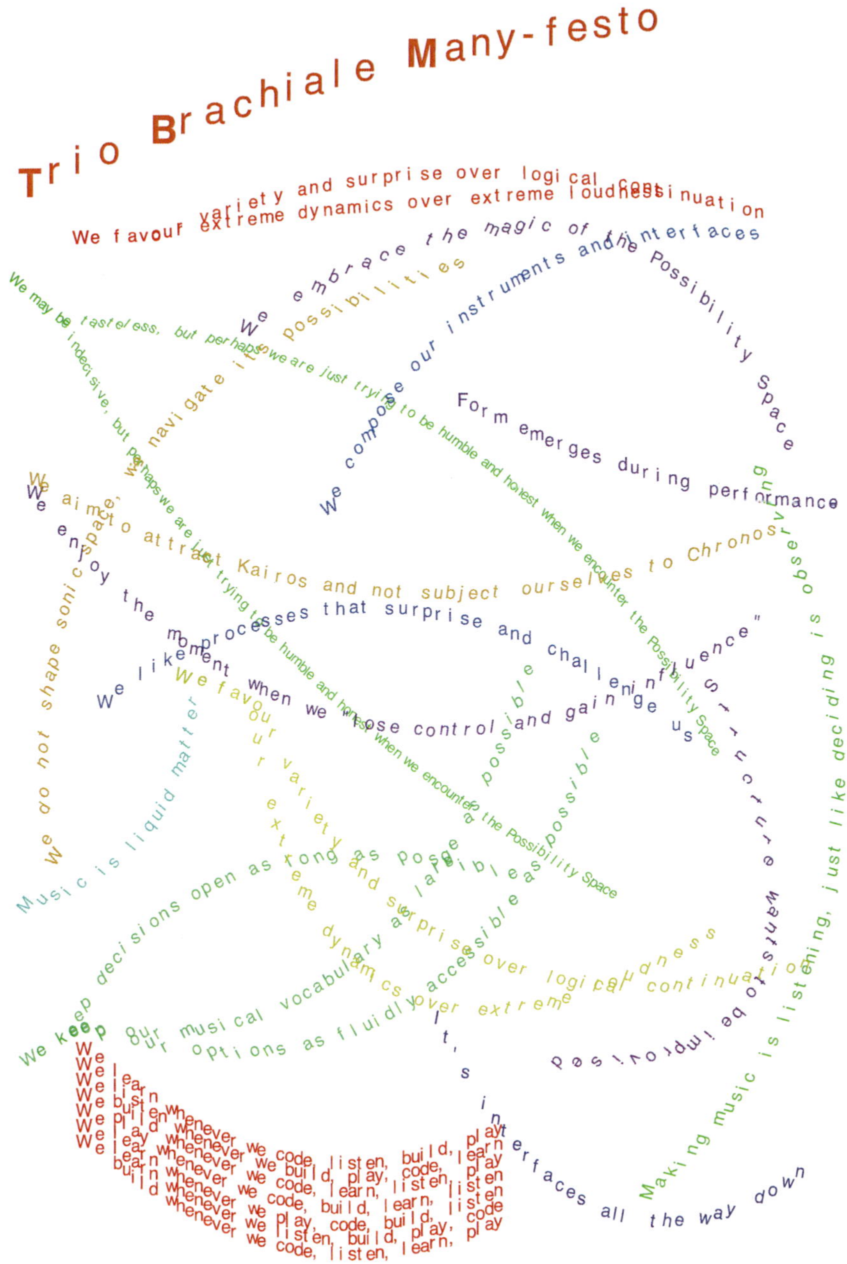
Trio Brachiale Many-festo
We favour variety and surprise over logical continuation
extreme dynamics over extreme loudness
We embrace the magic of the interfaces
We compose our instruments and the Possibility Space
We may be tasteless, but perhaps we are just trying to be humble and honest when we encounter the Possibility Space
indecisive, but perhaps we are just trying to be humble and honest when we encounter the Possibility Space
We navigate its possibilities
Form emerges during performance
We aim to attract Kairos and not subject ourselves to Chronos
We enjoy the moment
We like processes that surprise and challenge us
when we lose control and gain "influence"
We do not shape sonic space,
Music is liquid matter
Structure wants to be improvised
We keep decisions open as long as possible
We keep our musical vocabulary as accessible as possible
our options as fluid as possible
It's interfaces all the way down
Making music is listening, just like deciding is observing
We learn whenever we code, listen, build, play
We listen whenever we build, play, code, learn
We build whenever we code, learn, listen, play
We play whenever we listen, build, code, learn

2 Conceptual Contexts for Contemporary Instrumentality

2.1 *Music Making in Cybernetic Terms*

Music making is a complex social context in which relationships between its various actors can be fruitfully discussed from a cybernetic perspective. This allows us to introduce notions that are essential for our later discussion of concrete instrument designs, control schemes, and collaborative process.

A minimal list of elements in common Western music includes: musician, instrument, composer, instrument maker, audience, and performance site. Their roles can overlap: Composers may be also instrumentalists, audience members may also be musicians, musicians may also make their own instruments or setups.

Nonetheless, there is a division of roles in Western music practice: when a composer writes music for musicians, he (note typical gender) decides what the musicians should play on their instruments, and the musicians perform these instructions as exactly as possible. The composer's decisions in turn are limited by the instrument maker's choices for the instrument's possibility space. The instrument maker depends on properties of materials, and physical laws for vibration and acoustics. Now the resulting decision hierarchy appears to have this order.

physics ⟶ luthier ⟶ instrument → composer ⟶ musician ⟶ audience

On second thought, there are other dependencies: One can claim that the musician depends as much on the instrument maker as the instrument maker depends on the musician, her/his physical and mental abilities, and her willingness to study and play the instrument. So the order becomes.

musician ⟶ luthier ⟶ instrument → composer ⟶ musician ⟶ audience

Assuming that a majority of musical pieces composed are intended as popular music, one could claim that the audience is the real authority on what composers will compose. That puts the audience from the bottom to the top of the list.

audience ⟶ musician ⟶ luthier ⟶ instrument → composer ⟶ musician

Of course, these contradictions are easy to resolve by considering the mental model behind the relationships: The examples so far assumed hierarchical order, a "chain of command", modeled on classical causality. Simplified to three elements, here is a linear chain in music.

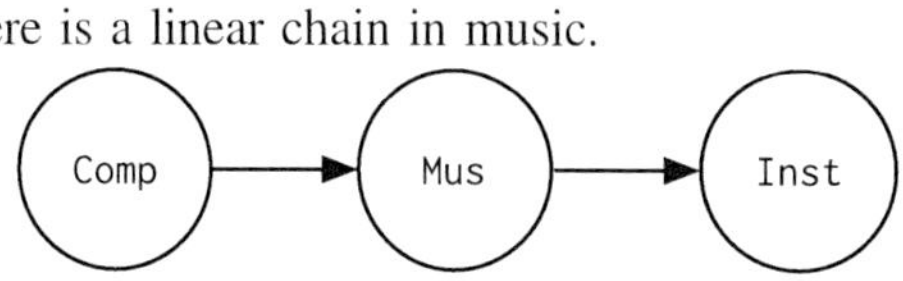

As soon as the composer listens to the music coming from the instrument, the chain becomes a loop—the early cyberneticians called this *feedback* and *circular causality*.[2] When a musician discusses with a composer how well which ideas work on the instrument based on "how the instrument feels" when playing, we have a *network of influences*. With more than three elements, such networks become dense very quickly. Drawing these influence paths at equal strength is a simplification, as their strength may vary depending on who interprets them.

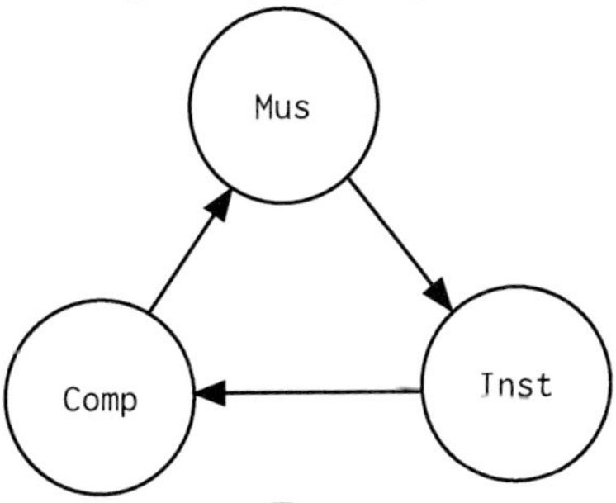

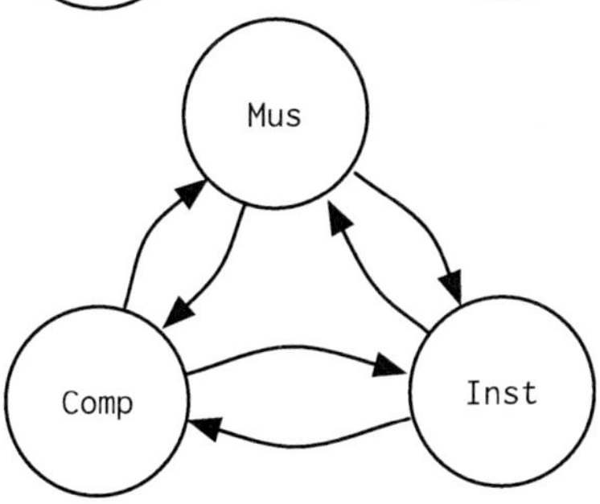

2.2 *Learning Music in Cybernetic Terms*

"Think back to the time you first touched an instrument. Remember the wondrous sound that came out? […] You were content to hear the sound come back to you. This was the unfolding of a natural process" (Werner 1998: p. 28).

Werner's portrait of the relationship between a child and an instrument corresponds with circularity, a key concept in systems theory. When playing an instrument, a musician performs bodily actions, and receives sensual feedback (auditive, haptic, visual, etc.) from the instrument. This feedback informs the musician and therefore influences her further interaction with the instrument.

[2]The Macy conferences which crystallised the cybernetics concepts in the 1940s and 50s were held under the title "Circular Causal and Feedback Mechanisms in Biological and Social Systems". The proceedings of the sessions from 1949–53 are reprinted in (Pias 2003).

"Stimulated by the sound, your curiosity about music could have grown from there. If you were left alone, you might have developed various relationships to the different sounds on that instrument. […] Perhaps we would have many more musical languages, creative techniques, ways of playing the instruments and even innovative fingerings if everyone had been left to their own devices for the first few years with an instrument" (ibid.).

As Werner continues, he sketches out the potentials that can unfold in such heterarchic relationships when they are not channeled, directed or guided: The child's playfulness and discovery are not limited by the musical systems of her culture and their conventions. Both players of this 'musical game' (Werner) are active, perceiving participants. Within this heterarchic interaction lies a highly complex understanding that goes beyond the perspectives of conventional music pedagogy.

"Unfortunately … Western rationalist tradition has fallen in love with the trivialisation operation. … Everything must be trivialised, even our children" (von Foerster 1999 (trans. AdC)).

While trivialisation is desirable in everyday contexts (in a bicycle, we want the pedaling, steering and braking operations to always work the same way), it is not desirable at all in nontrivial systems, such as human beings. Consequently, von Foerster proposed to only allow legitimate questions in education, i.e. questions to which the teacher has no expected answer. This would encourage independent thinking, as opposed to the reproduction of rote answers, which only prove that responder will not surprise anyone (and thus demonstrate having acquired triviality).

Rule systems are always constructions and agreed on by convention. However, choosing desirable rules will make a difference, as will being aware there is always more beyond them. No single system can ever claim exclusive correctness (except for monotheistic religions, and then only when seen from the inside); thus cultivating one's awareness of which system one is in, what is it like, what could be changed, and especially, what is outside of it, is good preparation for all kinds of creative involvement with the world.

As Heinz von Foerster put it, "The ethical imperative: Act always so as to increase the number of choices" (von Foerster 1999: p. 303).

3 Expanding Possibility Spaces in Art and Technology

"First we thought the PC was a calculator. Then we found out how to turn numbers into letters with ASCII—and we thought it was a typewriter. Then we discovered graphics, and we thought it was a television. With the World Wide Web, we've realized it's a brochure."

Douglas Adams

"If there is a sense of reality, and no one will doubt that it has a right to exist, then there must also be something one can call sense of possibility. Who has it does not say for example: Here this or that has happened, will happen, must happen; but he invents: Here [this or that] might, could or should happen; and if one explains to him that something is the way it is, then he thinks: Well, it could likely be different. Thus one could even define the sense of possibility as the ability to think everything that might as well be, and not to consider that which is any more important than that which is not."

Robert Musil, der Mann Ohne Eigenschaften (Trans. AdC)

Following Musil, we can attempt a working definition of a central concept for our work:

A *possibility space* is the variety of next interactions with the environment a human (or artificial) agent can imagine in the current situation.

Glancing sideways at music and art in the 20th century, one can claim that artists both discovered and explored new possibility spaces beyond the ones Douglas Adams mentions: creating new works not only within existing cultural conventions, but creating new systems of rules within which to create new pieces. This shifted attention of both artists and audience from product to process.

3.1 Technological Revolutions and Possibility Spaces

A series of technological revolutions changed societies worldwide throughout the 20th century. We discuss them here only from the context of musical practice, and only as instructive examples for how such developments expand possibility spaces, in order to put the new possibility spaces we are working with in historical context. In common engineering terminology, things have uses they are designed for, and then people discover other possibilities for them, which some may consider misuse, appropriation and the like, but in fact they are simply unforeseen affordances as soon as someone notices them.

The **electronic revolution** brought many changes where this happened:

Electromagnetic transmission and reception of audio signals allowed for transporting sound and music in space. While inventors, artists and theorists considered symmetrical communication by radio highly desirable, commercial applications of one-to-many communication quickly became the norm. This not only allowed the invention of new forms of highly effective propaganda, but also of new art forms like radio plays, created purely for experience by listening.

Electromagnetic recording on tape was designed for transporting sound and music in time, but some discovered the possibility for time-axis manipulation beyond fixing mistakes by cut-and-splice—extremely intricate tape music compositions became conceivable.

Electric amplification of voices and musical instruments was initially only intended for addressing larger audiences. The intimacy of close-miked voices allowed creating new vocal styles, and the new timbres of electric guitars, Hammond organs and electric pianos defined many of the popular music styles since 1950.

The simulation of physical vibration by electronic means first interested physicists, and later, engineers began modelling oscillations of musical instruments with their help. Already in the 1950s, pioneers like Louis and Bebe Barron experimented with circuits for sound synthesis informed by cybernetic ideas of circular causality.

The **digital revolution** in the arts started with recording and storing texts, sounds and images as numbers to make them "perfectly reproducible", and to be able to repair or improve them. In music production, the dawn of digital audio workstations (DAWs) mainly emulated previous technologies: tape machines, mixing desks and signal processing devices, as if Douglas Adams had needed another example.

Again, people discovered new possibilities: a great variety of time axis manipulations became constituting elements for emerging new styles of music, most prominently *sampling*. Genres like *glitch* or *clicks'n'cuts* even specialised in the new possibilities at the extreme ends of the originally intended affordance space.

The **algorithmic revolution** began with people noticing the possibility that, rather than recording and manipulating data material—sounds, images, texts—one could also instruct machines how to generate numbers such that they can be experienced as texts, images and sounds. This shift of attention from finished material product to the design of the generative processes themselves has changed the practical working methods of many artists dramatically, far beyond the specific domains of generative art and computational art. Composing both sounds and musical structures were central concerns in early Computer Music in the 1950s, as was creating both texture and organization of computer graphic artworks in the 1960s.

The **realtime revolution** which followed seemed at first simply a consequence of Moore's Law that computing power increases exponentially—affordable machines became fast enough to generate satisfyingly complex sounds and images in realtime. Musicians very quickly understood the expanding possibility space—they could interact with processes while they run and become entangled in system-observer interaction, spreading the concepts of second order cybernetics widely though often unwittingly.

The developments at STEIM since the 1970s are very much in this spirit, as well as laptop music like that promoted by MEGO in Austria. The creation of the SuperCollider audio programming language (McCartney 1996) represents a milestone: With its combination of efficient DSP and eclectic inclusion of multiple language paradigms, it continues inspiring artists (like ourselves) to create complex musical works. Realtime audio on general-purpose personal computers became available in the mid 1990s and was quickly widely adopted.

Since then, the idea of live coding has become possible—writing algorithms as a performance and changing them while they run, as promoted by TOPLAP (Toplap

2004–16). This flexibility in dealing with running processes shifts the time scales at which different decisions are feasible.

The current **autonomous agency revolution** has been quietly happening all along in the form of nontrivial machines, machines with idiosyncratic behavior, and other surprise generators: British cyberneticians like Gordon Pask, Ross Ashby and Stafford Beer are being rediscovered via the writings of Andrew Pickering, and in electronic music, a series of artists and designers including Louis and Bebe Barron, David Tudor, Don Buchla, Rob Hordijk, Peter Blasser, Jessica Rylan and others have created systems showing highly idiosyncratic autonomous behaviors. Of course, autonomy makes more sense when extended with sentience of the environment and memory of experience, and in the ongoing renaissance of artificial intelligence (under the name 'machine learning'), computing power is now sufficient for interesting realtime uses of such neural networks.

4 Undecidable Questions, Responsibility, and Collaboration

What do these ever-expanding possibility spaces mean? Is now everything equally possible and should we equally realize it all? Would doing so reduce the world to white noise? Or should we realize nothing? Would that reduce the world to silence? For these cosmic forms of option anxiety there is an elegant solution proposed by Heinz von Foerster:

"Only those questions that are in principle undecidable, we can decide" (von Foerster 2003a, b, c: p. 297).

While many questions can be solved systematically (which science proclaims to do), some questions cannot, and for those humans have to make choices, and accept the responsibility for their choices. A very simple example would be political questions: how much of which resources should a society spend on public health? transportation? infrastructure? There is no single correct answer; societies are responsible to find ways to negotiate between conflicting interests.

"When what is observed is observed by an observer, that observer is responsible for the observation, the sense he makes of it, and the actions he takes based on that sense. Since each observer is different, it is difficult to make general ethical points, because the responsibility belongs to each particular observer" (Glanville 2003).

The necessity to make choices and the responsibility for making them does not mean that humans have full understanding of the situation, and even less that they have full control of it. The world and its elements are complex and historical processes, so the same actions will not necessarily lead to the same consequences, which is exactly the definition of non-triviality. Based on incomplete knowledge, humans make decisions that seem reasonable at the time, and they may or may not work as desired. In short, control is an illusion, and at best there is influence toward what seems desirable.

In the more specific situation of humans collaborating in artistic endeavors, different models of control are in common use.[3] Ensembles may have a leader, who feels fully responsible for the musical outcome and ready to take all the decisions for it. While this may be satisfying for the leader, the other musicians may feel under-valued, and their particular abilities and creativity will have very little effect on the overall musical result.

"The aesthetical imperative: If you desire to see, learn how to act" (von Foerster 2003a, b, c: p. 303).

In our view, more symmetrical forms of collaboration are not only possible, but extremely desirable to put into practice. When we assume that control is an illusion, and there is only influence, it makes much more sense to welcome creative suggestions by collaborators, particularly those that seem unlikely or uncomfortable. This strategy can be mind-opening in many ways, from deconditioning one's artistic preferences to diving into ways of seeing the world that differ radically from one's own intellectual habits. Just like conversations that go in unexpected directions, the exchange of ideas by practical experimentation allows for very deep forms of unexpected learning, and potentially much more rewarding results than when following the model of the lone artist genius. This is a personal preference, of course.

"To repeat myself, the world may or may not be complex, but that is not my concern. My concern can only be with what I may know, or, more precisely, what I may understand about how I can understand my understandings. In that knowing I am always present" (Glanville 2007).

5 Limitations and Their Possibilities

Reading von Foerster's ethical imperative[4] in this context, the number of choices concerned is not that of an individual, but the number of choices a social network can access. This implies that limiting one's personal options may and often will increase the variety of possibilities of a collective one is part of.

"[…], that for any system to effectively control any other system, not restricting its possible outcomes a priori, that system must have at least as much variety as the other system, where variety is a measure of the number of possible states the systems may attain" (Glanville 2003).

Human cognition and its embodiment apparently introduce some limitations: There is "The Magical Number Seven, Plus or Minus Two" (Miller 1956), the number of distinct elements we can grasp co-instantaneously. Taking this literally

[3]Howard Becker's book *Art Worlds* is a brilliant and highly entertaining detailed study on how authorship and distribution of financial rewards are negotiated in a wide range of art disciplines (Becker and Howard 1982).

[4]"Act always so as to increase the number of choices." (von Foerster 2003a, b, c: p. 227).

for a moment, one might conclude that instruments ought to be designed for simplicity and predictability—in cybernetic language, they should be trivial machines, where the same input always yields the same output. When we put ourselves into such a machine, we get ourselves back. In other words, when we tell a machine everything it should do, and then it only does exactly that, we only get a mirror of our own thinking.

Luckily, musicians generally do not worry about Miller's magic number, and so they find out how to interact with, say, a violin which has many more variables than seven. The instrument exceeds them and, in playing it, they can exceed themselves and their rational grasp of things.

Likewise for creative machines: When they are non-trivial, they exceed us. With patience and respect, we can learn to influence them, thus putting ourselves into them, and through the interaction (Gordon Pask would call this "conversation"), exceed ourselves.

"We need few variables to create almost incalculable interactive possibilities. We merely need a few parts that will interact, producing the sorts of combinatorials that lead us to impossibly large numbers" (Glanville 2007).

Miller's magic number rule actually applies at multiple levels: human cognition tends to group elements into single conceptual chunks, and then 7 ± 2 applies to the list of chunks at the next level. This hierarchical nesting corresponds directly with the concept of MetaControl (see below). Combining Glanville's notion of few variables with nesting in magic number-sized groups of parameters/influence points may well be completely sufficient for very complex systems; and if we see this number as the gateway to a highly complex larger system (composer, player, machines, …), we arrive at a truly vast space of possibilities.

"Constraints! I love constraints" (Fredrik Olofsson).[5]

For ten years, Swedish audiovisual artist Fredrik Olofsson restricted the visual aspects of his works to using the color red exclusively, switching to green in 2014 (Olofsson n.d.). This can be read both as the exclusion of the question of color altogether and as a self-chosen constraint to encourage deep investigation of the specific aesthetic possibilities of one single color.

The artists in the Oulipo group (Ouvroir de littérature potentielle) devised an enormous variety of constraints for their writing practice. The best-known oeuvre in this style is George Perec's novel "La Disparition" (Perec 1969), which completely eschews the letter "e".

Analogously in digital music practice, one can choose constraints by delegating arbitrary parts of the processes and the decisions they require to machines. This liberates the musician from some responsibilities and frees resources for her to focus on other aspects. If we will, this converts the formerly minus two into a plus two in a different field or dimension.

Delegating decisions in this way requires a meta-level decision making process in which again responsibilities can be distributed between musician and machine.

[5]Personal communications, 2005–2016.

So the process of composing pieces or designing musical instruments more and more resembles designing processes. Making music then demands decisions within the different processes across the different levels. It is on this perspective that the operations of composing, playing music and making the instrument, traditionally assigned separately to musician, composer and instrument builder collapse into a single possibility space.

6 Second Order Virtuosity—A Definition

The term "virtuoso" is today mostly used in the context of music, and one typical definition is as "a performer of exceptional skill with particular reference to technical ability", which the second entry further clarifies: "a true virtuoso is both technician and artist" (OED 2012).

Its linguistic career began with generally referring to "a person very skilled or knowledgeable in something" in renaissance Italy, often in one of "the arts", which then included today's "science". The "English virtuosi", gentlemen who engaged in "natural philosophy" (which later became "science") and its experiments, were ridiculed in their time. They included brilliant amateur scientists Robert Hooke and Robert Boyle, who were later recognised as pioneers of modern science (see Nicholson 1973). In the 17th and 18th century, "virtuoso" was used to describe highly accomplished musicians in general, whose skills included composing. Only in the 19th century did the term narrow down to mean exclusively "excellent performer". We will try to expand on these historical definitions in light of the artistic practices discussed in this article.

When using computers and electronics for music beyond emulations of pre-existing technology, musicians encounter a new class of possibilities to explore: the invention of processes for performance.

"Since there is no more musical concreteness to the computer than there is in a CD player it is essential to think hard about the physicality of an instrument, how it should present itself to the performer. Since there is no physical given there is nothing to do but to invent one's own" (Ryan 1991: p. 6).

Thus, an expanded definition of a computer music virtuoso would include the invention of processes as a desirable ability to learn. The invention of possibilities before one can practically make use of them is a higher order cybernetic process and therefore best considered a conversational task (see Glanville as discussed earlier), or as Joel Ryan puts it for the context of electronic music:

"Each link between performer and computer has to be invented before anything can be played. But these 'handles' are just as useful for the development or discovery of the piece as for the performance itself. In fact the physicality of the performance interface helps give definition to the modeling process itself. The physical relation to a model stimulates the imagination and enables the elaboration of the model using spatial and physical metaphors. The image with which the artist works to realize his or her idea is no longer a phantom, it can be touched, navigated and negotiated with" (Ryan 1991: p. 5).

Let us consider this: Seen from the outside, an instrument is a physical interface, the audible sound produced, and the black box in between. The processes contained in the black box[6] receive raw sensor data from human actions and magically turn them into sound coming out of loudspeakers. For a computer music virtuoso, the black box consists of "interfaces all the way down": Modules do little tasks and inform other little modules via handles or interfaces, all of them contributing to the overall behavior of the system—its sound and feel as an "instrument". This ranges from hardware choices like sensors, computers, amplifiers and speakers to implementations of special objects in hardware or code. For example, one little module converts sensor data to more useful ones, maps them to a choice of processes and some of their parameters, while other parameters may come from the processes' internal states. Conceptually, this notion of modularity is common practice in computer science.

In audio software, such modularity is often organized in categories like mapping, sound synthesis, effects, modulators and mixers, which is a convenient entry point into the possibility space of computer music. It facilitates organizing and composing sound from familiar elements. But this familiar categorization can obscure the unique potentials computers offer as musical instruments: unorthodox cross-connections can create forms of musical expressivity that can only emerge within such idiosyncratic systems. Here, investigation means experimentation and navigating towards the unknown.

In our understanding, a computer music virtuoso fluently invents unique idiosyncrasies within computer-human-physical systems which can form temporary instrument-like musical identities. These are best explored in a conversational manner, playing them to understand their manifest and hidden qualities, informing future exploration and refinement. In this cyclic workflow, musicians continuously redesign their systems, creating new kinds of instruments that produce not only sound, but also compose, conduct, manipulate, mix, or distribute streams of signals. Learning by exploring and playing become undistinguishable, in private study and in public performance, and the eventual degrees of predictability versus surprise (or control versus *MetaControl*) one chooses for the constituent elements of one's performance setup become central artistic concerns.

In this sense, we consider software as a fluid medium in which the conscious use of handles wherever possible allows for making and remaking artistic decisions at every level of the processes involved. Languages designed for live-coding even allow exchanging every part of the code machinery at any time. Such interventions can then also be delegated to machines, to allow for rewiring and rewriting variants to occur that human actors would not think of. We call setup designs that allow such meta-control interventions meta-instruments, and at times they reach so far

[6]In many scientific domains, a black box is any device of interest that has inputs and outputs, and whose internals are not known a priori.

beyond simple experimental tools that one is willing to ascribe forms of nontrivial agency and emergent vitality to them.

All of these ideas inform our notions of modern instrumentality, which encompasses composing instruments, all the interfaces between humans, processes and subprocesses in machines, and composing the fluid, on the fly reconfigurations of such setups which current computer-based music systems afford. We would like to conclude by proposing a definition of *Second Order Virtuosity*:

> A *Second Order Virtuoso* is an artist of exceptional skills with regard to technical and mental ability to create, observe, and shape time-based art works. This entails dealing with a wide range of processes, from simple interfaces to the idiosyncratic entities that become possible in environments integrating physical objects, electronic circuits and computers equally. She is well-prepared to make meaningful decisions, both intuitive and well-considered, across time scales from preparing years ahead of time to composing in realtime.

7 Praxis—Advanced Instrumentalities in Electronic Music

This section discusses work by luminaries in our obscure field, who provide striking examples for the kind of idiosyncratic performance and design we find so inspiring that we incorporate them into our concepts and setups. We contrast them with aspects of our own design approaches, which aim to continue the tradition of creative re-invention we see ourselves in.

7.1 Sources of Inspiration in Performance and Design

7.1.1 Joker Nies: Circuit Bending and Performance

Joker Nies is a musician who understands the specific instrumentality of electronics very deeply. One of the pioneers of circuit bending in Europe, he has held numerous workshops for it, which gave many participants—including the members of Trio Brachiale—new creative options, and deeper understanding of some of the stranger possibility spaces one can find with electronic devices.

He is a virtuoso performer on "raw electronics"—on a heavily circuit-bent Suzuki Omnichord (giving this tragic failure of an accompaniment instrument a new lease on life), the TI Speak and Spell (a 1970s learning toy for children that has highly a bendable speech synthesis circuit and programs), and many others. For many years, he collaborated with Rob Hordijk, notably on the design of a chaotic synthesizer box called 24/7, steering it from an ingenious studio device to a unique instrument for live performance, which he plays brilliantly in contexts including jazz-club improvisation gigs, media art performance shows and New Music concerts (Fig. 1).

Fig. 1 Joker Nies with setup at New York Electronic Arts Festival 2009. Image © Joker Nies

7.1.2 Rob Hordijk: Autopoietic Synthesis and Chaos

Rob Hordijk's instrument designs embody sound processes that have an inner life. His creations, including the *Blippoo Box*, the DIY kit *Benjolin* and boutique modular synths, exhibit complex apparent *behavior* by self-modulating their own parameters, even when no human player is manipulating their controls. This autonomy gives players the necessary time to focus on listening/observing the current state of musical flow without being immediately active. Rob described one of his instruments to a musician who uses it as something that "plays the player" (101010oxo n.d.).[7]

One of Hordijk's most unique inventions is the "rungler", a finely tuned pattern generator at the edge between chaos and order: it has "states of balance" where it near-repeats sequences of waveforms (like a sequencer or LFO), and a propensity to lose this balance due to minute disturbances in the environment (e.g. thermic drift or user input). When that happens, it creates what Hordijk calls a "stepped havoc wave" (Hordijk 2009a), similar to a sample-and-hold circuit, before finding a new equilibrium in a different orbit. Hordijk traces this back to "Dutch composer Jan Boerman, who some fifty years ago had the idea to define a musical area between a pure sine as the simplest musical entity and pure noise as the most complex entity

[7]http://hordijk-synths.info/news/2015/05/07/nordvargr.html.

and explore the space in between these two. Chaotic systems seem to neatly fit in this area" (Hordijk 2009b).

Of course, this is merely one facet of the rich gamut of Hordijk's creations; there are many more highly interesting designs and inventions worth mentioning. For example, in "The Blippoo Box: A Chaotic Electronic Music Instrument, Bent by Design" (Hordijk 2009a) he describes the aesthetic rationale behind this instrument in great detail, explaining notions like well-tempered chaos and others. As composer Richard Scott puts it, "as well as being unique creative instruments, the depth of research and thinking he puts into them is such they may be seen as artworks or as compositions in their own right" (Scott 2013).

7.1.3 Peter Blasser: The Poetics of Synthesynthesis

Peter Blasser creates some of the most inspiredly idiosyncratic electronic instruments we know of. It's hard to do justice to the sheer scope of his inventiveness, from technical to the conceptual to the poetic language he uses for his work, so we only mention a few of his design concepts here that have been influential to our practice. For further reading, his website (Blasser n.d.), especially the "philosophical paperz" (Blasser 2016a, b), and his PhD thesis (Blasser 2015) are highly recommended (Fig. 2).

Fig. 2 Echo Ho performing a kitchen concert on a Blasser KitTenNetTik synthesizer kit assembled by Ralf Schreiber. The 100+ metal rod electrodes are the "androgynous nodes", connected by alligator clips to vegetables and body contacts. Image © Hannes Hoelzl 2006

- "Bent by design": A term also used by Hordijk, this is an invitation to users to modify instruments. Circuit benders usually "bend" used electronic devices such as toy keyboards by adding contacts or changing parts of their circuits in order to open the possibility space of these simple instruments far beyond the horizon of the designers' imagination. Blasser and Hordijk give hints where in their circuits further exploration might be fruitful, enabling their clients to personalize their instrument, e.g. by choosing the values of "hairy capacitors" such that the overall voicing is in the preferred range.
- "Androgynous nodes": Canonical electronics defines exactly which points in a circuit can be outputs and which inputs, enabling predictable connections between them. This allows different combinatorial configurations of modularized instruments. Several Blasser instruments, e.g. the KitTenNetTik series, contain many nodes that can act BOTH as input and output points.

"If you start not from "knowledge", but instead at any random and humble point within the aaji, you will see more than just arrows pointing in and out, but directionless flows, the stuff of simultaneity. No matter how hard you try, you cannot make an assemblage into male (only giving) or female (only receiving) […] The aaji gives and receives—rives like a river".[8]

This invention not only challenges one's perspective on electrical circuits, it also has radical effects on the playability of the instrument. Any arbitrary point in the circuit can be monitored as a sound output—the entire circuit becoming an electronic free jazz ensemble, and one can choose what to listen to. By sidestepping well-defined modularity, one can reconfigure the sound-producing circuit in near-endless combinations, radically extending the possibility space, and likely allowing creative emergence in a strict sense (Cariani 2008).

- Body contacts: Hard-wiring the androgynous nodes is just a special case of connections that may generally range from zero to infinite resistance. This is especially inviting hand-playing techniques, where the performer conducts changing portions of the electric flow through her body. By touching two or more nodes with varying pressure, one can delicately modulate the degrees of cross-talk between them. This technique, allows micromotoric subtlety in musical expression quite unrivalled by conventional controllers like pots, faders, and switches.

This little glimpse into the Blasser universe shows the possibility expanding, exploding nature of his work. What also fascinates us about his instruments, is that they are quite evidently designed such that the ambition to master them becomes meaningless. If the very inner workings of the instrument change with each cable that is dis-/connected, predictability is gone in the blink of an eye. As a consequence, learning and performing cannot be separated, so the audience can witness

[8]"Aaji" means circuitry, and "knowledge" means mainstream engineering thought, as in Shannon's information theory (Blasser n.d.).

the musician's learning process on stage[9]; arguably a defining feature of experimental art practice.[10] In terms of Second Order Virtuosity, what can be mastered here is the seamless integration of the learning process into the flow of musical performance.

7.2 *Trio Brachiale Concepts*

In the following three sections, each Trio member discusses one example research topic that has been integrated as a practical module or concept in his live setup.

7.2.1 Random Orbits—A hitchhiker's Guide to Navigation in Possibility Space (Hoelzl)

Random Orbits are a strategy for the exploration of high-dimensional parameter spaces, as they occur in complex synthesizers, and also in routines that sequentially produce musical events. With varying numeric input state—apart from the mappings and internal structure—they can produce output states (i.e. sounds) of extremely varying musical meaning. One very viable strategy for players to cope with the combinatoric explosion of possible states is to store favorite settings as presets, thus creating a historical archive of 'visited places' in the parameter space, much like marking named dots on a map.[11]

For exploration, random settings for all parameters can easily be produced by the machine, virtually beaming the player into distant uncharted territory she would not necessarily reach by intentional parameter changes. By exploiting a side effect of the methods computers use for creating random numbers, namely the ability to set seeds for the *pseudo-random number generators*, these seemingly unpredictable decisions become both fully deterministic and reproducible. The near-infinite amount of a computer's available numbers to use for random seeding offers us an equally huge number of places in possibility space that we can access reproducibly at any time—an archive of markers much like for presets, except this time pointing not into the past, but into the future.

Where sudden, white-noise-like parameter jumps are not desirable, one may limit the deviations from the previous state to a smaller range. Musically, this can be interpreted as variations of arbitrary strength, while conceptually, the brownian nature of this movement brings us back to notions of the historicity of processes (Fig. 3).

---

[9]Master improviser Robin Hayward reports that he deliberately modifies the mechanics of his tuba prior to performances, in order to push himself into a situation of re-learning on stage. (personal communication 2015).

[10]For a contemporary definition of musical experimentalism, see Emerson (2014).

[11]Compared to the total-recall convenience in music production environments, presets for non-trivial processes are not intended for full reproduction; they only put the process in a similar orbit.

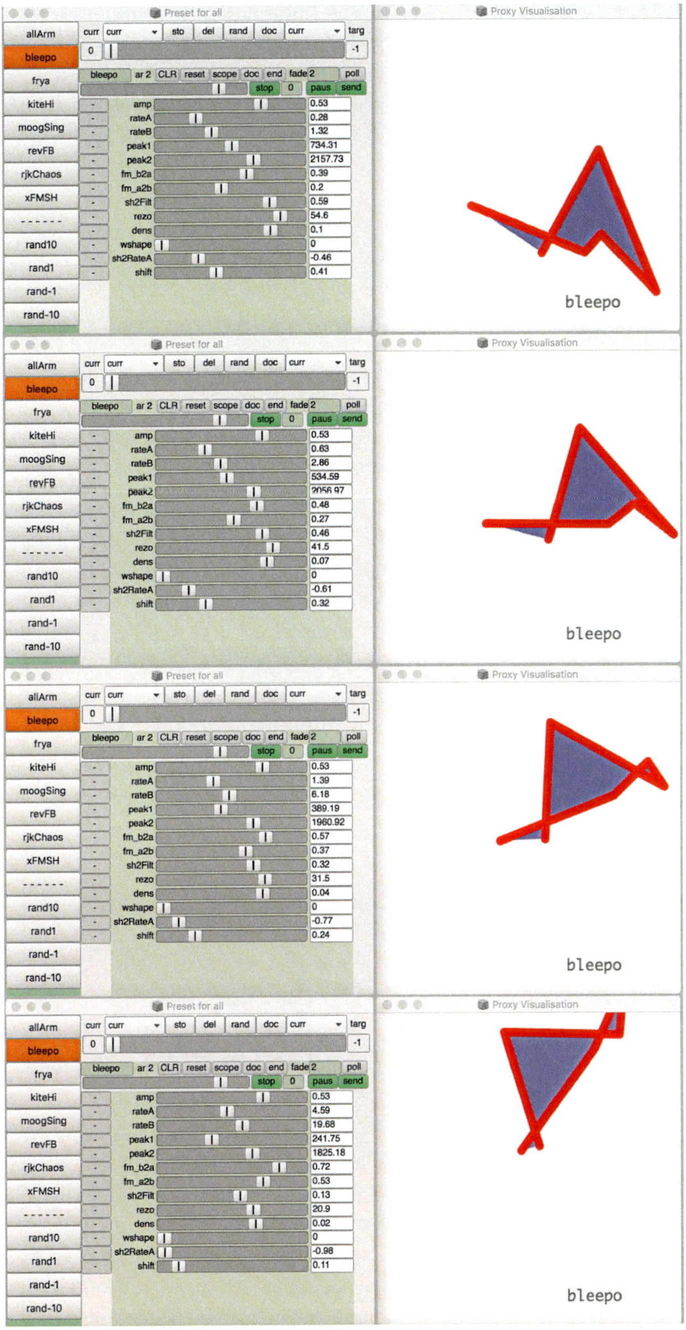

Fig. 3 A sound process in random orbit, 4 stages. The preset window on the left allows selecting a sound processes (*left*), access its presets (*top*) and edit its parameters (*right*). The *right* windows show a simple vector visualisation of the parameter state. Image © Hannes Hoelzl, 2016

Personally, I was struck by the effects that emerge when combining these three seemingly trivial techniques into what I call *Random Orbits*: starting from a given preset, seeded brownian moves enable a player to traverse the possibility space in steps along exactly reproducible trajectories that tend to converge after many repetitions. The metaphors of space and movement have wide ranging musical implications which we have only begun to study, and many others awaiting exploration.

Three examples:

- By mapping some keys of a keyboard to different random seeds and variation amounts, and others to presets, one can move from known departure points in a variety of zig-zag lines, with easily audible results, essentially establishing new poly-parametric scales of convincing musical precision and logic, akin to 'playing etudes in the pseudo-random scale'.
- The tendency to convergence often lets the state vector run into border zones of the possibility space. This has yielded many novel sounds, since it is rather uncommon in usual 'golden middle' playing styles to reach those outer zones. Several options are available for handling these approaches to the limits of each dimension. Currently, clipping is implemented, i.e. hitting the border of the numerical space like a hard wall. Alternatives include wrapping (as in old video games), bouncing/mirroring, or new developments like nonlinear compression near borders, and mapping to sinusoidal rotation.
- For most acoustic instruments, big steps in musical parameters are rare in the literature, because they become motorically increasingly harder to execute. Obviously, for electronic machinery no such graduation exists and step size can be arbitrarily played as a musical meta-parameter.

7.2.2 MetaControl—Lose Control, Gain Influence (de Campo)

The notion of MetaControl emerged for me while collaborating on the modality-toolkit, a software library based on the idea of highly modal interfaces/instruments, i.e. setups where a small number of physical interfaces can access and control a variety of processes in a great variety of ways (Modality Team 2015).

Like much other software, modality makes 'one to one mapping', where one interface element (such as a slider) controls one process parameter (such as speed), very easy, thus privileging it over its alternatives. I wondered what a polar opposite would be like, and sketched out the *Influx* class: here, the default mapping is that M continuous interface elements affect N continuous process parameters by freely definable sets of weights (Fig. 4).

When these weights are random (as they are by default), the physical control elements become really powerful: Moving a single control element traverses the parameter space along a multidimensional diagonal, changing every process

Fig. 4 Hybrid performance configuration used by Alberto de Campo with Trio Brachiale and Kairos Theory. Clockwise from *bottom left*: Manta controller by Snyderphonics, Cracklebox by STEIM, GamePad by Thrustmaster/Ferrari, Blippoo Box by Rob Hordijk, Apple Macbook Pro with SuperCollider, Faderfox UC-4, Saitek Cyborg X joystick, MOTU 1248 audio interface. Image © A. de Campo

parameter to some degree; moving a different control does the same along a very different axis. These experiments led to a deep insight (for me at least).

Not knowing the technical details of a mapping allows performers to fully concentrate on the experience of playing, learning to navigate intuitively in the parameter space of the process. (By comparison, having one controller per parameter places cognitive load on the performer, and allows interventions that may appear simplistic: changing only, say, the speed of the process and nothing else lets the audience hear a 'fader move').

These first experiments quickly suggested combinations with existing strategies, such as storing snapshots of parameter states as presets, and controller movements as control loops, both as material the evolving performance can refer back to in a multitude of ways.

Finally, following the notion of 'influence', a perspective that fascinates me *control polyphony,* where several (human or machine) sources influence one running process, and several processes receive influence from one or several sources. This approach creates shared networks of influence which jointly create a sound world that can genuinely surprise both the audience and the players themselves. As I tried to summarize it in (de Campo 2014), the concept of 'Lose Control, Gain Influence' (LCGI) is about 'gracefully relinquishing full control of the processes involved, in order to gain higher-order forms [of] influence on their behavior.'

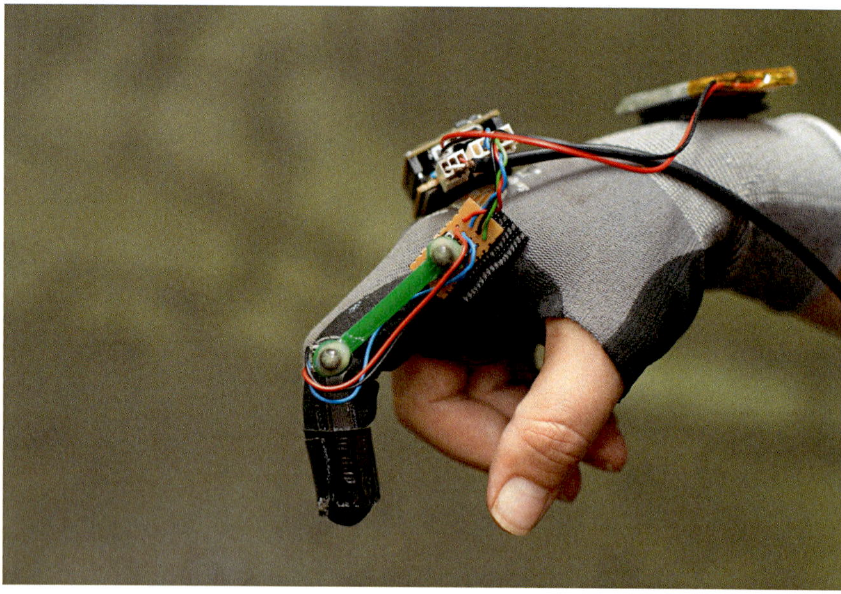

Fig. 5 *The Finger* by Dominik Hildebrand Marques Lopes. Image © Katharina Hauke

The software elements involved, *Influx, NdefPreset, TdefPreset,* and *KtlLoop*[12] have been used extensively by the Trio und many others in the context of Generative Arts Class at UdK Berlin, and further extensions (such as Nudge and NudgeGroups) are continuously extending the range of MetaControl possibilities.

As a practical example, please see the verbal score and hear a recording of **MetaControl Study No. 1,** my first piece that explores fundamental MetaControl notions, online (de Campo 2016).

7.2.3 The Finger and its Possibility Space (Hildebrand Marques Lopes)

The Finger is a wireless interface that I developed in 2010. It was intended to be a very inexpensive and fast to build solution for a gesture based controller to use in the one week workshop "All You Need is Gloves!" held by Hannes Hoelzl at UdK's Generative Arts Class. It evolved to become my main tool for playing concerts and exploring gesture based digital music for almost six years now (Fig. 5).

As soon as the first hardware prototype was working, continuous development of all software parts involved began, while the hardware functionality remained

[12]All published as SuperCollider extension libraries.

constant. Within the first years, I realized that even if I used synthesis patches I was well acquainted with and had formerly used with other controllers, they acquired a new, different quality or character when mapped to and played with *The Finger*. Often, the result was surprising and I learned that modeling interfaces, mappings and synthesis processes can be done separately, but that it is the combinatorics of all stages that form instrumental identity. If and how these identities work for me can only be explored by playing, and I rarely got what I expected while designing the individual parts. I really appreciated this. And without being able to consciously explain what I do (even to myself), I could rapidly form body knowledge to play my sound patches through this interface.

I started to design different mapping strategies by first focusing on finely tuned one-to-one mappings, traversing gesture recognition techniques and then working on complex MetaControl structures and pattern generators. How to implement switching between configurations of sound patches and mappings on stage turned out to be at least as important as the question of what sounds to play, and since I always wanted to get rid of the computer onstage, the instrument itself should suffice to do all the configuration switching.

Through playing more than 50 concerts in different constellations and musical styles ranging from experimental improvisation to club-like electronic music, in solo shows and ensembles, I figured out that less organization, less thinking about what to do next enabled me to be more focused and present in the moment. This is, at least retrospectively, the same reason why I like to improvise music in contrast to rehearsing and repeating phrases over and over to master playing certain compositions.

So, with the software I develop and use now, I try to explore how to share or entirely hand over the organization of different sounds and mappings to my instrument. In order to do this, I use strategies that are explained in the Random Orbits and MetaControl sections of this article, with respect to the idea that I want to create a heterarchic relationship between my instrument and myself. The instrument should become a playing partner with equal opportunities to make decisions. It can switch to different sound patches and redefine mappings, and it can also play little phrases of parameters that go into sound processes just as my sensor inputs do.

I have to fully concentrate to observe the current inner state of the instrument, learn how I can influence what is possible, while all of a sudden, everything can change and the game starts again. I have to play to find out what is possible to play, which means there are no known paths to follow—and thus no mistakes. By not moving my body, I can even let the instrument play by itself. The only parameter that is fully under my control with a fixed mapping is volume. This is the minimum of control needed to create musical structures with pauses and dynamics.

Pointing to the origin of these thoughts and ideas, which is second order cybernetics, I call this non-hierarchical approach *Conversational Communication*—controller and controlled are roles given by an observer.

8 Conclusions and a Future Agenda

We have proposed a new concept which we find essential for understanding musical practice and instrumentality in the 21st century: Second Order Virtuosity. It integrates insights from a large array of contexts, from applying second order cybernetics to music, the ethics of decisions and responsibilities, the stranger possibility spaces that computer technology opens, and from the uses of constraints in the arts.

Our main sources of inspiration clearly apply some of these insights in their practice. The discussion of the practical implementations of our ideas and the experience we haven been gathering with them hopefully provide a deep look into our idiosyncratic, but (we believe) quite consistent artistic universe. It embodies the insights that playing equals learning, autonomy of process is the most radical new possibility to emerge from current technology, and that artistic decisions affecting multiple time scales can be meaningfully made in the flow of performance.

At this point, the question that remains is: What to do next? Here is our future agenda:

Improvised music has long moved from uniform freedom to second order freedom—choosing the degrees of freedom freely for each piece, which for example Steve Lacy calls "poly-free music". We have only begun experimenting with this approach.

Networks of influence can naturally extend between the players' setups—and these influences again allow delegating some decisions to outside actors, so the individual can choose to restrict hand-made decisions to a smaller set of aspects.

Machines can become more equal partners in heterarchic networks of human and machine actors. Giving them similar access to decisions opens intriguing possibilities: They can provide a variety of "director's input", from oracular hints à la Oblique Strategies, concrete instructions about playing roles, to directly switching the selection and configuration of the modules the human players are currently playing with. Our ensemble Kairos Theory has begun exploring this.

Finally, we believe that the new possibility spaces for music call for, and will eventually lead to, new terms for the kinds of configurations and strategies used in contemporary musical practices like those in our group. Second order cybernetics and the existing new practices we reported on appear to be quite fruitful sources for such terms. We propose Second Order Virtuosity, MetaControl, and Random Orbits as small contributions to a terminology that extends received notions of instrumentality to accommodate new forms of artistic practice.

References

101010oxo. (n.d.). http://hordijk-synths.info/news/2015/05/07/nordvargr.html. Accessed August 5, 2016.

Becker, H. S. (1982). *Art worlds*. Univ. of California Press.

Blasser, P. (2015). *Stores at the Mall* (PhD Thesis). Wesleyan University. Available online at http://wesscholar.wesleyan.edu/etd_mas_theses/107/. Accessed August 05, 2016.
Blasser, P. (2016a). http://www.ciat-lonbarde.net/fyrall/index.html. Accessed May 27, 2016.
Blasser, P. (2016b). http://ciat-lonbarde.net/fyrall/paperz/index.html. Accessed May 27, 2016.
Blasser, P. (n.d.). http://ciat-lonbarde.net/. Accessed May 27, 2016.
Cariani, P. (2008). *Emergence and creativity*. Emoçao Art. ficial 4.0: Emergencia [Exhibition volume].
de Campo, A. (2014). Lose control, gain influence—Concepts for Metacontrol. In *Proceedings ICMC|SMC|2014*. Athens, Greece.
de Campo, A. (2016). *On MetaControl*. https://3dminalmanac.wordpress.com/?s=MetaControl. Accessed May 08, 2016.
Emerson, G. (2014). https://3dminalmanac.wordpress.com/2015/02/01/experimentalism. Accessed July 27, 2016.
Glanville, R. (2003). Second-order cybernetics. In F. Parra-Luna (Ed.), *Systems Science and Cybernetics, Encyclopedia of Life Support Systems, developed under the Auspices of the UNESCO*. Oxford: EoLSS Publishers, electronic. Available at http://cepa.info/2326. Accessed August 03, 2016.
Glanville, R. (2007). *"Designing complexity" performance improvement quarterly* (Vol. 20(2), pp. 75–96). Available at http://asc-cybernetics.org/systems_papers/Complexity%20finally%20reworked.pdf. Accessed August 03, 2016.
Hordijk, R. (2009a). The Blippoo Box: A chaotic electronic music instrument, bent by design. *Leonardo Music Journal, 19*, 35–43.
Hordijk, R. (2009b). *What the $#%$ is a rungler*? http://electro-music.com/forum/post-281728.html. Accessed August 05, 2016.
McCartney, J. (1996). SuperCollider: a new real time synthesis language. *ICMC Proceedings*. Online at http://www.audiosynth.com/icmc96paper.html. Accessed August 05, 2016.
Miller, G. A. (1956). The magical number seven, plus or minus two: Some limits on our capacity for processing information. In *The Psychological Review* (Vol. 63, pp. 81–97). Online at http://cogprints.org/730/1/miller.html. Accessed August 05, 2016.
Modality Team. (2015). Website http://modality.bek.no. Accessed May 20, 2016.
Nicholson, M. J. (1973). Virtuoso. In *Dictionary of the History of Ideas* (Vol. IV, p. 486). New York, NY.
OED. (2012). Virtuoso. In *The Oxford Dictionary of Music* (6th ed.). Oxford University Press.
Olofsson, F. (n.d.). http://www.fredrikolofsson.com/pages/overview-green.html
Perec, G. (1969). *La disparition*. Paris: Denoël.
Pias, C. (2003). *Cybernetics|Kybernetik. The Macy-Conferences 1946–1953. Bd. 1 Transactions/Protokolle*. Diaphanes.
Roads, C. (2001). *Microsound*. MIT Press.
Ryan, J. (1991). *Contemporary music review*. (Vol. 6, Part 1, pp. 3–17).
Scott, R. (2013). *Rob Hordijk's Blippoo Box, Benjolin and active matrix mixer*. http://richard-scott.net/research/blippoo-box/. Accessed May 25, 2016.
SuperCollider team. (n.d.). *Quarks extensions* are available at https://github.com/supercollider-quarks. Accessed May 20, 2016.
Toplap Team. (2004–2016). http://toplap.org. Accessed May 25, 2016. This is the main organization promoting and documenting live coding.
Trio Brachiale. (2016). http://www.earweego.net/TrioBrachiale/. Contains general info, some browser-based animated versions of the many-festo, and materials for the examples in this article under the link "instrumentalities".
von Foerster, H. (1999). *2 × 2 = Grün, Bemerkungen zum Verhältnis des Menschen zu seinen Maschinen* (Audio track, at minute 8:58). supposé Edition 3 (trans. AdC).
von Foerster, H. (2003a). *Understanding understanding: Essays on cybernetics and cognition*. Springer Science & Business Media.

von Foerster, Heinz (2003b). *On constructing a reality* (pp. 211–227). Understanding understanding. New York: Springer.
von Foerster, H. (2003c). *Ethics and second-order cybernetics* (pp. 287–304). Understanding understanding. New York: Springer.
Werner, K. (1998). *Effortless mastery*. Jamey Aebersold Jazz.

Part IV
Listen Perceive Feel

Mapping, Causality and the Perception of Instrumentality: Theoretical and Empirical Approaches to the Audience's Experience of Digital Musical Instruments

Gina Emerson and Hauke Egermann

Abstract Digital musical instruments (DMIs) rarely feature a clear, causal relationship between the performer's actions and the sounds produced. Instead, they often function simply as *controllers*, triggering sounds that are or have been synthesised elsewhere; they are not necessarily sources of sound in themselves (Miranda and Wanderley 2006). Consequently, the performer's physical interaction with the device frequently does not appear to correlate directly with the sonic output, thus making it difficult for spectators to discern how gestures and actions are translated into sounds. This relationship between input and output is determined by the mapping, the term for the process of establishing relationships of cause and effect between the control and sound generation elements of the instrument (Hunt et al. 2003). While there has been much consideration of the creative and expressive potential of mapping from the perspective of the performer and/or instrument designer, there has been little focus on the experience of those receiving DMIs. How do spectators respond to the perceptual challenge DMIs present them? What influence do mapping and other aspects of instrument design (e.g. the type of controller used and the sound design) have on the success of an instrument when considered from the spectator's point of view? And to what extent can (and should) this area of artistic exploration be made more accessible to audiences? This article aims to consider these questions through providing a critical review of the existing theoretical and empirical work on DMI reception.

G. Emerson (✉)
Audio Communication Group (Sekr. EN-8), Technische Universität Berlin,
Einsteinufer 17c, 10587 Berlin, Germany
e-mail: gina.emerson@tu-berlin.de

H. Egermann
Department of Music, University of York, Heslington, York YO10 5DD, UK

T. Bovermann et al. (eds.), *Musical Instruments in the 21st Century*,
DOI 10.1007/978-981-10-2951-6_23

1 Introduction

The existing discourse on digital musical instruments (DMIs) has tended to focus on the instrument designer and performer's perspective, evaluating DMIs in terms of their expressivity and playability. However, more recently, there have been calls to place more of an emphasis on those in the position of receiving DMIs. Audience members represent important stakeholders in the DMI development process (O'Modhrain 2011); their responses can contribute much to the evaluation of designs as successes or failures and it is their expectations of what an instrument should be, their concepts of *instrumentality*, that designers and performers are ultimately confronted with. One facet of DMI design that can have a strong impact upon reception is the perceptibility of a causal link between the performer's interaction with the device and the sonic output produced (hereafter gesture-sound causality). If the mapping design is not easily perceptible, spectators can struggle to discern how gestures and actions are translated into sounds. This essay provides a critical summary of the existing literature on audience experiences with DMIs, centering on the perception of gesture-sound causality. Two strands of relevant theoretical and empirical research are considered: (1) existing work on the relationship between perceptible gesture-sound causality and the degree of *liveness* a performance has; and (2) the connection between causality and audience understanding. The relationship between these areas and audiences' views on instrumentality will be highlighted throughout. The concluding section turns to the consequences for DMI designers and the issues surrounding making DMI performances more accessible to audience members.

2 Causality and Liveness

The term *liveness* was introduced by Philip Auslander (1999, 2009), who uses it to refer to the cultural value that live performance is commonly accorded. Live musical performance has *authenticity*; it involves effort and skill and is thereby often considered to be experientially different, and in part more valuable, than listening to a recording (Auslander 1999, pp. 73–128). The notion of liveness invokes many related concepts that are relevant when considering audience perceptions of DMIs and has been taken up in DMI-focused research most prominently by John Croft in his 2007 essay, *Theses on Liveness*. Here, Croft breaks down the concept of liveness into two different kinds, *procedural* liveness and *aesthetic* liveness. The former is applicable whenever live sound is produced and manipulated in real time. Aesthetic liveness, in contrast, is a level above procedural liveness, in which some kind of input makes 'aesthetically meaningful' differences to the audio output in a real-time context, by which a significant, detectable musical change or development over the course of the performance is meant (Croft 2007, p. 61). This aspect of an identifiable relationship between input and output from the spectator's perspective is central to Croft's overall concept of what can legitimately be considered a live performance:

> Thus the onus of justification of liveness is shifted to the causal link between the performer's action and the computer's response. It is a question of the specificity of the relation: if many perceptibly different inputs generate outputs with no pertinent differences […], then the liveness is merely procedural and not aesthetic—pre-recorded sounds would do the job as well or better. (ibid.)

On the basis of this, Croft formulates eight conditions for the relationship between input and output in DMIs and similar technologies that, if fulfilled, would allow for greater aesthetic liveness in performances and also grant such devices a stronger claim to instrumentality (ibid., pp. 64–5). These can be summarised into two underlying principles: that the performer's gestural input should always be met with an immediate and fitting response from the instrument (e.g. a large input gesture should result in a loud or accented sound) and that the instrument should itself have some degree of internal consistency that makes it possible to learn how to control the system and makes its sound recognisable. By drawing on the spectator's point of view in order to better inform artistic practice, Croft's essay represents an important step in the theoretical work on DMI reception and allows for the development of more concrete audience-centred approaches to evaluation (see also Schloss 2003).

A small collection of studies has taken up the issues of liveness and causality in empirical investigations. Berthaut et al. (2015), for instance, explicitly investigated Croft's link between liveness and causality. They showed participants video recordings of performances with three DMIs that had been created for the experiment and designed specifically to make gesture-sound causalities difficult to perceive: the first featured a temporal delay between input gesture and sounded result, the second was mapped so that discrete or short gestures resulted in sustained sounds (and vice versa) and the third instrument was only partially under the performer's control, with some audio features at times produced automatically by the computer (ibid., p. 385). The videos were presented both with and without animated pointers that were colour-coded to represent the musical parameter being controlled at any one time (pitch, timbre, loudness and musical pattern). These 'visual augmentations' were designed to clarify the mapping for the participants and were tested in an experiment for their ability to achieve this. Participants were simply asked to rate the extent to which they thought the performer's gestures influenced the music (described here as a rating of performer *agency*, i.e. their level of activity and the extent to which this is effective or has an impact) and how confident they were about this. As predicted, the ratings for perceived agency and the confidence ratings were higher in the conditions that featured the visual augmentations, which suggests, as the authors conclude, that providing audiences with a clearer understanding of the mapping causality can establish a greater sense of performer agency (and therefore liveness) in performances with DMIs (ibid., pp. 385–6).

A recent study by the present authors provides further evidence of the link between causality and liveness. Participants rated video clips of performances with DMIs with causal and acausal mapping designs in their original version and in a manipulated audiovisual mismatch version in which a section of video from one performance was presented together with a section of the audio recording of a different performance (Emerson et al. forthcoming). It was hypothesised that the causal and original DMI performances would be rated more positively in terms of emotional response and as more stimulating and interesting, with the manipulation having a stronger effect on the ratings for the causal DMIs. The results indicated that a lack of perceptible causality does have a negative impact on ratings of DMI performances. The instruments in the causal group were viewed as considerably more interesting and more successful at holding the participants' attention than the acausal instruments and in comparison to their manipulated versions. The acausal group received no significant difference in ratings between original and manipulated clips. This latter result, that the participants gave similar ratings for the original and manipulated versions of the acausal instruments, implies that they did not perceive much of an aesthetic difference between them, echoing Croft's comments about the possibility of a recording equalling the impact of a live performance in case of some DMIs (see p. 2 above). If a manipulation was able to create almost the same impression upon an audience as the original performance, this furthermore shows how a lack of perceptible gesture-sound causality could thwart performers' attempts at communicating an artistic or expressive goal to spectators and could decrease the perceived instrumentality of the DMI.

3 Causality and Audience Understanding of DMIs

The most extensive empirical work on audience understanding of DMIs comes from A.C. Fyans, Michael Gurevich and Paul Stapleton, who have explored, via a series of studies, the extent to which the skill displayed by the performer is a factor in audience evaluations (Fyans and Gurevich 2011), as well as in how far audience members are able to notice errors made in DMI performances (Fyans et al. 2010). In Gurevich and Fyans' later article (2011), these investigations are combined into a single paradigm. Participants were asked to watch recordings of performances with a theremin and a *Tilt Synth*, a DMI developed specifically for the experiment, and were then interviewed on their perceptions of instrumentality, skill and error.

The performers were not experts in their instruments and, therefore, several errors occurred (eight in total by the theremin player, five by the Tilt Synth player, as identified by the performers themselves). Overall, only two out of twenty-seven participants correctly identified errors in the Tilt Synth performance; it was generally held that such an instrument was 'error-free', as it did not appear to allow for any fine-grained control of pitch, in contrast to the theremin (ibid., p. 172). The participants' descriptions of how they thought the instruments functioned, especially for the Tilt Synth, were largely vague and inaccurate. The language used by participants furthermore suggested that the two instruments were quite differently perceived. When talking about the theremin, they tended to focus more on the instrumentalist's gestural control, its relationship to the sonic output and made positive comparisons to acoustic instruments, whereas for the Tilt Synth, their general lack of understanding of its functioning prevented this (ibid., p. 173). This also implies that perceived skill, which arises from greater understanding, is still an important category in the evaluation of a performance, as has been suggested elsewhere (e.g. Auslander 2009, pp. 603–4).

Emerson et al. (forthcoming) conducted a survey at a concert featuring performances with four new DMIs, which was designed to collect spectators' responses to DMI performances in an explorative manner. Forty-nine audience members were asked to note down what they paid attention to during the performances and to describe their overall impression of each one. The majority of answers mentioned having paid attention to the gesture-sound causality. The responses frequently included such terms as 'gesture' (n = 7), 'movement' (n = 21), 'relation' (n = 8) and 'interaction' (n = 16), with some audience members reporting having directed their attention to interpreting the movements of specific parts of the body (e.g. 'the hands', n = 25). Further to this, fourteen respondents mentioned that they focused on trying to understand the relationship between the performer's interaction with the device and the sounds produced:

I was trying to figure out how the moving of the cubes corresponds to the changes in sound.
(Respondent 16)
[I was] trying to understand the logic of relation between [the] players, relating the blocks as causes to effects.
(Respondent 9)
[I focused on the] hands/hand movements and [their] connection with the sounds.[1]
(Respondent 21)
How are the sounds produced?[2]
(Respondent 20)
I wanted to understand how the sounds are produced, so I paid attention to the movements.[3]
(Respondent 11)
It was a shame that there was a large gap between movement and interaction—no direct feedback [was] detectable.[4]
(Respondent 49)

[1]Original German: , […] Hände/Handbewegung und Zusammenhang mit Sound'.

[2]Original German: 'Wie entstehen die Klänge?'

[3]Original German: 'Ich wollte verstehen, wie die Klänge entstehen, also habe ich auf die Bewegungen geachtet.'

[4]Original German: 'Ich fand es schade, dass zwischen Bewegung und Interaktion eine starke Lücke klaffte—keine direkte Rückkoppelung spürbar.'

[It was] unclear for me, what was made by [by the] laptop and what not.
(Respondent 30)

Such comments suggest that mapping configuration can be the source of some confusion for audience members. Indeed, two respondents mentioned that a brief explanation of how the presented instrument works would have been helpful to them (Nos. 4 and 11). In addition to this, there were eight responses that questioned whether 'instrument' was the correct word to use for the devices displayed, which shows how audiences can become engaged with issues of instrumentality when confronted with new DMIs:

Is it even an instrument if you can't control it and the sounds are not predictable?[5]
(Respondent 6)
The instrument and the computer aren't separated enough to understand properly the way the instrument works.
(Respondent 31)
[It is] more of a sound-instrument than a musical instrument.[6]
(Respondent 40)

This reveals how audience members' experiences with DMIs often revolve around trying to understand the process of music-making, which is often done through comparing DMIs to their existing notions of what a musical instrument is and how it should be interacted with. It appears that problems related to deciphering the gesture-sound causality of a DMI can then come to hinder audience members' aesthetic appreciation.

4 Conclusion: From User Experience to Audience Experience

The literature discussed here provides useful insights into the various concepts that are involved in an audience's reception of a DMI performance, namely the perception of agency and liveness, overall comprehension and the perceived level of skill displayed. It has been emphasised that the perceptibility of the gesture-sound causality can impact how further aspects of the performance are viewed. As illustrated in Fig. 1, we propose that a clear gesture-sound causality is the foundation for understanding the basic functioning of the instrument, which then underlies such higher-level evaluative concepts as perceived liveness and skill (Juslin 2013; Leder et al. 2004).

How can designers and performers approach the task of incorporating insights from this body of research into DMI practice? It seems most important to simply

[5]Original German: 'Ist es überhaupt ein Instrument, wenn man es nicht kontrollieren kann und die Sounds sind nicht vorhersehbar?'

[6]Original German: 'Eher ein Klang-Instrument als ein Musikinstrument.'

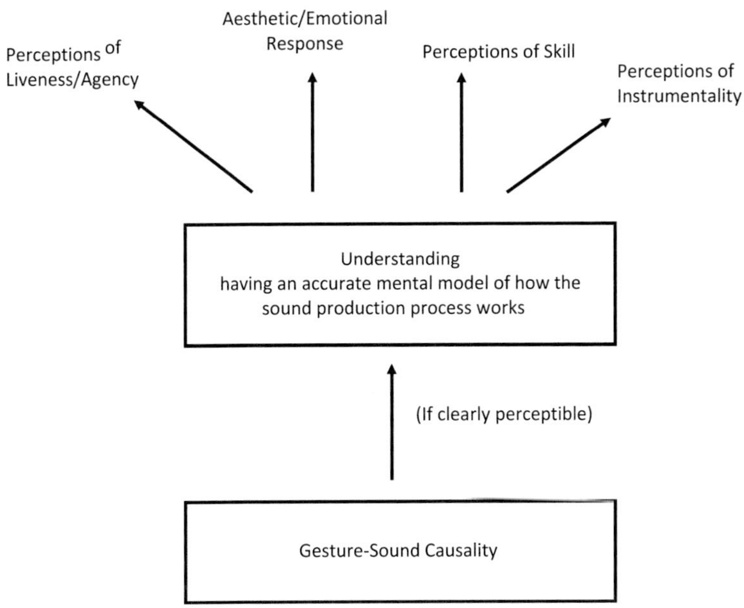

Fig. 1 A model of the DMI reception process

consider DMIs from the audience's perspective, drawing on spectators' feedback in the design process where possible (Barbosa et al. 2012), considering ways in which to provide more information during performances (Berthaut et al. 2013) and granting more attention to sound design (Jordá 2004). In this way, by shifting the focus from the user to the audience, the project of creating new DMIs could result in richer perceptual experiences that interest and provoke spectators, encouraging their engagement with questions of instrumentality and new ways of producing electronic music.

References

Auslander, P. (1999). *Liveness: Performance in a mediatized culture*. London and New York: Routledge.

Auslander, P. (2009). Lucille meets Guitarbot: Instrumentality, agency, and technology in musical performance. *Theatre Journal, 61*(4), 603–616. doi:10.1353/tj.0.0277

Barbosa, J., Calegario, F., Teichrieb, V., Ramalho, G., & McGlynn, P. (2012). Considering audiences' view towards an evaluation methodology for digital musical instruments. In *NIME 2012: Proceedings of the 12th International Conference on New Interfaces for Musical Expression* (pp. 403–408). Ann Arbor, MI. Retrieved from http://www.nime.org/proceedings/2012/nime2012_174.pdf

Berthaut, F., Marshall, M., Subramanian, S., & Hachet, M. (2013). *Rouages*: Revealing the mechanisms of digital musical instruments to the audience. In *NIME 2013: Proceedings of the*

International Conference on New Interfaces for Musical Expression (pp. 165–169). Daejeon, South Korea. Retrieved from http://nime.org/proceedings/2013/nime2013_51.pdf

Berthaut, F., Coyle D., Moore, J. W., & Limerick, H. (2015). Liveness through the lens of agency and causality liveness through the lens of agency and causality. In *NIME 2015: Proceedings of Conference on New Interfaces for Musical Expression* (pp. 382–386). Baton Rouge, LA. Retrieved from http://www.nime.org/proceedings/2015/nime2015_272.pdf

Croft, J. (2007). Theses on liveness. *Organised Sound, 12*(01), 59. doi:10.1017/S1355771807001604

Emerson, G., Förstel, G., & Egermann, H. (in revision). Gesture-sound causality from the audience's perspective: Investigating the influence of mapping perceptibility on the aesthetic perception of new digital musical instruments. *Psychology of Aesthetics, Creativity and the Arts.*

Fyans, A. C., Gurevich, M., & Stapleton, P. (2010). Examining the spectator experience. In *NIME 2010: Proceedings of the 10th International Conference on New Interfaces for Musical Expression* (pp. 451–454). Sydney, Australia. Retrieved from http://www.nime.org/proceedings/2010/nime2010_451.pdf

Fyans, A. C., & Gurevich, M. (2011). Perceptions of skill in performances with acoustic and electronic instruments. In *NIME 2011: Proceedings of the 11th International Conference on New Interfaces for Musical Expression* (pp. 495–498). Oslo, Norway. Retrieved from http://www.nime.org/proceedings/2011/nime2011_495.pdf

Gurevich, M., & Fyans, A. C. (2011). Digital musical interactions: Performer–system relationships and their perception by spectators. *Organised Sound, 16*(2), 166–75. doi:10.1017/S1355771811000112.

Hunt, A., Wanderley, M., & Paradis, M. (2003). The importance of parameter mapping in electronic instrument design. *Journal of New Music Research, 32*(4), 429–440. doi:10.1076/jnmr.32.4.429.18853

Jordà, S. (2004). Instruments and players: Some thoughts on digital lutherie. *Journal of New Music Research, 33*(3), 321–341. doi:10.1080/0929821042000317886

Juslin, P. N. (2013). From everyday emotions to aesthetic emotions: Towards a unified theory of musical emotions. *Physics of Life Reviews, 10*(3), 235–266. doi:10.1016/j.plrev.2013.05.008

Leder, H., Belke, B., Oeberst, A., & Augustin, D. (2004). A model of aesthetic appreciation and aesthetic judgments. *British Journal of Psychology, 95*(4), 489–508.

Miranda, E. R., & Wanderley, M. M. (2006). *New Digital Musical Instruments: Control and Interaction Beyond the Keyboard* (Vol. 21). Computer Music and Digital Audio Series. Middleton: AR Editions.

O'Modhrain, S. 2011. A framework for the evaluation of digital musical instruments. *Computer Music Journal, 35*(1), 28–42. doi:10.1162/COMJ_a_00038

Schloss, W. A. (2003). Using contemporary technology in live performance: The dilemma of the performer. *Journal of New Music Research, 32*(3): 239–242. doi:10.1076/jnmr.32.3.239.16866

Western Orchestral Instruments in the Foreground: What Features Make an Instrument More Attractive for a Solo Role in Concertos?

Song Hui Chon

Abstract Music often consists of multiple instruments and parts. Some serve a more foreground role (such as carrying a melody) whereas others offer background support (namely, as accompaniment). Musical solos are probably the clearest example of the foreground usage. What factors then make a specific instrument attractive for performing a solo? For example, an instrument might be preferred for a solo function if there are many virtuoso musicians. Or listeners might request a solo of a rare instrument. In this chapter, we examine the popularity of an instrument to play a solo role using four factors: pitch, loudness, timbre, and performer pool size. We focus on the concerto repertoire in Western classical music, since the titles bear a clear designation of the solo instrument(s). Our hypothesis is that an instrument will be attractive for a solo if it can produce high pitches and loud sounds, has a salient timbre, and has many skilled performers available to play it. Correlation and multiple regression results were mostly in agreement with the hypothesis; an instrument is more likely to serve in a solo role when it has a higher median pitch, a highly salient timbre, and there are a larger number of trained musicians.

1 Introduction

In terms of sales, some musical instruments are more popular than others. The U.S. sales statistics published in the music industry census report (Music Trades 2014) show the guitar (both acoustic and electric) to be the most purchased instrument (2,472,000 units were sold in 2013), followed by ukuleles (966,340 units), and portable keyboards (912,500 units, which is comparable to a total of 911,400 units of brass/woodwind/string instruments for schools sold in 2013). Among various

---

S.H. Chon (✉)
Department of Electrical, Computer, and Telecommunications Engineering Technology, Rochester Institute of Technology, One Lomb Memorial Drive, Rochester, NY 14623-5603, USA
e-mail: songhui.chon@rit.edu

T. Bovermann et al. (eds.), *Musical Instruments in the 21st Century*,
DOI 10.1007/978-981-10-2951-6_24

factors one might expect to influence decisions to purchase and learn an instrument are price, portability, musical expressiveness, and ease of playing. On the other hand, other aspects might influence the choice of an instrument to listen to. For example, we might well prefer listening to an instrument that is rare, expensive, and difficult to play, though we may not want to learn how to play it ourselves.

Composers do not necessarily need to master the instruments they write for. Although influenced by other factors like cultural and economic ones, a composer's musical decisions regarding instrumentation might be motivated more by the sound or timbre of the instrument and its capacity for musical expressiveness. Acoustically, instruments differ in a variety of ways, including their range of pitch, loudness, timbre, and their ability to blend with other instruments (Sandell 1995; Chon and McAdams 2012; Tardieu and McAdams 2012; Lembke and McAdams 2015). These acoustic characteristics in turn affect the function or role that an instrument might fulfill. Huron et al. (2014) showed that different instrument properties are predictive of that instrument's capacity to convey sadness. For example, it is difficult to play sad music on a banjo, because the rapid decay of plucked strings renders slow performance difficult. Conversely, it is difficult to perform rapidly on a harmonium, making that instrument better suited to somber expressions. In short, different instruments offer different musical affordances (Huron and Berec 2009).

Johnson (2011) carried out an empirical study of instrumentation patterns that began with exploratory research followed by hypothesis testing to find different functions for various instruments. Specifically, Johnson assembled a sample set of individual orchestral sonorities representing the 19th-century art music repertoire. For each sampled work, he randomly selected a single sounded moment and coded which instruments were present. Using cluster analysis, he found three overarching clusters for 19 orchestral instruments. He interpreted these groupings as suggestive of three functions reflecting different instrumentation purposes, dubbed *Standard*, *Power* and *Color* (SPC). *Standard* instruments (including violin, viola, cello, contrabass, flute, oboe, clarinet, and bassoon) are the most commonly used instruments and tend to sound throughout the course of musical works. *Power* instruments include trumpet, trombone, French horn, tuba, piccolo, and timpani, which are often used for loud energetic passages; lastly, *Color* instruments introduce timbral contrast, variety and novelty. Bass clarinet, contrabassoon, cornet, harp, and English horn belong to the *Color* cluster. Following a similar but longitudinal approach using instrumentation patterns found in orchestral music in 1701–2000, Chon et al. (in preparation) also found three clusters of orchestral instruments in moderate agreement with Johnson's SPC model; nonetheless they were interpreted as reflecting different compositional functions: four string instruments (violin, viola, cello, and contrabass) form a cluster (a.k.a. the *Strings* cluster), constituting the *skeleton* of a musical scene. Flute, oboe, clarinet, bassoon, and French horn form the *Core Winds* cluster. These instruments have been consistently included in orchestral compositions more often than other wind instruments. The biggest cluster, the *Effects* cluster, includes the rest of instruments—bass clarinet, contrabassoon, cornet, harp, and English horn, trumpet, trombone, tuba, piccolo, and timpani, which is a mixture of instruments from Johnson's *Power* and *Color* clusters providing extended ranges of pitch, timbral effects, and loudness.

2 Solo Use of Western Orchestral Instruments

Perhaps the most unambiguous case of a solo is in the designation of solo instruments in concerto titles. Therefore, we operationalize an instrument's likelihood to be used in a foreground role as the number of concertos featuring that instrument as (one of) the solo instrument(s) in the title, such as Schumann's *Piano Concerto in A minor, Op. 54*. Since thousands of concertos have been written, and since titles are relatively easy to sample, there is practical merit in using concerto titles as an operationalization for frequency of instrument foregrounding. If an instrument is commonly used in solos, it is reasonable to assume that it is also likely to have many concertos written for it. Of course, we do not expect that the concerto count for an instrument will be perfectly representative of its general appearance as a featured or foreground instrument. Nevertheless, we might expect a reasonably strong correlation. Accordingly, for the purposes of this study, we will make use of the number of concertos for various instruments as our dependent measure.

We might suppose that some instruments are better suited to a solo function than others. For example, we might expect a flute or a cello to feature as a solo instrument more often than a piccolo or a contrabass. This raises the motivational and focal question of our study: "What makes some instruments better suited to function in a solo capacity?"

3 Possible Factors for the Choice of Solo Instruments

3.1 Perceived Loudness Capacity

One factor that might be presumed to influence the choice of a solo instrument is the perceived loudness of the instrument. Instruments are typically capable of playing a wide range of dynamic levels. Although the average dynamic level is of interest, the principal consideration is the capacity of the instrument to play loudly. In Western orchestral music, the norm is a multi-part texture in which several instruments sound concurrently. Such situations will inevitably cause auditory masking, where one sound would partially or wholly obscure another sound. Masking is related to loudness (Fletcher and Munson 1937), and typically the louder of two sounds will mask the quieter sound more than vice versa (Zwicker and Fastl 2007, p. 61). Accordingly, one might predict that those instruments capable of producing a louder sound would be favored in a solo function.

3.2 Pitch Height

In contrapuntal music of the Baroque period (1600–1750), each musical line carried equal importance. As the Baroque style declined, more emphasis came to be placed on the higher-pitched line. Since then, the most important musical lines (e.g., melody) tend to be in the upper-most part (Stauffer 2006, p. 43). Marie and Trainor (2012) reported empirical evidence for the *high voice superiority effect for pitch*, where auditory streams of higher pitch are perceptually more prominent than those of lower pitch. A subsequent study by Trainor et al. (2014) found that this increased prominence was also noticeable in computational auditory models. Chon and Huron (2014a, 2015) have raised a possible link between the *high voice superiority effect for pitch* and masking: even though a lower-pitched pure tone is known as a better masker than a higher-pitched pure tone, between two complex tones a higher-pitched one is more effective in masking a lower-pitched tone. This phenomenon, then, might set a biological disposition toward a greater emphasis on streams of higher pitch.

Whatever the underlying cause of this *high voice superiority effect for pitch* might be, the empirical observation remains that the highest pitches in a complex sonority are likely to be more perceptually important (Chon et al. 2013a, b). Note that this effect relates to relative rather than absolute pitch. Conceptually, a high pitch can always be made to sound less prominent by concurrently including an even higher pitch. Accordingly, we might predict that those instruments capable of playing the highest pitches will tend to be favored whenever a composer seeks a solo or prominent function.

3.3 Timbre Salience

Chon (2013) proposed the concept of timbre salience as the attention-capturing quality of timbre. Her conjecture was that instruments differ in their capacity to capture listeners' attention, which in turn might have affected the way instruments have been used in orchestration practices. Chon and McAdams (2012) further found that a highly salient timbre would not blend well with others. In effect, a salient instrument tends to *stand out* from its surroundings.

In a perceptual experiment, Chon and Huron (2015) measured the timbre salience of 15 orchestral instruments defined as the correct identification rate of an instrument in a unison dyad, while controlling for pitch, loudness, and effective duration. Specifically, they provide estimates for clarinet, English horn, flute, French horn, harp, harpsichord, marimba, oboe, piano, trombone, trumpet, tuba, tubular bells, cello, and vibraphone, which will be used for analyses in this study.

3.4 *Performer Pool Size*

Apart from the sonic factors mentioned above, we might expect that the skill level of a performer would also play a role. A composer may be motivated to write for an instrument, if there is a great performer of the given instrument. For example, a prominent horn player of the late 18th century named Joseph Leubgeb inspired Mozart, and Michael and Joseph Haydn to compose horn concertos (Heartz 1995, p. 277). As it would be easier to find a good musician from a larger group of candidates, the availability of talented soloists would be influenced by the size of the pool of pertinent performers. Consequently, it is possible that the preference for some instrumental solo might be correlated with the number of available performers.

4 Aims

Our current study seeks to predict how frequently various instruments are used in a solo or foreground capacity. In light of the preceding discussion, we analyze the possible role of four factors: loudness capacity, pitch height, timbre salience, and performer pool size. Other factors such as the age of an instrument undoubtedly play a role. However, in the absence of any reliable way of operationalizing these other concepts, we will focus on the above four factors. Formally, our hypothesis may be stated as follows:

H. The number of concertos written for an instrument is positively correlated with the instrument's loudness capacity, pitch height, timbre salience, and performer pool size.

5 Method

5.1 *Concerto Data*

We obtained a list of concerto titles from allmusic.com using the search term *concerto* in January 2014. This website offers a large online database of songs, albums, artists, and compositions that can be easily searched. The query returned the titles of over several thousand compositions. Many of these specified a single solo instrument in the title; some include more than one, whereas some others do not mention any. We reviewed each concerto title to ensure that it identifies a solo instrument (or more) and that it belongs to the categories of both *orchestral* and *concerto*, which were encoded at the time of data entry. The titles that did not meet the above criteria were excluded, such as *Concerto for Orchestra, Op. 12* by Bartók, which featured the orchestra as one hyper instrument for solo, or *Concerto pour piano seul* by Alkan, which is not an orchestral but an instrumental composition.

Note that some concertos are not unique to a single instrument. For example, J.S. Bach wrote *Concerto for flute or recorder, violin, strings and continuo in D major, BWV 1050a*, which features a solo violin and either a flute or a recorder solo. In such cases, a single composition was coded as multiple compositions featuring each of the different instruments as serving solo or foreground functions. The tallies for each of the corresponding instruments were individually incremented.

In order to carry out this study, we need some guideline by which to accommodate possible variants. Since the purpose of this study is to determine which factors influence the choice of an instrument for foreground functions, the overarching criterion for including alternate names should be the degree to which that instrument resembles the common modern instrument. For example, the posthorn bears a close resemblance to the French horn with regard to timbre, loudness, and (to a lesser degree) pitch height. Accordingly, we might include Johann Beer's *Concerto à 4 for posthorn, corno da caccia, 2 violins and continuo in B flat major* in our tallies for French horn. In addition, instruments are sometimes referred to by a variety of names (such as violin and fiddle). In other cases, a similar instrument may have slightly different names, such as the *oboe* and the *oboe d'amore*. In some other cases, a rather general term might be found, such as the *Clavier*, which means a keyboard instrument as a whole. One might also argue that since harpsichord is an ancestor of piano, the concertos featuring a harpsichord should be tallied with those for a *piano*, as the latter would be more general of the two instruments. However, they are kept separate, as harpsichord reappeared as a solo instrument in concertos in the 20th century. The same reason was applied to the decision to keep recorder and flute separate. Table 1 identifies the modern instrument equivalents used in this study.

Table 1 List of historic instruments and their modern equivalents

Historic instruments	Modern (equivalent) instruments
Chalumeau, Basset Horn, Basset Clarinet	Clarinet
Transverse flute, Flauto d'amore	Flute
Oboe d'amore, Oboe da silva, Oboe da caccia	Oboe
Hunting horn, Corno da caccia	French Horn
Cornet	Trumpet
Baritone horn, Euphonium	Trombone
Fiddle	Violin
Viola da gamba	Cello
'Keyboard' or 'Clavier' before 1780	Harpsichord
'Keyboard' or 'Clavier' after 1780	Piano

5.2 Predictor Variables

The dependent variable (DV) to predict is the number of concertos written for various instruments featured for solo. In order to test our hypothesis, we need to characterize each instrument in terms of the four independent variables (IVs): perceived loudness capacity, pitch height, timbre salience, and performer pool size.

5.2.1 Perceived Loudness Capacity

There exist a few reports on the average loudness of various instruments (Miśkiewicz and Rakowski 1994; Moore 2010; Klonari et al. 2011). However, we were unable to find any comprehensive measurement that covers all 18 instruments identified in the list of concertos. Subjective judgments concerning instruments' perceived loudness capacity were deemed appropriate for the purposes of our study. Accordingly, the loudness estimates represent subjective ratings by experienced musicians rather than objective acoustical measurements.

We conducted a perceived loudness capacity survey in two parts. First, participants were asked to rank order the instruments from most loud to least loud. The idea here was to establish mental anchors that would guide and facilitate the second task. Then, participants were given a physical line (nine points along the line) on paper for each instrument and asked to place a mark (X) corresponding to the instrument's deemed capacity for loudness. Our index of perceived loudness capacity is simply the median rating for a given instrument across all 11 experienced musicians who participated. The results shown in Table 4 showed a high correlation with the peak levels of the instruments reported in Moore (2010), $r(8) = 0.64$, $p < 0.05$, implying that our participants' judgments are a realistic estimate of the perceived loudness capacity of these instruments.

5.2.2 Pitch Height

Each instrument produces a range of pitches. Depending on its physics, an instrument's timbre might be rather homogenous (such as with string instruments) or have distinctive registers (such as with flute or clarinet). Table 2 summarizes the conventional pitch ranges of the 18 instruments. The highest and lowest pitch information was obtained from Adler (2002), except for the recorder whose information is from Campbell et al. (2004, p. 140). We then calculated the pitch range in semitones, as well as its middle point (*median pitch*) for each instrument. When a pitch range in semitones is an even number, the middle point is between pitches, in which case the lower pitch was chosen as the median.

Table 2 Pitch distribution by instrument

Instrument	Conventional range (Adler 2002)			
	Highest pitch	Lowest pitch	Median pitch	Range in semitones
Violin	B7	G3	A5	53
Piano	C8	A0	E4	88
Flute	D7	C4	G5	39
Cello	E6	C2	D4	53
Oboe	G6	A$^{\#}$3	D5	34
Clarinet	G6	D3	A$^{\#}$4	42
Harpsichord	F6	F1	B3	61
Trumpet	C6	F$^{\#}$3	A4	31
French horn	F5	G1	F$^{\#}$3	47
Viola	A6	C3	A$^{\#}$4	46
Organ	C7	C2	F$^{\#}$4	61
Bassoon	D$^{\#}$5	A$^{\#}$1	G3	42
Guitar	E5	E2	A$^{\#}$3	37
Harp	G$^{\#}$7	B0	D4	82
Trombone	F5	E2	A$^{\#}$3	38
Recorder (soprano)	D7	C5	C$^{\#}$6	27
Saxophone (alto)	A$^{\#}$5	C$^{\#}$3	F4	34
Contrabass	G4	C1	A2	44

In light of the *high voice superiority effect for pitch*, we could conjecture that an instrument capable of producing higher pitches lends itself to more solos, as a higher-pitched stream tends to carry a higher cognitive importance. On the other hand, an instrument with a narrow pitch range might not be very attractive for a solo role, as its expressive capability is limited. Therefore, an instrument that can produce high pitches across a wide pitch range could be the best candidate for a solo role.

As these four pitch-related parameters are not independent of one another, adding all four of them to the regression model will not be very useful. Hence, correlation was calculated (Table 3). The median pitch was chosen to represent the pitch information, as it shows the highest correlation with other factors.

5.2.3 Timbre Salience

Chon and Huron (2014b) reported the average identification rates of 15 instruments in unison dyads, which is taken as a measure of timbre salience. Unfortunately, only ten of the 15 instruments were included in the list of the 18 most popular

Table 3 Correlation of pitch-related parameters

	Highest pitch	Lowest pitch	Median pitch	Pitch range
Highest pitch	$r = 1.00$	$r = 0.21$	$r = 0.72^{**}$	$r = 0.53^{*}$
	$p < 0.001$	$p = 0.41$	$p < 0.001$	$p < 0.05$
Lowest pitch	$r = 0.21$	$r = 1.00$	$r = 0.83^{**}$	$r = -0.72^{**}$
	$p = 0.41$	$p < 0.001$	$p < 0.001$	$p < 0.001$
Median pitch	$r = 0.72^{**}$	$r = 0.83^{**}$	$r = 1.00$	$r = -0.21$
	$p < 0.001$	$p < 0.001$	$p < 0.001$	$p = 0.41$
Pitch ranges	$r = 0.53^{*}$	$r = -0.72^{**}$	$r = -0.21$	$r = 1.00$
	$p < 0.05$	$p < 0.001$	$p = 0.41$	$p < 0.001$

* and ** denote a 0.05- and 0.001-level significance, respectively

Table 4 18 most popular solo instruments and the corresponding number of concertos

Instrument	Number of concertos	Loudness capacity	Median pitch	Timbre salience	Performer pool size
Violin	1,894	5.59	A5	N/A	1,191
Piano	1,169	6.77	E4	0.58	4,337
Flute	606	3.76	G5	0.32	976
Cello	597	5.50	D4	0.82	682
Oboe	436	5.27	D5	0.19	557
Clarinet	330	4.73	A$^{\#}$4	0.30	817
Harpsichord	270	1.84	B3	0.85	413
Trumpet	263	8.39	A4	0.16	885
French horn	214	6.98	F$^{\#}$3	0.16	685
Viola	212	5.59	A$^{\#}$4	N/A	670
Organ	175	9.00	F$^{\#}$4	N/A	1,127
Bassoon	170	4.28	G3	N/A	499
Guitar	138	3.06	A$^{\#}$3	N/A	1,354
Harp	101	2.18	D4	0.20	220
Trombone	95	8.21	A$^{\#}$3	0.12	631
Recorder (soprano)	96	1.84	C$^{\#}$6	N/A	102
Saxophone (alto)	84	5.94	F4	N/A	702
Contrabass	58	5.64	A2	N/A	545

solo instruments in Table 4, hence this factor was excluded from the multiple regression analysis. Nonetheless, we calculated the correlation between the number of concertos and the timbre salience values for the common instruments.

5.2.4 Performer Pool Size

In order to estimate the number of performers, the number of full-time university-level teachers in the United States and Canada was obtained for various instruments from the College Music Society directory (Short 1994). The presumption is that the number of teachers for a given instrument will correlate with the number of performers, which will also reflect the instrument's popularity to serve in a foreground role. The directory includes an index that conveniently groups together faculty by instrumentation, so the total number of professors was counted for each instrument. The results are reported in the last column of Table 4. Even though these numbers do not match the exact numbers of proficient musicians at the time of compositions, as there exist no better measures to the best of our knowledge, we will include these counts as a factor for analysis.

6 Results

The resulting sample set included 6,559 concerto titles featuring 138 unique solo instruments or units. Among the 138 instruments, the most popular were violin, piano, and flute, which is hardly surprising. Also included were ensembles identified as the solo unit rather than specific individual instruments, such as those for a string quartet or for a brass quintet. The list contained many solo instruments that were featured in only a handful of concertos, such as didgeridoo, celesta, alphorn, banjo, glass harp, Ondes Martinot, and Theremin. In fact, 51 instruments were featured in only one concerto; 21 were featured in only two concertos; 104 instruments were featured in less than 10 concertos. Table 4 lists the 18 most popular solo instruments in the decreasing order of the number of concertos, along with each instrument's perceived loudness capacity, median pitch, timbre salience, and performer pool size. The first row in Table 4 indicates that there are 1,894 concertos featuring a violin solo. This number does not necessarily mean that the violin is the only solo instrument; 1,894 concertos could include concertos that feature a violin and other instrument(s) such as Beethoven's *Concerto for violin, cello and piano in C major, Op. 56*. The two most popular solo instruments, piano and violin, account for 46 % of the sampled concertos ((1,894 + 1,169 − 52)/6,559 = 0.46; 52 concertos were subtracted because they featured both a violin and a piano as solo instruments).

To evaluate our hypothesis, we calculated correlation between the concerto counts and each of the four factors: median pitch in midi number, perceived loudness capacity, performer pool size, and timbre salience. Moderate positive correlations were observed for median pitch, $r(16) = 0.42$, $p = 0.08$, and performer pool size, $r(16) = 0.53$, $p < 0.05$, and timbre salience, $r(8) = 0.46$, $p = 0.18$, but not for loudness capacity, $r(16) = 0.10$, $p = 0.69$. These results are mostly consistent with the hypothesis, leading to the interpretation that an instrument is frequently used for solo or foreground purposes when it is relatively high-pitched, with a salient timbre, and there are relatively large numbers of players on it. The loudness capacity turned out to have very little correlation with the concerto counts, suggesting that it may not have been an important factor to affect a composer's decision of a solo instrument.

A multiple linear regression has been carried out to examine if the three IVs (median pitch in midi number, loudness capacity, and performer pool size) could predict the number of concertos for the 18 instruments. Timbre salience values were not included due to the limited number of samples. DV was the log transformation of the number of concertos rather than the concerto counts because of the normality assumptions for regression. The model proved to be significant, $F(3, 14) = 4.49$, $p = 0.02$, $R^2 = 0.49$, $R^2_{adjusted} = 0.38$. However, when the effect of individual factors was compared, only median pitch and performer pool size were significant, $\beta = 0.46$, $p = 0.03$ for median pitch, and $\beta = 0.53$, $p = 0.02$ for performer pool size. Loudness capacity failed to achieve a significant effect, $\beta = 0.02$, $p = 0.92$. By removing loudness capacity, the new regression model performed slightly better in terms of the adjusted R^2 value, $F(2, 15) = 7.20$, $p < 0.01$, $R^2 = 0.49$, $R^2_{adjusted} = 0.42$. Between the two IVs, performer pool size alone could account for 28 % of variance in data, $F(1, 16) = 6.34$, $p = 0.02$, $R^2 = 0.28$, $R^2_{adjusted} = 0.24$. Median pitch explained another 20 % of variance, $F(1, 16) = 4.10$, $p = 0.06$, $R^2 = 0.20$, $R^2_{adjusted} = 0.15$.

7 Discussion

In this study, we aimed to explain what factors make an instrument more attractive for a solo role in concertos. Our hypothesis was that an instrument would be featured in a solo more often if it can play higher pitches, has a salient timbre, is capable of playing loudly, and there exists a large pool of musicians who play that instrument. In other words, a positive correlation was predicted between the concerto counts and each of the four factors. The number of concertos written for the 18 most popular instruments was analyzed using correlation and multiple regression with three factors, excluding timbre salience.

All four factors showed a positive correlation with the concerto counts of the 18 instruments, which is consistent with our prediction. Performer pool size turned out to be the best predictor of the number of concertos written for instruments. It explained 28 % of variance in data by itself, suggesting a close relationship between the number of skilled performers on an instrument and the number of compositions featuring that instrument in a solo capacity. The next successful factor was median pitch, explaining another 21 % of variance. This indicates that an instrument that can play higher pitches tends to be featured more often in a foreground role, which is consistent with the *high voice superiority effect for pitch*. Loudness capacity failed to predict the concerto counts, suggesting that loudness has probably not been very influential in a composer's choice of a solo instrument.

It is interesting to see performer pool size to be most effective in predicting the number of concertos. But when we try to interpret its meaning, we run into the classic *chicken-and-egg* problem: Have many concertos have been written for an instrument, say a violin, because there were many talented performers? Or did many musicians choose to learn the violin because there were many foreground roles already available for the instrument? Surely, there must be some form of circular interaction between these two factors, and therefore it may be impossible to distinguish causality.

The factors investigated in this study should not be considered exhaustive. There must be other factors that undoubtedly influence the choice of instrumentation in Western classical concerto works. Some instruments such as lutes were rarely included in orchestral music (Lambord 1916, p. 87), even though they have been used in many other forms of music making. Also, we did not consider the historical factor of the instrument age. The violin has been available for a much longer period of time than a clarinet; hence more concertos are expected to feature a violin solo than a clarinet solo. In addition, there could be factors of traditional and normative practice. For example, although the viola has been used in Western art music ensembles for practically as long as the violin, until recently it was rarely featured as a solo instrument for a concerto. There may be acoustic reasons accounting for this, but it might have been simply a matter of tradition. Furthermore, the instrumental training of composers themselves is apt to bias the choice of instruments. Mozart, for instance, was a highly trained pianist, so it should not be surprising that he wrote many piano concertos, with few concertos for other instruments. Moreover, in earlier times when composers were employed by the wealthy or noble, concertos were often written with the solo part for the employer or his wife or daughter, just like Quantz wrote flute concertos for his employer Frederick the Great of Prussia (MacDonogh 1999).

The popularity of high-pitched instruments for solo use is in agreement with the *high voice superiority effect for pitch* that the auditory stream of higher pitch tends to carry more cognitive importance. Perhaps the human auditory cognition system assigned greater importance on higher pitches to begin with, which might have led to more foreground roles for high-pitched instruments that was evident in many concertos written for them. This foreground prominence of higher-pitched instruments may then have drawn more musicians to study them, among who were gifted performers to inspire composers to write more foreground roles for these instruments.

References

Adler, S. (2002). *The study of orchestration* (3rd ed.). New York, NY: W.W. Norton & Company.

Campbell, M., Greated, C., &, Myers, A. (2004). *Musical instruments: History, technology, and performance of instruments of western music*. Oxford University Press.

Chon, S. H., & McAdams, S. (2012). Exploring blending as a function of timbre saliency. *Proceedings of the 12th International Conference of Music Perception and Cognition (ICMPC)* (pp. 222–223). Greece: Thessaloniki.

Chon, S. H. (2013). *Timbre saliency, the attention-capturing quality of timbre* (Unpublished Ph.D. dissertation). McGill University, Canada.

Chon, S. H., Huron, D., & DeVlieger, D. (in preparation). Exploratory study of Western orchestration—Patterns through history.

Chon, S. H., & Huron, D. (2014a). Does auditory masking explain high voice superiority? In *Proceedings of the 13th International Conference of Music Perception and Cognition (ICMPC)* (pp. 185–188). Seoul, Korea.

Chon, S. H., & Huron, D. (2014b). Instrument identification in concurrent unison dyads: The effect of timbre saliency. In *Proceedings of the 13th International Conference of Music Perception and Cognition (ICMPC)* (pp. 289–292). Seoul, Korea.

Chon, S. H., & Huron, D. (2015). Effects of pitch, intensity and timbre on frequency masking. *Journal of Acoustics Society of America, 137*(4), 2406–2406.

Chon, S. H., Schwartzbach, K., Smith, B., & McAdams, S. (2013a). Effect of timbre on voice recognition in two-voice counterpoint music. In *Proceedings of the Biennial Meeting of the Society for Music Perception and Cognition (SMPC) 2013* (pp. 31–31). Toronto, Canada.

Chon, S. H., Schwartzbach, K., Smith, B., & McAdams, S. (2013b). Effect of timbre on melody recognition in three-voice counterpoint music. In *Proceedings of the Stockholm Music Acoustics Conference (SMAC)—Sound and Music Computing Conference (SMC) 2013* (pp. 54–61). Stockholm, Sweden.

Fletcher, H., & Munson, W. A. (1937). Relation Between Loudness and Masking. *Journal of the Acoustical Society of America, 9*(1), 78–78.

Heartz, D. (1995). *Haydn, Mozart and the Viennese School, 1740–1780*. New York, NY: W.W. Norton & Company.

Huron, D., & Berec, J. (2009). Characterizing idiomatic organization in music: A theory and case study of musical affordances. *Empirical Musicology Review, 4*(3), 103–122.

Huron, D., Anderson, N., & Shanahan, D. (2014). 'You can't play a sad song on the Banjo': Acoustic factors in the judgment of instrument capacity to convey sadness. *Empirical Musicology Review, 9*(1), 29–41.

Johnson, R. (2011). The standard, power, and color model of instrument combination in romantic-era symphonic works. *Empirical Musicology Review, 6*(1), 2–19.

Klonari, D., Pastiadis, K., Papadelis, G., & Papanikolaou, G. (2011). Loudness assessment of musical tones equalized in a-weighted level. *Archives of Acoustics, 36*(2), 239–250.

Lambord, B. (1916). The orchestra and orchestral music. In D. G. Mason (Ed.), *The Art of Music: A Comprehensive Library of Information for Music Lovers and Musicians*. New York, NY: National Society of Music.

Lembke, S. A., & McAdams, S. (2015). The role of spectral-envelope characteristics in perceptual blending of wind-Instrument sounds. *Acta Acustica united with Acustica, 101*(5), 1039–1051.

MacDonogh, G. (1999). *Frederick the great: A life in deed and letters*. New York, NY: Macmillan.

Marie, C., & Trainor, L. J. (2012). Development of simultaneous pitch encoding: Infants show a high voice superiority effect. *Cerebral Cortex*, bhs050.

Miśkiewicz, A., & Rakowski, A. (1994). Loudness level versus sound-pressure level: A comparison of musical instruments. *Journal of the Acoustical Society of America, 96*(6), 3375–3379.

Moore, A. (2010). *Student musicians' perception of loudness and how it correlates to the measured level. Independent studies and capstones.* Paper 604. Program in Audiology and Communication Sciences, Washington University School of Medicine. http://digitalcommons.wustl.edu/pacs_capstones/604

Music Trades. (2014). *The music industry census 2014.* Englewood, NJ: Music Trades Magazine. www.musictrades.com

Sandell, G. J. (1995). Roles for spectral centroid and other factors in determining "blended" instrument pairings in orchestration. *Music Perception, 13*, 209–246.

Short, C. R. (Ed.). (1994). *Directory of Music Faculties in Colleges and Universities: US and Canada, 1992–94.* Binghamton, NY: College Music Society.

Stauffer, G. B. (2006). The modern orchestra: A creation of the late eighteenth century. In J. Peyser (Ed.), *The orchestra* (pp. 41–72). Milwakee, WI: Hal Leonard.

Tardieu, D., & McAdams, S. (2012). Perception of dyads of impulsive and sustained sounds. *Music Perception, 30*(2), 117–128.

Trainor, L. J., Marie, C., Bruce, I. C., & Bidelman, G. M. (2014). Explaining the high voice superiority effect in polyphonic music: Evidence from cortical evoked potentials and peripheral auditory models. *Hearing Research, 308*, 60–70.

Zwicker, E., & Fastl, H. (2007). *Psychoacoustics: Facts and Models* (Vol. 22). Springer Science & Business Media.

Instruments Unheard of: On the Role of Familiarity and Sound Source Categories in Timbre Perception

Kai Siedenburg

Abstract Musical timbre has traditionally been treated as a sensory phenomenon, that is, as a "surface feature" that resides in the musical moment. The role of familiarity with sound source categories and instrument families has remained unexplored. The current chapter takes a dedicatedly cognitive view on timbre and argues that long-term familiarity and knowledge about instrument categories affect even such supposedly low-level tasks as dissimilarity ratings. As a background, the chapter provides a conceptual framework for the notion of timbre, as well as an outline of basic results from timbre dissimilarity ratings and instrument identification. Results from a previous study on the role of sound source categories in timbre dissimilarity ratings are then discussed in depth (Siedenburg et al. in Frontiers in Psychology 6, 2016b). This study collected timbre dissimilarity ratings for tones from acoustic musical instruments as well as for novel, digitally transformed tones. The main pieces of evidence to be discussed come from rating asymmetries and a regression model. It is argued that timbre perception is characterized by an interplay of sensory and categorical representations, reflecting acoustic facets and learned sound source and instrument categories of musical instruments. Implications for the design of novel digital musical instrument design are being discussed.

1 Introduction

How would a symphony sound like if you had never heard an orchestral instrument before? Would it all be "one strange flavor"? In other words, to which extent do memory, familiarity, and learned instrument categories play a role in the perception of musical timbre? Probing the nature of timbre cognition in such ways is of particular musical relevance today, given that the design of novel instruments has become a central concern of contemporary musical practice.

K. Siedenburg (✉)
Department of Medical Physics and Acoustics—Signal Processing Group,
University of Oldenburg, D-26111 Oldenburg, Germany
e-mail: kai.siedenburg@uni-oldenburg.de

T. Bovermann et al. (eds.), *Musical Instruments in the 21st Century*,
DOI 10.1007/978-981-10-2951-6_25

In the field of music cognition, auditory schemata acquired through long-term experience have mainly been studied for pitch, harmony, and rhythm (see e.g., Krumhansl 1990; Huron 2006). On the contrary, timbre has traditionally been treated as a primarily sensory phenomenon independent of cognition that resides in the musical moment. Accordingly, timbre perception should not be subject to long-term familiarization. Neurophysiological studies on the auditory processing of timbre have started to provide evidence for the contrary position by showing that instrumentalists exhibit altered neural responses to sounds from their own instrument (cf., Pantev et al. 2001; Shahin et al. 2008; Strait et al. 2012). However, it remains unclear whether these results reflect conscious perceptual experience.

In the current article, I discuss timbre perception from the angle of sound source categories and sound familiarity. I will begin by providing a short recapitulation of the concept of timbre and a review of basic findings in timbre dissimilarity ratings and instrument identification. I will then discuss the role of sound source categories in timbre dissimilarity ratings along the lines of a previous study (Siedenburg et al. 2016a, b). The main hypothesis will be that timbre perception is characterized by an interplay of sensory and categorical representations, reflecting acoustic facets and learned instrument categories of musical instruments and their corresponding semantic associations. In closing, implications for instrument design are being discussed.

2 The Concept of Timbre

2.1 Background on Timbre Perception

Timbre is here understood as an umbrella term that denotes the bundle of auditory features (other than pitch, loudness, duration) that contributes to sound source identity and sound quality. These features are based on acoustic cues such as the spectral envelope distribution, attack sharpness, spectrotemporal variation or modulation, roughness, and noisiness, in addition to features that may be idiosyncratic to certain instruments (see e.g., McAdams 2013). In order to circumvent some of the confusion that the term is sometimes associated with (cf., Krumhansl 1989; Hajda et al. 1997), it is important to distinguish three facets of the concept.

First, timbre is an umbrella term for auditory attributes, i.e., features of a perceptual representation in the mind of a listener. Timbre does not denote the physical properties of a sound event, nor any of its representations via waveforms, spectrograms, audio descriptors, etc. This position does not deny the necessity to investigate the perceptual representation of timbre by using acoustic models and descriptors (e.g., Peeters et al. 2011; Siedenburg et al. 2016a, b), but regards timbre as a primarily psychological entity, similar to other auditory attributes such as pitch and loudness (which should similarly not be confused with fundamental frequency and intensity, respectively).

Second, timbre has a qualitative and a categorical face. The bundle of sensory features required to perceptually categorize a sound source may not be identical to that used for the assessment of qualitative dissimilarity between sounds. Whereas in the former case it is likely that listeners rely on a sub-set of most reliable cues (which likely differs across individual sound sources), the latter case calls for features that are shared between a number of stimuli, such that these can be compared with to each other. More generally speaking, perceptual systems are highly adaptive and there might not be one "hard-wired" perceptual representation of timbre, but the set of most salient features may depend on the specific perceptual task at hand (cf., Shamma and Fritz 2014).

Third, there are multiple scales of different timbral granularity. Whereas it is reasonable to speak about timbral differences of sounds from different musical instruments or families of instruments, even a single instrument may yield a whole palette of distinct timbres. The properties of these covary with playing style (articulation), pitch register, and dynamics (playing effort). Note that it is likely that this pattern of covariation between auditory attributes is used as a cue for the identification of sound sources, although to date only little research investigates this question empirically (cf., Marozeau et al. 2003; Handel and Erickson 2004). One could yet think of a finer scale of timbral granularity, for instance as emerging from the timbral differences (e.g., "coloration") that arise between different audio renderings of the same underlying audio material (e.g., Lindau et al. 2014). For these reasons, it is also misleading to suggest that one sound-producing object or instrument yields exactly one timbre. Contrary to parlance of "the timbre of the clarinet", there is no single timbre that *fully* characterizes the clarinet. The timbre of a clarinet tone depends on pitch, playing effort, articulation, fingering, etc. In terms of a biological analogy, a single type of sound-producing object or sound-synthesis algorithm yields a "timbral genus" that may encompass various "timbral species". These species may differ along various parameters, such as playing technique, covariance with pitch and loudness, or expressive intent. Genera group into "families" (e.g., string vs. brass timbres) and at some point into "kingdoms" (timbres related to, say, acoustic vs. electronic means of sound generation).

2.2 *Timbre Dissimilarity*

Most cornerstones of empirical research on the perceptual representation of musical timbre are based on dissimilarity ratings: Two tones are presented in succession per experimental trial, and listeners rate their degree of dissimilarity, such that the task does not require any verbal labeling of sounds. Starting with the early work of Plomp (1970), Wessel (1973), and Grey (1975), multidimensional scaling (MDS, see Kruskal 1964; Winsberg and De Soete 1993) has been the most important tool for the analysis of the resulting dissimilarity data. Its basic idea is to yield a spatial configuration of the rated stimuli, the *timbre space*, in which spatial distance corresponds to rated dissimilarity. The space is spanned by the rating data's latent dimensions which can be interpreted psychophysically by correlation with continuous acoustic descriptors. For example, McAdams et al. (1995) presented a three-dimensional solution including values for dimensions or features specific to each sound, as well as weights on shared dimensions and specificities for latent classes of subjects. The first spatial dimension correlated with (log-) attack time (AT), the second with the spectral center of gravity (SCG), the third with spectral variation over time ("spectral flux"). SCG and AT have been confirmed to be perceptually salient in a number of studies (Lakatos 2000; Halpern et al. 2004; Caclin et al. 2005). Recently, Elliott et al. (2013) used high-dimensional modulation spectra that represent a signal's joint spectro-temporal variability in order to provide an acoustic basis for the five-dimensional MDS space they had obtained. They observed that the approach has similar predictive power compared to an acoustic description based on scalar audio descriptors (including measures such as spectral and temporal center of gravity).

2.3 *Instrument Identification*

In comparison to dissimilarity ratings, there are fewer empirical studies that investigated the acoustic features utilized by humans in musical instrument identification. For that reason, there is not yet a detailed empirical account on the matter (see McAdams 1993, for a review of theoretical models). In principle, it is important to note that the cues used for dissimilarity ratings must not necessarily be identical to those used for instrument identification (cf., Siedenburg et al. 2016a). For instance, Agus et al. (2012) demonstrated that neither solely spectral nor solely temporal properties can account for results obtained in perceptual timbre categorization, although these properties usually play the most salient role in dissimilarity judgments. Task sensitivity also seems to be an important component in reconciling apparent divergences between studies showing that listeners implicitly recognize the subtlest variations in frozen noise sounds (Agus et al. 2010), or discriminate between subtle changes in spectrotemporal behavior (McAdams et al. 1999), but appear to be insensitive to such subtleties in timbre dissimilarity ratings. At the same time, there are experimental situations in which listeners appear to be sensitive to certain mechanical properties of sounding objects carried by acoustic cues when making dissimilarity ratings, but will use only a subset of those for source material identification, i.e., the ones that are most reliable for the specific experimental task (McAdams et al. 2010).

Laying out a general framework for the study of psychological similarity, Tversky (1977) noted that different perceptual features may contribute to different tasks: "Our total data base concerning a particular object (e.g., a person, a country, or a piece of furniture) is generally rich in content and complex in form. It includes appearance, function, relation to other objects, and any other property of the object that can be deduced from our general knowledge of the world. When faced with a particular task (e.g., identification or similarity assessment) we extract and compile from our data base a limited list of relevant features on the basis of which we perform the required task" (p. 329). Similar to furniture, musical timbre varies along a variety of features that enable identification and discrimination. Yet, if confronted with the task of similarity assessment, subjects could rely on the perceptually most salient properties according to which musical tones can be most easily compared with one another. Spectral center of gravity and attack time may be dimensions well suited for such comparative tasks. For absolute identification or classification, however, it would be a non-optimal strategy not to make use of all other available features that reduce ambiguity between stimuli. This may be particularly important for a perceptual parameter such as timbre, for which the *'perceptual consequences of the multiplicity of cues created by the sound production process are varied. […] Any single cue will provide some level of identification performance, and combinations of cues usually will produce better performance than a single one. Moreover, the effectiveness of any cue will vary across contexts'* (Handel 1995, p. 433).

3 Sound Source Categories in Timbre Perception

This section discusses results from a study that explored the ways in which sound source categories of familiar acoustic tones affect timbre dissimilarity ratings (Siedenburg et al. 2016b). The described experiments compared the ratings of tones from orchestral instruments with those of unfamiliar, "unheard" synthetic counterparts that were selected via a first experiment as to less readily evoke familiar instrument families or sound source categories.

4 Stimuli and Sound Transformation

The stimuli used in the experiment consisted of 14 recordings of single tones from common musical instruments and a set of transformed sounds. The acoustic tones included common instruments from the Western orchestra (bass clarinet (BCL), bassoon (BSN), flute (FLT), harpsichord (HCD), horn (HRN), harp (HRP), marimba (MBA), piano (PNO), trumpet (TRP), bowed violoncello (VCE), violoncello pizzicato (VCP), vibraphone (VIB), bowed violin (VLI), and violin pizzicato (VLP)). Additionally, 70 transformed tones were derived by means of a digital signal transformation from the 14 acoustic tones. All sounds had a fundamental frequency of 311 Hz (E♭4) and a duration of 500 ms.

For the synthetic transformations, the spectro-temporal signal envelopes and temporal fine structures of two recorded sounds were purposely mismatched, a procedure that has been shown to yield "chimæric" perceptual properties in speech synthesis (cf., Smith et al. 2002). Spectrotemporal amplitude envelopes (ENV) were extracted by filtering the signals with a 24-channel Gammatone-filterband (Patterson et al. 1992) and subsequent half-wave rectification and low-pass filtering of each subband, yielding time-varying estimates of the amplitude envelope of each Gammatone subband. The signals' temporal fine structures (TFS) were obtained by dividing each subband by its respective ENV. By mismatching TFSs and ENVs from the 14 original sounds, a large set of novel sounds was generated. 70 sounds were selected from this set to be used in the first experiment of the study (see the original publication for more details on the sound synthesis and selection process).

5 Instrument Identification

The first experiment tested how well musicians could identify recordings and transformations and how they rated the familiarity of stimuli. 15 musicians listened to single presentations of these tones and chose an identifier from a list of eight possible options. The list consisted of six musical instrument names that included the correct label and five randomly chosen labels from the remaining set or the labels of the two original sounds involved in the transformation process. For instance, if a transformation was derived from the TFS of a piano and the envelope of a violin, then both instrument names, piano and violin, would be part of the list. The list further contained the two options *unidentifiable* and *identifiable but not contained in list*. If the latter option was selected, participants were asked to provide a short description in writing. They then heard the sound a second time and were asked to rate their perceived familiarity on an analog-categorical scale (not discussed in detail here).

By construction, correct responses for the identification task only existed for the recordings. Here, correct identification rates ranged from 0.46 (BCL and BSN) to 1.0 (TRP). The mean identification rate for all 14 recordings was 0.73 (SD = 0.180) with chance baseline equal to 1/8 = 0.125. The bass-clarinet (BCL) was the only recording for which an alternative category, *unidentifiable*, was selected most often (0.53). This means, musicians were able to identify instruments on the basis of tones of 500 ms duration from a single presentation. At the same time, the data exhibited considerable variance in the percentage of correct identifications across different instruments (ranging from 46 to 100 %), which parallels the divergent estimates of identification accuracy in the literature (Srinivasan et al. 2002).

From the remaining 70 transformations, 29 were most often identified as other musical instruments (i.e., the category that was selected by the majority of subjects) with average selection rates of 0.47 (SD = 0.12). From these 29 transformations, the category chosen most often for 23 sounds was an instrument that was involved in the transformation either with its ENV or TFS. This may highlight the capabilities of musicians to associate sounds with their sources, although chimæras were quite dissimilar to any of their (ENV or TFS) generators.

Thirteen transformations were most often selected as *unidentifiable* with stimulus-wise mean selection rates of 0.55 (SD = 0.21). Twenty-eight transformations were selected as *identifiable, but not in the list* with mean selection rates of

0.55 (SD = 0.16). If subjects had selected the latter category, they were asked to briefly describe what they had heard in a written response. Three different types of responses appeared most often here: 41 % of these responses mentioned single orchestral instruments; 37 % mentioned a mix of multiple instruments (e.g. "piano and trombone in unison"); 16 % mentioned electronic means of sound synthesis; 6 % of responses were hard to categorize (e.g., participant 7: "Ahh yes patch 87: plucking a frog."). The fact that around twice as many transformations were perceived to be *identifiable but not contained in the list* as opposed to simply *unidentifiable* highlights that listeners tend to frequently commit false positives, i.e., tend to misattribute sources rather than abstaining from sound source identification.

6 Modeling Dissimilarity Ratings

In a second and third experiment, we collected pairwise dissimilarity ratings for a set of 14 recorded acoustic tones, and those 14 transformations rated as least familiar in the previously described experiment, as well as a mixed set that contained seven recordings and seven transformations. Several characteristics of the data highlighted the pertinent role of sound source categories in dissimilarity ratings.

A first line of evidence concerned the asymmetry of ratings. The timbral dissimilarity of sound A followed by sound B must in principle not be equal to the reverse order of presentation. In shorthand, d(A, B)/ = d(B, A). The difference d(A, B) − d(B, A) then measures the asymmetry of the pair (A, B). If there was a subset of ratings such that its average pairwise difference deviated from zero, this would indicate asymmetric rating tendencies for this subset. This is what was observed in the data: Distributions of differences did not deviate from zero for the sets of recordings and transformations. However, there were significant deviations from symmetry for the subset of across-stimulus-type comparisons: dissimilarities of recordings followed by transformations were smaller than in the reverse order of presentation. In short, d(rec, trans) < d(trans, rec). This finding was confirmed in an altered experimental design with a different group of subjects. No simple acoustic effect can plausibly account for this effect of directionality. For that reason, these findings suggest that there are situations in which category membership plays an important role in dissimilarity ratings and participants appear to take into account more than just acoustical cues. It may be that the first sound of the pair is taken as a reference, according to which the second is compared. In that case, the dissimilarity of the prototype (the acoustic sound) to the variant (the transformation) is larger than the dissimilarity of the variant to the prototype. In this sense, familiar instruments may here act as "timbral hubs" that "twist" the topology of a perceptual dissimilarity space. Such an interpretation would be coherent with the classic work of Tversky (1977), laying out a psychological framework for the treatment of psychological dissimilarity that not necessarily always coheres with the metric axioms. Among other things, Tversky argued that the symmetry axiom is particularly problematic in many cases (e.g., "North-Korea is more similar to Red China than Red China is to North-Korea.").

A second line of evidence stems from regression modeling of the rating data from the set of recordings. In these acoustic tones, instrument category and acoustic qualities of course coincide to a large extent (cf., Giordano and McAdams 2010), although not completely. Take the difference between the piano and the harpsichord or the vibraphone and marimba; the members of both pairs may feature quite different acoustic qualities although they belong to the same instrument family: keyboard and mallet instruments, respectively. Using an exploratory regression analysis, we thus set out to quantify the degree to which musicians relied upon acoustic and categorical types of stimulus representations in their timbre dissimilarity ratings.

A large set of 34 scalar audio descriptors from the Timbre Toolbox (Peeters et al. 2011) was used, encoding spectral, temporal, and spectro-temporal properties of the acoustic signal. These were employed in conjunction with a partial least-squares regression model (PLSR). The latter is particularly suited to deal with collinear predictors by creating latent components that act as regressors in the final model (Wold et al. 2001). In contrast to principal component analysis followed by multivariate linear regression, PLSR optimizes the latent decomposition such that the covariance with the dependent variable of interest is maximized. Dependent variables were the corresponding dissimilarity ratings averaged across subjects and the order of presentation. Predictor variables were formed by stacking absolute differences of descriptor values for the respective pairs, such that there were 34 independent variables of size 91 ($=14 \cdot 13/2$).

Figure 1 (panel A) displays the predicted and observed dissimilarities for the model based on the acoustic descriptors. Although there is generally a good fit, the plot highlights two outliers (annotated as 1 and 2 in the plot). Point 1 stems from the marimba-vibraphone pair for which the acoustic model overestimated the dissimilarity rating, and point 2 from the harp-trumpet pair, for which ratings were underestimated on acoustic grounds alone. This once more suggests that listeners

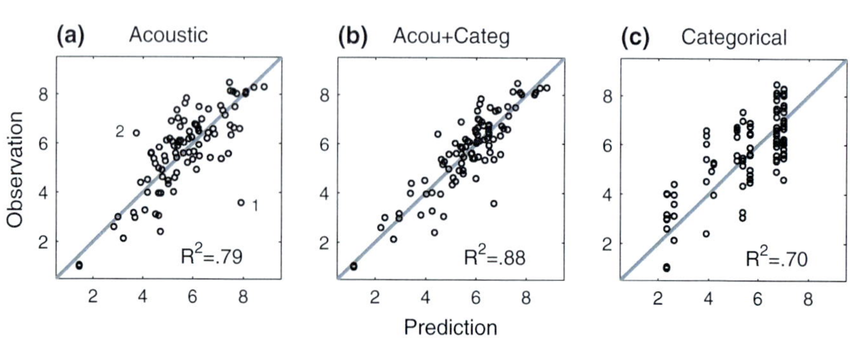

Fig. 1 Mean pairwise dissimilarity ratings for the acoustic tones (observations; y *axis*) and predictions based upon acoustic descriptors **a**, audio and categorical predictors combined **b**, and category membership of the instruments **c**. Data points 1 and 2 in the *left* panel are discussed in the text. Graphic reproduced from Siedenburg et al. (2016b)

not solely based their ratings on acoustic information, but also took into account categorical information such as instrument families: Because the marimba and the vibraphone are both percussion instruments, they were rated as more similar than their acoustic dissimilarity would have predicted. The reverse may have been at play for the harp and the trumpet, members of the string and brass families, respectively.

In order to provide a quantitative footing for this intuition, we considered four additional categorical predictors of dissimilarity related to instrument families and the mechanics of sound production. These categories were not based on continuous acoustic descriptions of the audio signal, but may have been inferred perceptually and thus could have influenced the dissimilarity ratings. We thus updated the model by including four novel predictors which encoded (i) instrument family membership (woodwind, brass, keyboard, string, percussion), (ii) source resonator (string, air column, bar), (iii) general source excitation (continuous, impulsive), and (iv) specific source excitation (blown, bowed, struck, pluck). Each predictor (i–iv) was encoded as a binary variable, i.e., yielding a dissimilarity of 0 if a pair of sounds shared the respective feature, and 1 otherwise. For instance, categorical predictor (ii) would yield a 0 for the pair marimba-vibraphone, because the resonators of both instruments are bars. Figure 1 (panel B) displays the resulting prediction of the model that includes both acoustic and categorical predictors. The inclusion of categorical predictors significantly improved the correlation between predicted and observed values, and the resulting prediction shared almost 90 % of the variance with the observed dissimilarity rating data (compared to an $R^2 = 0.79$ for the solely acoustic model). It is also visible that the inclusion of the categorical predictors improved the quality of the prediction for the two poorly predicted pairs discussed above (marimba-vibraphone, harp-trumpet). Finally, the relevance of categorical representations is emphasized by the fact that the four categorical predictors alone already shared 70 % of the variance with the observed rating data (panel C).

Finally, it is to be mentioned that a different series of experiment (not further discussed here) used the same set of 14 recordings and 14 transformations in a short-term recognition task. In every trial participants were required to match a probe sound to a preceding sequence of sounds (Siedenburg and McAdams 2016). Results suggested that timbres from familiar acoustic instruments were easier to match compared to the timbres of novel synthetic tones. Because categorical and semantic representations may be more readily available for the acoustic recordings, yielding a more multifaceted and robust cognitive representation, these sounds may be more easily held in short-term memory.

7 Discussion

At the beginning of this article, I suggested that traditional musical instruments yield whole palettes of different timbres, depending on the multifaceted ways in which they are excited which also covary with pitch and dynamics. In order to identify instruments in complex musical contexts, it seems likely that the auditory system makes use of this structured variety of acoustic cues. Furthermore, both auditory sensory representations and learned sound source categories appear to play a role in timbre dissimilarity ratings. The latter hypothesis was supported by the discussion of findings on asymmetric dissimilarity relations which cannot be explained on mere acoustic grounds, as well as the beneficial contribution of categorical predictors in a dissimilarity model (Siedenburg et al. 2016b). Results from a short-term memory task pointed in a similar direction (Siedenburg and McAdams 2016). In effect, the listening brain represents, simultaneously, the "sound" of an instrument and the "idea" of how that sound was generated. It seems likely that this duality—here derived from the perspective of perception and cognition—constitutes an important factor in the design of new digital musical instruments (DMIs).

A tentative lesson from this perspective could be that novel DMIs should be created in ways that afford for the rapid association of the gestural means of sound generation, the visual appearance of the instrumental interface, and the timbral features of its sound synthesis design, even if the latter is in principle unconstrained. This essentially calls for sound design that only uses a highly constrained portion of the timbral possibilities available in the digital realm—aiming for a well-defined "timbral genus"—but at the same time calls for great variability on a more fine-grained scale of timbral detail. This may help to provide a structured array of auditory cues that listeners can exploit as a means to learn to associate a specific timbral genus with the visual and semantic information provided in musical performance. Such a learned association may yield the basis for the perception of *causality* between gesture and timbral microvariation, perhaps a necessary condition for acoustically grounded expressivity that allows performers of DMIs to sculpt their sounds and communicate their sonic intentions.

At the same time, one should keep in mind that these considerations represent a traditional approach to DMI design which takes the causality inherent to acoustic instruments as a strict model. Although we know that this is a good starting point in order to make novel DMIs accessible and apprehensible, there may also be aesthetic potential in the opposite, "anti-causal" approach that might pursue the attempt to partially trick the tight coupling of acoustics and cognition.

Acknowledgments Portions of this article are based on two previous publications (Siedenburg et al. 2016a; Siedenburg et al. 2016b). I would like to thank my co-authors Stephen McAdams, Ichiro Fujinaga, and Kiray Jones-Mollerup for their help.

References

Agus, T. R., Suied, C., Thorpe, S. J., & Pressnitzer, D. (2012). Fast recognition of musical sounds based on timbre. *The Journal of the Acoustical Society of America, 131*(5), 4124–4133.

Agus, T. R., Thorpe, S. J., & Pressnitzer, D. (2010). Rapid formation of robust auditory memories: Insights from noise. *Neuron, 66*, 610–618.

Caclin, A., McAdams, S., Smith, B. K., & Winsberg, S. (2005). Acoustic correlates of timbre space dimensions: A confirmatory study using synthetic tones. *The Journal of the Acoustical Society of America, 118*(1), 471–482.

Elliott, T., Hamilton, L., & Theunissen, F. (2013). Acoustic structure of the five perceptual dimensions of timbre in orchestral instrument tones. *The Journal of the Acoustical Society of America, 133*(1), 389–404.

Giordano, B. L., & McAdams, S. (2010). Sound source mechanics and musical timbre perception: Evidence from previous studies. *Music Perception, 28*(2), 155–168.

Grey, J. M. (1975). *An exploration of musical timbre* (Unpublished doctoral dissertation). CCRMA, Stanford University.

Hajda, J. M., Kendall, R. A., Carterette, E. C., & Harshberger, M. L. (1997). Methodological issues in timbre research. In I. Deliège & J. Sloboda (Eds.), *Perception and cognition of music* (pp. 253–306). New York, NY: Psychology Press.

Halpern, A. R., Zatorre, R. J., Bouffard, M., & Johnson, J. A. (2004). Behavioral and neural correlates of perceived and imagined musical timbre. *Neuropsychologia, 42*(9), 1281–1292.

Handel, S. (1995). Timbre perception and auditory object identification. In B. C. Moore (Ed.), *Hearing* (Vol. 2, pp. 425–461). San Diego, CA: Academic Press.

Handel, S., & Erickson, M. L. (2004). Sound source identification: The possible role of timbre transformations. *Music Perception, 21*(4), 587–610.

Huron, D. (2006). *Sweet anticipation: Music and the psychology of expectation*. Cambridge, MA: MIT Press.

Krumhansl, C. L. (1989). Why is musical timbre so hard to understand? In S. Nielzén & O. Olsson (Eds.), *Structure and perception of electroacoustic sound and music* (Vol. 846, pp. 43–53). Amsterdam, The Netherlands: Excerpta Medica.

Krumhansl, C. L. (1990). *Cognitive foundations of musical pitch*. Oxford, UK: Oxford University Press.

Kruskal, J. B. (1964). Multidimensional scaling by optimizing goodness of fit to a nonmetric hypothesis. *Psychometrika, 29*(1), 1–27.

Lakatos, S. (2000). A common perceptual space for harmonic and percussive timbres. *Perception and Psychophysics, 62*(7), 1426–1439.

Lindau, A., Erbes, V., Lepa, S., Maempel, H.-J., Brinkman, F., & Weinzierl, S. (2014). A spatial audio quality inventory (SAQI). *Acta Acustica United with Acustica, 100*(5), 984–994.

Marozeau, J., de Cheveigné, A., McAdams, S., & Winsberg, S. (2003). The dependency of timbre on fundamental frequency. *The Journal of the Acoustical Society of America, 114*(5), 2946–2957.

McAdams, S. (1993). Recognition of sound sources and events. In S. McAdams & E. Bigand (Eds.), *Thinking in sound: The cognitive psychology of human audition* (pp. 146–198). Oxford, UK: Oxford University Press.

McAdams, S. (2013). Musical timbre perception. In D. Deutsch (Ed.), *The psychology of music* (3rd ed., pp. 35–67). San Diego, CA: Academic Press.

McAdams, S., Beauchamp, J. W., & Meneguzzi, S. (1999). Discrimination of musical instrument sounds resynthesized with simplified spectrotemporal parameters. *The Journal of the Acoustical Society of America, 105*(2), 882–897.

McAdams, S., Roussarie, V., Chaigne, A., & Giordano, B. L. (2010). The psychomechanics of simulated sound sources: Material properties of impacted thin plates. *The Journal of the Acoustical Society of America, 128*(3), 1401–1413.

McAdams, S., Winsberg, S., Donnadieu, S., De Soete, G., & Krimphoff, J. (1995). Perceptual scaling of synthesized musical timbres: Common dimensions, specificities, and latent subject classes. *Psychological Research, 58*(3), 177–192.

Pantev, C., Roberts, L. E., Schulz, M., Engelien, A., & Ross, B. (2001). Timbre-specific enhancement of auditory cortical representations in musicians. *Neuro Report, 12*(1), 169–174.

Patterson, R. D., Robinson, K., Holdsworth, J., McKeown, D., Zhang, C., & Allerhand, M. (1992). Complex sounds and auditory images. *Auditory Physiology and Perception, 83*, 429–446.

Peeters, G., Giordano, B. L., Susini, P., Misdariis, N., & McAdams, S. (2011). The Timbre Toolbox: Extracting audio descriptors from musical signals. *The Journal of the Acoustical Society of America, 130*(5), 2902–2916.

Plomp, R. (1970). Timbre as a multidimensional attribute of complex tones. In R. Plomp & G. F. Smoorenburg (Eds.), *Frequency analysis and periodicity detection in hearing* (pp. 397–414). Leiden, The Netherlands: Suithoff.

Shahin, A. J., Roberts, L. E., Chau, W., Trainor, L. J., & Miller, L. M. (2008). Music training leads to the development of timbre-specific gamma band activity. *Neuroimage, 41*(1), 113–122.

Shamma, S., & Fritz, J. (2014). Adaptive auditory computations. *Current Opinion in Neurobiology, 25*, 164–168.

Siedenburg, K., Fujinaga, I., & McAdams, S. (2016a). A comparison of approaches to timbre descriptors in music information retrieval and music psychology. *Journal of New Music Research* (in press).

Siedenburg, K., Jones-Mollerup, K., & McAdams, S. (2016b). Acoustic and categorical dissimilarity of musical timbre: Evidence from asymmetries between acoustic and chimeric sounds. *Frontiers in Psychology, 6*(1977). doi:10.3389/fpsyg.2015.01977

Siedenburg, K., & McAdams, S. (2016). The role of long-term familiarity and attentional maintenance in auditory short-term memory for timbre. *Memory*, published online. doi:10.1080/09658211.2016.1197945.

Smith, Z. M., Delgutte, B., & Oxenham, A. J. (2002). Chimaeric sounds reveal dichotomies in auditory perception. *Nature, 416*(6876), 87–90.

Srinivasan, A., Sullivan, D., & Fujinaga, I. (2002). Recognition of isolated instrument tones by conservatory students. In *Proceedings of the 2002 International Conference on Music Perception and Cognition, Sydney* (pp. 17–21), July 17–21.

Strait, D. L., Chan, K., Ashley, R., & Kraus, N. (2012). Specialization among the specialized: Auditory brainstem function is tuned into timbre. *Cortex, 48*(3), 360–362.

Tversky, A. (1977). Features of similarity. *Psychological Review, 84*(4), 327–352.

Wessel, D. L. (1973). Psychoacoustics and music: A report from Michigan State University. *PACE: Bulletin of the Computer Arts Society, 30*, 1–2.

Winsberg, S., & De Soete, G. (1993). A latent class approach to fitting the weighted Euclidean model. *CLASCAL. Psychometrika, 58*(2), 315–330.

Wold, S., Sjöström, M., & Eriksson, L. (2001). PLS-regression: A basic tool of chemometrics. *Chemometrics and Intelligent Laboratory Systems, 58*(2), 109–130.

What If Your Instrument Is Invisible?

Dafna Naphtali

Abstract As an electronic musician I am largely occupied with capturing and manipulation of sound in real time—specifically the sound of instruments being played by other musicians. Also being a singer, I've found that both of my instruments are often perceived as "invisible". This article discusses various strategies I developed, over a number of years, in order to "play" sound manipulations in musically reactive ways, to create a live sound-processing "instrument". Problems were encountered in explaining what I do to other musicians, audience, and audio engineers about what I do, technically and musically. These difficulties caused me to develop specific ways to address the aesthetic issues of live sound-processing, and to better incorporate my body into performance, both of which ultimately helped alleviate the invisibility problem and make better music.

1 Motivation

As an electronic musician and vocalist, I am largely occupied with the capture and manipulation of sound in real time—the sounds of my voice and of instruments being played by other musicians. Often, both of these instruments are "invisible" to the audience. Over 20 years I have used terms like "real-time live sound-processing", "audio machinations", or "{kaleid-o-phone}" to describe what I do. I grappled with explaining to other musicians, my audience and audio engineers what it is, exactly, that I am doing on stage. I found the technical explanation to be difficult enough, but explaining myself aesthetically and musically has been as interesting and challenging a task. With each new project or collaboration, new issues of musical and acoustic aesthetics arose, which I studied, played with, and sometimes held onto, as part of my growing library and palette for performance and communication. In this text, I will explain why my audio setup feels so consistently to be an *instrument* rather than a pile of audio effects.

D. Naphtali (✉)
New York, USA
e-mail: post@dafna.info

T. Bovermann et al. (eds.), *Musical Instruments in the 21st Century*,
DOI 10.1007/978-981-10-2951-6_26

2 Guidelines

In the mid-1990s, I wanted to experiment and find ways to play sound manipulations in musically reactive ways, as an "instrument", and especially in improvised contexts.

I first started, by dragging a Macintosh Plus computer with MIDI-controllable effect units to clubs and venues—and gigging meant carrying a whole lot of equipment. I did not know what to call what I was doing with all the sound processing, but I knew I wanted to do it, and was compelled by what I was discovering in this unchartered territory.

I was lucky to start with a group of very open-minded instrumentalists: with drums, bass, saxophone, a multi-instrumentalist (brass, reeds), a live video artist, and me, singing (in a group sometimes called "Collison" or "Free Jazz Video Collision"). I put microphones on everyone. I routed the sound I brought in through a mixer into an effects unit controlled by a Max[1] patch I wrote. In this way, I created tiny and constantly changing loops, wild mood swings with comb filters, pitch-shifting madness, accompanied by a bit of my singing in between.

Back then, although there were a few instrumentalists in New York bringing computer-based electronics into small club environments, the aural transformations I was experimenting with were more likely heard on the concert stage of a university, in academic computer music.

The musicians I played with both appreciated and supported what I was doing, and reacted very musically to my electronics, experimenting along with me. After our shows, however, people would often say "… fabulous, but you really should do more!"; they thought I was only *doing something* when I was singing. My sound manipulations were not attributed to me because the audience could neither visually identify these sounds as coming from me, nor identify them aurally as my contribution. I realized that I created a complex combination of sounds, that were either being ascribed to the other musicians (who were my source material), or that was so strange as to be unidentifiable, and certainly not created by the singer.

Through this experience, and supported by fellow musicians, I came to discover that I wanted and needed to create a separate sonic identity with my live processing, one that would be distinct from the other players' sounds. I would have to synthesize these ideas myself from my own musical influences and technical background.

Over time, I came up with two guidelines to help me find a sonic identity and tackle my invisibility problem. They are primarily about respecting the sound and independence of the other musicians and help me to make good music:

[1]I started using Max in 1992 (v2.0 Opcode) at a time when commercial/affordable versions of Max could not yet do live signal processing. Real-time signal processing using MSP was added in 1997. Max/MSP is now developed and maintained by Cycling '74.

1. Never interfere with a musician's own musical sound, rhythm or timbre.
2. Be musically identifiable to both co-players and the audience.

The challenge in finding ways to be musically identifiable to an audience, however was somewhat more complex. I found that to *show* what I was doing in a non-ostentatious, and genuinely musical way was only possible if I followed my guidelines. But in addition, the organization of my performance setup, my *instrument*, needed to follow a third guideline:

3. The performance setup must incorporate my body and allow for physical interaction between the technology and myself.

The setup would have to be flexible and physically responsive, just like the acoustic instruments of the musicians.[2]

3 My Practice

My audio system is meant for performance. In this sense it is reliable and consistent enough for me to think of it as an *instrument*. It can however also be wildly unpredictable, even chaotic, in extremely fun and musical ways: at times I work very close to feedback and find resonances with filters, pitch-shifters and delays. I create complex musical textures with these effects, creating responses and statements that are intrinsically related to, and simultaneously derived from, the music around me. My work, as such, is extremely subject to the acoustics of the performance space. Therefore, the sonic character of my contribution is subject to the physical realities of temperature, humidity, and acoustics of the performance space.[3] My source sounds are unpredictable, since they are being created by other people. My processed versions of these sounds, which I try to harness/control into being something I can "play", are always subject to environmental conditions. Due to these ephemeral qualities, my live performances require long set up times and long sound checks.

No matter how well rehearsed, each and every performance, even those with the same instrumentation, music, and performance space, will be inescapably different. As a performer, I feel as if I were trying to build a small fire on stage, fanning it all

[2]Many of the following ideas were presented as a workshop "Live sound-processing strategies" (at Harvestworks in New York, May 2012), and as a six-day intensive course "Aesthetics of Live Sound Processing" (at UniArts Sound Art Summer Academy in Helsinki, August 2014). The demonstration performances at the 2012 workshop with were with Robert Dick (flute) and Satoshi Takeishi (percussion).

[3]Invited by pianist Kathleen Supové to process her playing "Phrygian Gates" by John Adams. I used many of the techniques described in this article, including feedback, and discovered radical variations in my sound in each of ten performances we did in the same venue! The patches and processing were nearly identical in each case, were well rehearsed. Only the weather changed, (and the people in the audience) and therefore the way in which my feedback processes worked in that space.

along, building it and keeping it going. But unlike an actual fire, mine is also an interactive system that I can control rhythmically and quickly. This has been the key in creating my instrument.

4 Invisibility, Performance Practice and Gender Trouble[4]

At first I was very occupied with trying to figure out what to do technically: to avoid sounding like a simple "effect", and to fit into an ensemble sonically using my new ideas. There was no time to worry if anyone else could actually understand what I was doing.

As an example—imagine seeing a saxophonist perform a rhythmic figure that is looped and filtered, but one that is controlled by someone standing on the other side of the stage (who is also singing). By seeing and hearing, it is easy to identify what the saxophone player does, but there is no visual cue or connection with the sound processor/singer. Due to this visual incongruity, the sounds I created were often attributed to the sax player, to other musicians holding instruments, or even to the sound engineer in the back of the room—but were not attributed to me.

I did not worry about this lack of audience understanding (yet), because I was too busy explaining myself to the sound engineers at the clubs where I played. Without going into great detail here about the invisibility that still existed for a woman working with technology at that time, let me simply state that the combination of an unusual technical setup and my gender led to the absolute necessity for me to allocate an extra 15–20 min to *every* setup and load-in, for every gig. It was necessary to plan for enough time for a long "chat" with the sound engineer; to explain what I was doing in detail, even to justify my methods. I had to convince the engineer, usually male, that I did actually know what I was doing even if it was unorthodox, and that I did know how to handle the equipment, in spite of my gender. I knew from experience that without this chat, as soon as the engineer heard feedback (even if intentional and part of my sound), he would turn it down. He would likely mix my electronic sounds so as to be quieter than all the other instruments, assuming my contribution to be merely sound effects (and singing).

What a relief, that nowadays I rarely need to have these conversations. It is often not even necessary to explain my live sound-processing in detail at all, as it is more commonplace. Nor is it necessary to prove myself as competent, as there are many more women working with music technology, both as performers and engineers. These days, a musician performing with a laptop is rather ordinary, and many musicians and engineers are familiar with the idea and practice of live sound-processing for a variety of musical styles, even pop music, and in all kind of

---

[4]In writing this article, I struggled with whether I should first be describing my technical solutions or my aesthetic ones. I have opted for the chronological explanation since it was difficult to solve my aesthetic problems without the technical ability to do so. Over time I have come to understand these as distinct problems, yet more intertwined than expected.

venues. These days, when I explain my practice to an audio engineer, I hear a lot more "Oh yes, someone was in here last night doing that".

With more experience, I saw the need, and crafted a careful approach for sound processing, one that is clearer to my fellow musicians. Once I had refined the way I use my sounds, I developed a "body-oriented" live electroacoustics practice. My hope is that this approach enables even a more general, uninitiated audience, to understand that in my performances, the "singer" is also an electronic musician, shaping her instrument out of sound.

5 Making the Invisible More Understood: Using My Body

My initial musical training was as a jazz guitarist and, later I also studied classical voice. My teachers were very strict about wanting me to keep perfectly still while playing or singing, so as to focus my energy entirely into my instrument. In recent years, I have had to undo this training, to some degree, for the simple reason that electronic instruments do not generally involve acoustic vibration.

These missing vibrations are what would normally allow me to feel physically connected to my (acoustic) instrument or voice, even when I am standing perfectly still. With electronic instruments, the physical gestures I make take the role of the vibrations in creating this connection. Using gesture controllers helps me to connect with my audio, in the moment, and not just passively as a listener, post facto, through my ears. Moving and connecting my body to sound/music/rhythm is of paramount importance. This is how I have turned what I do into a performable instrument.

I must move, and I must touch something while I'm playing. I shape the music using actions that require physical effort corresponding to the sound. The best controllers (ones that I can either hit hard enough or that are sensitive enough to allow for subtle control), help me to use my entire body, and lessen the "invisible instrument" problem.

Over the years, I've performed and experimented with many different "controllers-of-the-moment" and interfaces as part of my instrument. At first, I did wonder whether the use of controllers—Wii Remote, MIDI fader boxes, iOS devices, LEAP motion—was intended more to facilitate my playing or rather to promote audience engagement. If these controllers were for my audience's benefit, then was this to "show them" that I am performing? That felt awkward to me. If the controllers were meant for me to make my performing easier, then my challenge would be to find ways to make each physical gesture analogous to the musical gestures, avoiding overuse of simple correlations (such as moving hands up and down to control volume or pitch.)

In recent years I started getting positive feedback about this from fellow musicians—that I do move while performing and actively engage my body. One collaborator[5] wrote I bring "the energy and physicality of free jazz to experimental

[5]Pianist/composer Gordon Beeferman, personal communication 19 January, 2016.

electronic music" which I took as a high compliment. It has turned out that all the dance classes I took growing up, and Dalcroze Eurhythmics taken as an undergraduate, have all informed my electronic music practice.

Looking back to the mid-90s—to when experimental video artist Kristin Lucas joined many of our performances in "Collision", it is clear that her practice at that time served both as a contrast and counterpart to my own performance identity in the group. Lucas performed using a toy guitar as a video controller, standing on stage amongst the musicians; cameras attached to her and to the other musical performers. With both a projector and a camera mounted on her bicycle helmet, Lukas projected her intriguing images, created in the moment, onto the the musicians/instruments and the entire room, in the same sort of appropriation of the immediate environment that I was experimenting with. (She collected and processed images, I collected and processed sounds). Yet, Lucas' wonderfully prescient "video instrument", one normally not visible to the audience, was intended to be seen, and perceived as connected to the music, while my "musical/audio" instrument remained "invisible", unless I sang and processed my own voice.

It would be simple to solve the invisibility problem, if all that was needed was to use the right gestural controller, or perform in a more outwardly visible or extroverted way as Lucas did. However, it turned out that much more than a visual identification was necessary—my instrument had to be *aurally* identifiable too. Hence my approach is more complex and involves an aesthetic approach to listening and acoustics, as well as active engagement in my body.

6 Sound Decisions

This aural identification, can be broken down into several simple parameters of electroacoustic music, and specific approaches I have taken using these parameters specific to playing with live sound-processing.

6.1 Envelope/Dynamics

Being able to shape the volume and subtleties of my sound is an important part of my musical expression and my instrument, as it is for all other instruments. Shaping the envelope and dynamics of my live-processed sounds is central to my performing, and it is the first thing I teach my students. *If I cannot control volume, I cannot do anything else described in this article. I use volume pedals and other interfaces, as well as compressor/limiters for constant and close control over volume and dynamics.*

6.2 *Rhythm*

I cannot express strongly enough how important control over rhythm is to my entire concept. It is what makes my system feel like an instrument. My main modes of expression are expressed via timbre and rhythm. Melody and direct expression of pitch using electronics are slightly less important to me, though the presence of pitches is never to be ignored. I choose rhythm as my common ground with other musicians. It is my best method to interact with them.

Nearly every part of my system allows me to create and change rhythms by interactively altering delay times, or by tapping in the desired pulse. Tapping in pulses has helped me put my body into my performance, and therefore helped me with my invisible instrument problem. We need our bodies involved!

I am strongly attracted to polyrhythms, which is not surprising, my family is Greek, so there was lots of dancing in odd time signatures growing up. Because it is so prevalent in my music, I implemented a mechanism that allows me to tap delay times and rhythms that are complexly related to what is happening in the ensemble at that moment.

Pianist Borah Bergman once explained a system he though I could use for training myself to perform complex rhythms, and I created a Max patch to facilitate this. Later I implemented this polyrhythmic metronome, to control the movement between any two presets quickly, creating polyrhythmic electroacoustics. Other rhythmic control sources I have used included Morse Code as rhythm, algorithmic processes, and a recreation of features influenced by North Indian Classical Tala.[6]

7 Delays and Filtering

As stated in my first guideline, I avoid interfering with another musician's own musical sound, rhythm or timbre. If I were to play in a purely "transformative" manner (Rowe 1993), or merely mirror what they do, I worry that I might interrupt their own thought process about rhythm and timbre. Instead of "piggy backing" my sound on theirs, I'd rather aurally transform and reflect on their sound-making, creating a statement of my own, based on their sound.

To achieve this, my primary tools are delays and filters. Filters change timbre. Yet when my filtered version of a performer's sound is mixed with their own direct live sound, my contribution might not be heard as an entirely separate gesture. The

[6]I have learned some patterns related to North Indian Classical Tala through self-study and private study with other musicians/collaborators, some quite accomplished, who were willing to help me find ways to use them in my work. These patterns, which I use in my live sound-processing work are merely reflections of these encounters and collaborations. For example of the patterns, see those listed at https://www.ancient-future.com/theka.html. In my programming, I assigned the various syllables, each to a particular preset in my patches and the result reflects the Tala-inspired patterns in the live sound processing.

filtering that I do might be heard merely as an effect. To avoid this, I take extra care when changing certain parameters. It all comes down to filters, delays and time. They are all interrelated.

By modulating delay time, it is possible control everything from timbre to rhythm to musical form. In Karlheinz Stockhausen's lectures "Four Criteria of Electronic Music" (1972), he described a continuum of timbre-rhythm-form—inherent perceptual qualities determined by the rate of notes: from very fast pulses to very, very slow rhythms. Perceptually, rhythmic pulses faster than 16-20 Hz become pitches of various timbres. Rhythmic pulses slower than this range might be heard simply as relating to traditional rhythms and tempo. If pulses/rhythms are played at slowest end of this continuum, they are perceived as pertaining to musical form.[7]

By modulating the amount of feedback in a delay line, I am also able to control sustain, and overlap of sound. In many cases, this also allows me to alter timbre, in that the overlap of many copies of a signal, effectively cancels out or emphasizes timbre-rhythm-form frequencies, creating something akin to a comb filter. The delay times and overlap relate directly to the timbre-rhythm-form continuum described by Stockhausen.

With interpolating delay lines, I am able to adjust the length of the delays in real-time without clicks or distortion. A welcome side effect is a Doppler effect, resulting in inadvertent pitch changes when increasing or decreasing the delay time. This is something fun and useful in forging a new sound out of what is being played by another musician.

7.1 Delay Time/Length/Interaction

The length of a delay in relation to an incoming sound is the key feature to be considered in order predict what will happen in a "musicianly" way.

I use long delays (i.e., longer than the captured phrase) to create rhythmic structures. This use of delay/repetition is a time-honored musical gesture, employed successfully for centuries in many kinds of music and heard today rather ubiquitously in loop-based electronic music.

To create a separate voice for myself, I must undermine the expectations set up by this structure. I could change the delays slowly over time as a long-form musical gesture, but I prefer to change my delays in a clearer musical gesture, in response to what I hear from the ensemble. Working with Doppler, with feedback levels, with various delay times and other effects are all effective to this end. But it also works to simply change delay parameters to create overlapping sounds or "not-overlapping" sounds.

[7]According to him, this happens with periods longer than 8 s between the single attacks.

If a delay is shorter than the length of its input phrase, it results in overlapping copies of the sound. This overlap causes filtering and resonance, especially when the delay effect has a high feedback level. The feedback creates even more overlap. By shortening a delay time to be shorter than the sound I'm delaying, I create instant resonance, or with slightly longer delay times, resonant textural density.

With delays of less than 30 ms (approximately), any incoming sound with sharp transients will turn into pitches that I can use as sound gestures. The frequency of the sounding pitch is inversely proportional to the length of my delay.[8] Conversely, sustained sounds and long tones resonate at the same frequencies, but they sound rather like they are being comb filtered rather than delayed because of the overlap. When I use this procedure on someone else's sound, I contradict my first guideline, and so, to maintain independence, I am careful how I use resonance on long tones.

These phenomena all relate, of course, to physical modeling and periodicity pitch. But all technical thoughts aside, I know that playing with delay length and feedback (overlap vs. non-overlap) gives me the ability to rapidly transform a sound into something seemingly unrelated to the sound I started out with and that I am processing.

7.2 Filter Strategies

Julius O. Smith, a longtime researcher at the Center for Computer Research in Music and Acoustics (CCRMA), wrote: "Any medium through which a signal passes can be considered a filter" (Smith 1985). This implies that all parts of my system—my delays, the room I am playing in, the speakers—all have an effect on my sound.

But how can *I* be identified, when I'm filtering another performer's sound? My solution is essentially that when I use effects to alter a source sound, I almost always make an active choice to do something *different* than what my "source musicians" are doing. To *be different*, I *listen first*. *Only then, do I react*—in one of the following ways.

7.3 Temporal Shift

I could play my filtered "response" later (in time) by feeding my filtered sound through one or more long delays. The delay applied to the filtered sound makes it

[8]In an example of "periodicity pitch"—a 1 ms delay line with high feedback resonates at 1000 Hz, 2 ms at 500 Hz, 3 ms at 250 Hz. This pattern continues until around the pitch is out of hearing range, (sub audio) and it actually starts to sound like delay.

evident that sound processing is happening and separate from my source. The temporal shift makes it clear that my processed output is not merely part of the original sound, but rather that it is something *I* am doing and contributing. Of course, I select my delay times keeping aware of any other rhythms and tempi played at that moment.

7.4 Sweep

I could modulate parameters such as Q factor, center frequency, bandwidth, and/or feedback. I like to perform these modulations rhythmically and in a way that differs from whatever my source musician does pitch-wise or rhythmically. This approach works best on sonically rich sounds.

If I observe my source musicians are playing quickly while using using a wide pitch range, I will choose to sweep center frequencies, or modulate other parameters, more slowly and create a contrast. If the other musicians are moving slowly, I may choose to move more quickly. If they create fast or complex rhythms, I do not compete, and stay fixed in a position. If they play drones or with a limited note range, I have the option to either articulate pitches by selecting a high Q factor and resonance for my filters or to perform fast rhythmic gestures, e.g., by quickly moving the center frequency around. In other words, the key to maintaining a separate identity when using filters is *contrast and/or temporal displacement.*

7.5 Pitch-Shifting and Other Manipulations

Another way to distinguish my sound is to use pitch-shifting. As a composer and singer/instrumentalist, I am keenly aware of the implications of all pitches that are being played by all the musicians, and do not pitch-shift sounds without considering the sonic and harmonic mess I might create if by disregarding stylistic and harmonic rules. I can choose to break these rules, of course, but I am always aware of what new pitches I've added to the overall sound, just as if I were playing these pitches/notes on a traditional instrument.

These considerations aside, I otherwise follow the same guidelines I use for filtering (regarding contrast and temporal displacement). If I pitch-shift a sound without delay or modulation, as with filtering, my contribution will likely be heard simply as an effect.

7.6 Reversed Sounds/Speed Changes

When reversing real-time sampled sounds, it is difficult, to make useful generalizations regarding the outcome, because the resultant sounds are so tied to the individual envelope of the sound. For the most part though I cannot predict an outcome unless I know in advance what the envelope of the sound will be, or unless I first sample and repeat the sound myself. There is some predictability and usefulness in resonant percussion sounds played backwards with their lovely sudden cutoffs, and how we all know how enticingly beautiful backwards piano sounds, but most instruments have significant variation in the types of envelopes they can produce, especially with skilled players. There are also a great many sounds that do not benefit from being reversed and others that are even humorous because of spectral distortions, and so the outcome is as varied as the input. This lack of ability to truly generalize on this subject was first noticed by with Pierre Schaeffer in the 1940s (Schaeffer 2012, p. 8).

Reversing a sound, however, does have its usefulness help however me create the distinction and otherness between my sound source and my own sound gestures, especially when combined with pitch-shifting or other processing.

Similarly, the speed of playback is not that interesting for me to generalize about beyond to say that some sounds are quite interesting when slowed down to reveal their internal rhythmic structures and hidden resonance. They can be interesting when sped up to create new rhythms. But how they actually sound is also very dependent on the algorithm that is used for the speeding up and slowing down, and whether or not it is accompanied by pitch changes (since this can now be decoupled). However, looping the rhythmic structures, or speeding up and slowing down, especially in conjunction with pitch-shift, are all useful and simple musical statements that I can make and use often. They are useful, because they nearly always will sound different than my source audio.

8 Independence Day—Feedback as Solution

After a few years of performing, progressing and learning about my instrument, (around 1998) I saw the need to become less dependent on others for my sound sources without relying on samples and pre-recorded sounds. The end to this dependency on others for sound came spontaneously during a recording session. At some point someone pointed at me to play a solo, and I realized that I could not play solo because all my sounds depended on someone else playing! To get around the dependency problem and play my solo that day, I innovated and improvised. By routing my effects processor back into itself I created *no-input* feedback, and also built audio feedback using the room tone. These impromptu solutions soon became a regular part of my instrument. Since then, I have found feedback to be an immensely rich source of sonic material that does not require anyone else nor even my voice to work.

I like to use quick changes and modulation of feedback in my delay systems to quickly increase or decrease density of my sound, and (as mentioned above) I also find it useful for creating sustained pitched sounds. This is why a critical part of my technical setup is the compressor/limiter on my output. The compressor is set with a fast attack and high compression ratio. This enables me to get dreadfully close to sound system feedback levels without actually hurting anyone's ears or losing control.

Using feedback as a musical source was central to the "Ha!" trope in my trio *What is it Like to be a Bat?* (1996, with Kitty Brazelton). In one of our pieces, I would yell/sing "Ha!" into the microphone and then trigger a sequence of enormous, feeding-back, swiftly changing delays resulting in Doppler effects. The result sounded *very* big and I used it in a duet with the drummer. Although it was at first challenging to consistently repeat the effect, I eventually found a way and the "Ha!" turned into a signature element of the piece "Sermonette". (Brazelton and Naphtali 2003)

9 Location, Location, Location

My earliest experiences performing as an electronic musician were at a time when I questioned all my roles as a performer. Nearly all of the electronic musicians I knew at the time were computer musicians trained in academia. My (mostly male) colleagues would participate in performances of their pieces, often from the middle of the audience, sitting at the sound board or a computer, playing from the vantage point of an audio engineer rather than as a performer from the stage.[9]

These considerations led to us questioning the role of our audio engineer in "What is it like to be a Bat?".[10] We considered the engineer to be a performing member of the group, using electronics from the desk. At one point we even had him sing a chorale with us, conducted from the stage.[11]

9.1 I Came to Ask Myself

When performing with sound utilizing the acoustics of the entire space as part of the musical gesture, at which position should I perform? Does standing on stage make me a performer? Does standing in the audience makes me a technician? How does the utilization of space (as part of my created instrument) differ from the attention given to the sound of a room by any trained acoustic musician? Do I want to perform on stage as a musician, or run my pieces off-stage, recognized as a high-end technician/composer/audio engineer? I discovered that I was uncomfortable being

[9]Two notable exceptions are George Lewis, and Mari Kimura, whose performances from the stage as composer/performers and improvisers with computers were very inspiring to me at that time.

[10]What is it Like to be a Bat?, originated as a "digital punk" trio (with co-composer Kitty Brazelton, and drummer Danny Tunick) back in 1997. Computer music, live processing, electroacoustic sound "tectonic plates", electric guitar, electric bass, drums and 2 multi-octave voices (Kitty and Dafna) (see What is it like to be a Bat? CD released 2003. http://www.tzadik.com/index.php?catalog=7707).

[11]Unfortunately, this, and perhaps some gender bias, resulted in some reviewers assuming that the audio engineer, or the drummer did all of the live electronics.

in the engineer's position. Although the sound was better sitting in the middle of the room, as a performer, I always preferred to be on stage, interacting with the other musicians. For all my projects, I therefore decided to be on stage, even when my role did not involve any singing at all.

10 Creating the Invisible Instrument So It Could Be Understood by Musicians[12]

10.1 Instrument Control

To turn my setup into an instrument, the first thing I did was to code up every available parameter I could control. What started to interest me was how the parameters moved in relation to each other. But I quickly realized that I could not effectively control very many parameters at the same time. I simply do not have that many fingers! Also, I needed to be able to control my instrument quickly, because otherwise I could not control rhythm, nor react rhythmically.

So, next I began to do two things: grouping parameters of my delays/pitch-shift/comb filters to control many things at once (which also expanded my palette), and creating algorithmic strategies to rhythmically control these parameter groupings. Grouping parameters into meta-parameters is similar to what is done in instrumental or vocal music, where many individual actions cause a vibrato on a violin or a glissando on a trombone. Vibrato and glissando are meta-parameters that group those actions.

The algorithmic controls allow me to interact and lock-in both rhythmically and poly-rhythmically with other musicians.

10.2 Presets/Starting Points

Arranging parameters I want to control into *presets* has been of tremendous value, because it helps me gain speed in my reactions. These presets are both starting points and safe zones that assure me of a known position or sound to which I can return, allowing me to experiment more freely and without reservation.

Augmenting this control, I set it up so that I can control my presets and transitions by manually sequenced them or triggering them algorithmically, or in short

[12]Some examples can be seen in a video of a performance for "Live Processing Strategies" a talk I gave at Harvestworks in New York City in May 2012. Joining me were Robert Dick (bass flute) and Satoshi Takeishi (percussion). I am processing both instruments and my voice in many of the ways described in this section. More information and examples can also be found at: http://dafna.info/instrument/. (Naphtali et al. 2012)

poly-rhythmic patterns. This rhythmic control over my "presets" (parameter groupings and their transitions) are another important and evolving part of what constitutes my instrument. I can create new presets on the fly and reuse them during performance.

10.3 Giving up Control Over Time = Automated Processes

Sometimes it is useful for some of my sound processes to work on their own. This way, I can create a richer sound world, especially when I'm playing solo. It also injects bits of unpredictability, surprises and aleatoric possibilities to my performances. Examples of these autonomous processes are: (1) random variations around the current value of parameters, and slowly drifting towards new values which can be set manually at any time; (2) totally randomizing parameters with the only control being the range of random numbers, offset and the speed of newly created values; (3) "Mutate" which randomizes given parameters around loop points of my live-recorded samples.

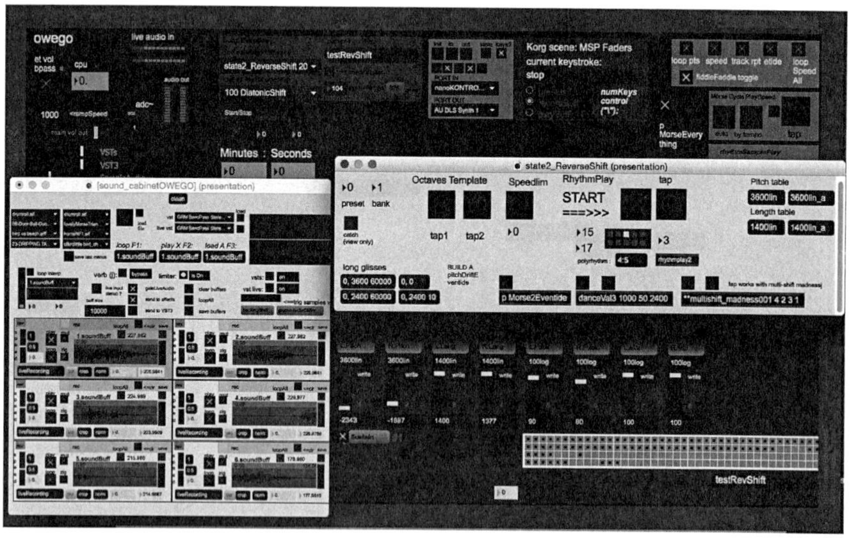

11 Voice Activated Controls

More recently, I added a set of "Voice-Activated" controls to my instrument: using my incoming audio signals as my only control source. These programming routines are meant to unfold in the background, autonomously. They were created originally for my voice in solo performances, but I have since used them with other instruments, projects and compositions.

In one routine, I connect pitch-tracking with a moving, sweeping subtractive synth. What feels like a "comet tail" becomes attached to the incoming sound, loosely following its pitch slightly weighted and behind schedule. Another control routine recognizes attack transients to decrease or increase the speed of a looping audio sample.

Though musically useful, these controls are invisible too. As much as I like them, they do not particularly help my audience with understanding what I'm doing on stage. Nevertheless, I like the outcome, and so I continue experimenting with it.

12 Outro

Building an electronic instrument that makes use of sounds that are not generated by the electronic musician-performer represents several challenges. The sounds, collected from elsewhere, and so one challenge is visually connecting those sounds to the electronic musician's body and person. If the sound source is another musician on stage, and also audible to the audience, there are further challenges in creating aural distinctions between the source and the electronic instrument's output. These challenges make creating such a live sound-processing instrument a difficult task. It is inherently *invisible* to the audience, although it might be present and recognizable to the musicians on stage.

These challenges can be addressed by carefully combining an awareness of acoustics, some basic psychoacoustics and good listening. The combination of these elements establish an aesthetic and predictive approach to live sound processing informed by an awareness of musical styles and rhythm. The electronic musician's own performance must furthermore be connected in some way to her body, through creative use of gestural controllers and other input devices, mapped to salient groupings of musical parameters in physically meaningful ways. This approach creates the musical distinction between the output by the electronic musician and her sources. Gestural controllers properly connected to these distinguished sounds create "control intimacy" (Moore 1988) noticed by both fellow performers and by the audience. With training and practice, this instrument will eventually become less "invisible" and better understood by all.

13 Learning from the Masters

For some further reading and research—here are some books and articles I recommend.

- Stockhausen's London lectures—particularly Lecture V: "Four Criteria of Electronic Music" available *as DVD or on UbuWeb*. Stockhausen discusses a continuum from pulses to rhythm to form (a sped up a rhythm can become a spectral quality, slowed way down it can be conceived of as form.) and many other such insights.)
- Pierre Schaeffer "Solfège d'Objet Sonore" (book/3 CDs)
- Pierre Schaeffer *In Search of a Concrete Music*. U of California, 2012
- Sonic Experience: *A Guide to Everyday Sound*. Edited by Jean-Francois Augoyard, Henri Torgue, McGill-Queens University Press 2005

References

Brazelton, K. & Naphtali, D. (2003). "Sermonette- Ha!" (track 4). What is it like to be a bat? Tzadik Records, CD.

Four Criteria of Electronic Music (KONTAKTE), Part 1. 1972. Prod. Robert Slotover. Perf. Karlheinz Stockhausen. Allied Artists, London. *UbuWeb Film & Video/Lecture 5*. http://ubu.com/film/stockhausen_lectures5-1.html. Accessed June 1, 2016.

Moore, F. R. (1988). The Dysfunctions of MIDI. *Computer Music Journal, 12*(1), 19–28.

Naphtali, D., Dick, R., & Takeishi, S. (2012). Live sound processing strategies. https://youtu.be/x9hWMSzMdTI. Accessed June 1, 2016.

Rowe, R. (1993). Interactive Music Systems: Machine listening and composing. pp. 6–8. Cambridge, MA: The MIT Press.

Schaeffer, P. (2012). *In Search of a Concrete Music*. (Trans. C. North & J. Dack Trans.). pp. 8–9. Oakland, CA: U of California.

Smith, Julius O. (1985). Fundamentals of digital filter theory. *Computer Music Journal, 9*(3), 13–23.

Printed in the United States
By Bookmasters